T0134390

Automotive Chassis Engineering

David C. Barton · John D. Fieldhouse

Automotive Chassis Engineering

David C. Barton
School of Mechanical Engineering
University of Leeds
Leeds
UK

John D. Fieldhouse
School of Mechanical Engineering
University of Leeds
Leeds
UK

ISBN 978-3-030-10200-5 ISBN 978-3-319-72437-9 (eBook)
https://doi.org/10.1007/978-3-319-72437-9

© Springer International Publishing AG 2018
Softcover re-print of the Hardcover 1st edition 2018
This work is subject to copyright. All rights are reserved by the Publisher, whether the whole or part of the material is concerned, specifically the rights of translation, reprinting, reuse of illustrations, recitation, broadcasting, reproduction on microfilms or in any other physical way, and transmission or information storage and retrieval, electronic adaptation, computer software, or by similar or dissimilar methodology now known or hereafter developed.
The use of general descriptive names, registered names, trademarks, service marks, etc. in this publication does not imply, even in the absence of a specific statement, that such names are exempt from the relevant protective laws and regulations and therefore free for general use.
The publisher, the authors and the editors are safe to assume that the advice and information in this book are believed to be true and accurate at the date of publication. Neither the publisher nor the authors or the editors give a warranty, express or implied, with respect to the material contained herein or for any errors or omissions that may have been made. The publisher remains neutral with regard to jurisdictional claims in published maps and institutional affiliations.

Printed on acid-free paper

This Springer imprint is published by Springer Nature
The registered company is Springer International Publishing AG
The registered company address is: Gewerbestrasse 11, 6330 Cham, Switzerland

Preface

A common concern of the automotive industry is that new recruits/graduates are more than able to operate the modern computer-aided design packages but are not fully aware or knowledgeable about the basic theory within the programmes. Because of that lack of basic understanding, they are unable to develop the commercial package(s) to suit the company's needs nor readily appreciate the output values. Even more important, as time progresses and that basic knowledge becomes rarer within companies, the reliance on commercial software suppliers increases, along with costs. There is a continuing need for companies to become self-sufficient and be in a position to develop bespoke design 'tools' specific to their needs.

The advances in electric vehicle technology and move towards autonomous driving make it necessary for the engineer to continually upgrade their fundamental understanding and interrelationship of vehicle systems. The engineers in their formative years of training need to be in a position to contribute to the development of new systems and indeed realise new ones. To make a contribution it is necessary to, again, understand the technology and fundamental understanding of vehicle systems.

This textbook is written for students and practicing engineers working or interested in automotive engineering. It provides a fundamental yet comprehensive understanding of chassis systems and presumes little prior knowledge by the reader beyond that normally presented in Bachelor level courses in mechanical or automotive engineering. The book presents the material in a practical and realistic manner, often using reverse engineering as a basis for examples to reinforce understanding of the topics. Existing vehicle specifications and characteristics are used to exemplify the application of theory. Each chapter starts with a review of basic theory and practice before proceeding to consider more advanced topics and research directions. Care is taken to ensure each subject area integrates with other sections of the book to clearly demonstrate their interrelationships.

The book opens with a chapter on basic vehicle mechanics which indicates the forces acting on a vehicle in motion, assuming the vehicle to be a rigid body. Although this material will be familiar to many readers, it is a necessary prerequisite to the more specialist material that follows. The book then proceeds to a chapter on steering systems which includes a firm understanding of the principles and forces involved under both static and dynamic loading. The next chapter provides an appreciation of vehicle dynamics through the consideration of suspension systems—tyres, linkages, springs, dampers, etc. The chassis structures and materials chapter includes analysis tools (typically FEA) and design features that are used on modern vehicles to reduce mass and to increase occupant safety. The final chapter on Noise, Vibration and Harshness (NVH) includes a basic overview of acoustic and vibration theory and makes use of extensive research investigations and test procedures as a means to alleviate NVH issues.

In all subject areas, the authors take account of modern trends, anticipating the move towards electric vehicles, on-board diagnostic monitoring, active systems and performance optimisation. The book contains a number of worked examples and case studies based on recent research projects. All students, especially those on Masters level degree courses in Automotive Engineering, as well as professionals in industry who want to gain a better understanding of vehicle chassis engineering will benefit from this book.

Leeds, UK David C. Barton
John D. Fieldhouse

Acknowledgements

The origins of this book lie in a course of the same name delivered to Masters level Automotive and Mechanical Engineering students at the University of Leeds for a number of years. The authors are grateful to those who have contributed to the design and development of the course, especially the late Professor David Crolla, Professor David Towers, Dr. Brian Hall and Dr. Peter Brooks, as well as to previous research students who have developed some of the case study material.

Contents

Chapter 1
Vehicle Mechanics

Abstract Before embarking on the focus of this book it was felt necessary to provide a basic understanding of the dynamic forces experienced by any road vehicle during normal operation. This chapter introduces such forces on a vehicle when considered as a rigid body. It discusses the source of each force in some detail and how they may be applied to predict the performance of a vehicle. It extends the normal straight-line driving to include non-steady state cornering and the case of car-trailer combinations. Each section generally includes typical problems with detailed solutions.

1.1 Modelling Philosophy

Most of the analyses of vehicle performance rely on the idea of representing the real vehicle by mathematical equations. This process of *mathematical modelling* is the cornerstone of the majority of engineering analyses. The accuracy of the resulting analysis depends on how well the equations (the mathematical model) represent the real engineering system and what assumptions were necessary in deriving the equations.

A vehicle is a complex assembly of engineering components. For different types of analysis, it is reasonable to treat this collection of masses differently. For example, in analysing vehicle acceleration/deceleration, it may be appropriate to lump together all the masses and treat them as if the vehicle were one single body, a lumped mass, with the mass acting at an effective centre of mass, commonly termed the "centre of gravity". For ride analyses however, the unsprung masses would typically be treated separately from the rest of the body since they can move significantly in the vertical direction relative to the body. Also, for an internal combustion engine (ICE) vehicle, the engine mass may be treated separately to represent its relative vertical motion on the engine mounts. For driveline analyses, the masses and inertias of the rotating parts in the engine, gearbox, clutch, drive shafts etc. may be separated from the rest of the vehicle mass.

© Springer International Publishing AG 2018

D. C. Barton and J. D. Fieldhouse, *Automotive Chassis Engineering*,
https://doi.org/10.1007/978-3-319-72437-9_1

This lumped mass approach is extremely useful for modelling the gross motions of the vehicle, i.e. in the longitudinal, lateral or vertical directions. The lumped masses are assumed to be rigid bodies with the distribution of mass throughout the body characterised by the inertia properties. Of course, no engineering component is strictly a rigid body, implying infinite stiffness, although it will in many cases be a perfectly adequate assumption to treat it as such. The vehicle body, typically made of pressed steel sections and panels spot-welded together is fairly flexible—a typical torsional stiffness for a saloon car is around 10 kNm/degree. For other types of analysis, e.g. structural properties or high frequency vibration and noise properties, the vehicle body would be treated as a distributed mass, (i.e. its mass and stiffness properties distributed around its geometric shape) and typically a finite element approach would be used for the analysis.

Using the lumped mass approach to a vehicle dynamics problem, the governing equations of motion can usually be derived by applying Newton's Second Law of Motion or its generalised version when rotations are involved which are usually called the Rigid Body Laws. The approach, which is the preferred method for tackling the majority of dynamics problems is:

(a) Define an axis system
(b) Draw the Free Body Diagram (FBD)
(c) Apply the Rigid Body Laws
(d) Write down any kinematic constraints
(e) Express forces as functions of the system variables
(f) The governing set of equations then come from combining (c), (d), and (e).

1.2 Co-ordinate Systems

There is a standard definition embodied in an SAE standard (SAE J670 Vehicle Dynamics Terminology) for a co-ordinate system fixed in the vehicle and centred on the vehicle centre of gravity (C.G.) as shown in Fig. 1.1. Note that the rotational motions of the vehicle body—roll, pitch and yaw—are defined in the figure. The vehicle fixed co-ordinate system which, therefore, moves with the vehicle is useful for handling analyses. For the analyses of vehicle performance in this chapter, however, a simple ground fixed axis system is appropriate and the analyses are restricted to two dimensions involving longitudinal, vertical and pitch co-ordinates.

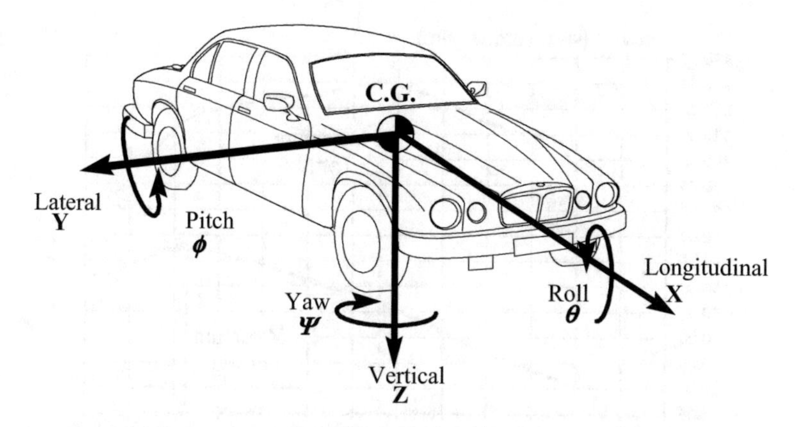

Fig. 1.1 The vehicle co-ordinate system as detailed in SAE J670 vehicle dynamics terminology

1.3 Tractive Force and Tractive Resistance

Static and dynamic calculations require an understanding of the dynamic forces and loads involved during motion. These may be referred to as tractive forces and resistances. The following sections discuss these loads and associated tyre properties.

1.3.1 Tractive Force or Tractive Effort (TE)

The tractive effort (TE) is the force, provided by the engine or electric drivetrain, available at the driven axle road/tyre interface to propel and accelerate the vehicle. For a conventional ICE vehicle, TE is given by:

$$TE = \frac{T_e \times n_g \times n_d \times \eta}{r} \tag{1.1}$$

where:

T_e engine torque
n_g gearbox ratio
n_d final drive (differential) ratio
η overall transmission efficiency
r effective rolling radius of tyre.

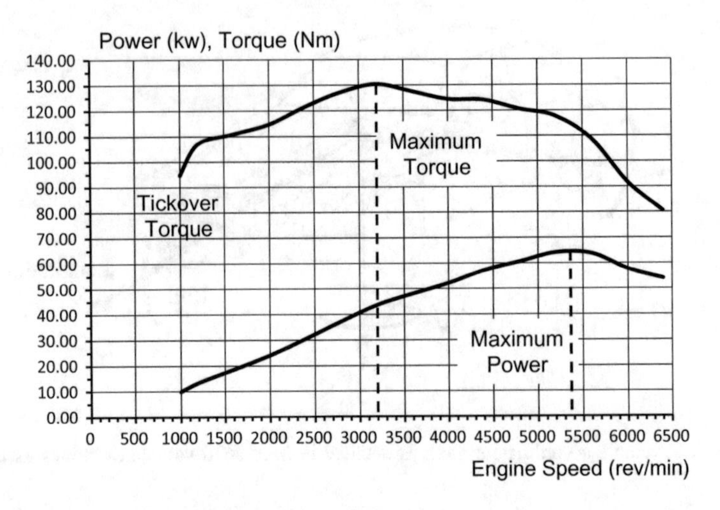

Fig. 1.2 Typical internal combustion engine characteristics

For a conventional ICE vehicle, the vehicle speed (v) is given by:

$$v = \frac{N_e \times 2\pi r}{n_g \times n_d} \tag{1.2}$$

where:

N_e Engine speed (rev/s).

A typical internal combustion engine power and torque characteristic plotted against engine speed is shown in Fig. 1.2. Note that the maximum torque and maximum power occur at different engine speeds. An electric or hybrid electric drivetrain would have different torque and power characteristic curves. However drivetrains are not the focus of the current book so are not discussed further.

1.3.2 *Tractive Resistances (TR)*

A vehicle's resistance to motion is due to three fundamental parameters: gradient resistance, aerodynamic drag and rolling resistance (with slow speed manoeuvres, turning resistance is also important).

1.3.2.1 Gradient Resistance (G_R)

If the vehicle is progressing up a gradient, G_R is the proportion of a vehicle's weight acting down a gradient—the "mg sin θ" component as indicated in Fig. 1.3. If the vehicle is progressing down the gradient, the component would be assisting the vehicle in which case the force would be termed "gradient assistance". G_R can be represented as a single force acting at the C.G. of the vehicle and parallel to the road surface:

$$\text{Gradient Resistance (Assistance)} = \pm\, \text{mg} \sin\, \theta \tag{1.3}$$

Note:

- Gradient Resistance/Assistance is proportional to vehicle weight (mg).
- The force normal to the road is always "mg cos θ" and, as θ tends to zero, the total normal force at the tyre/road interface approximates to mg, the weight of the vehicle.

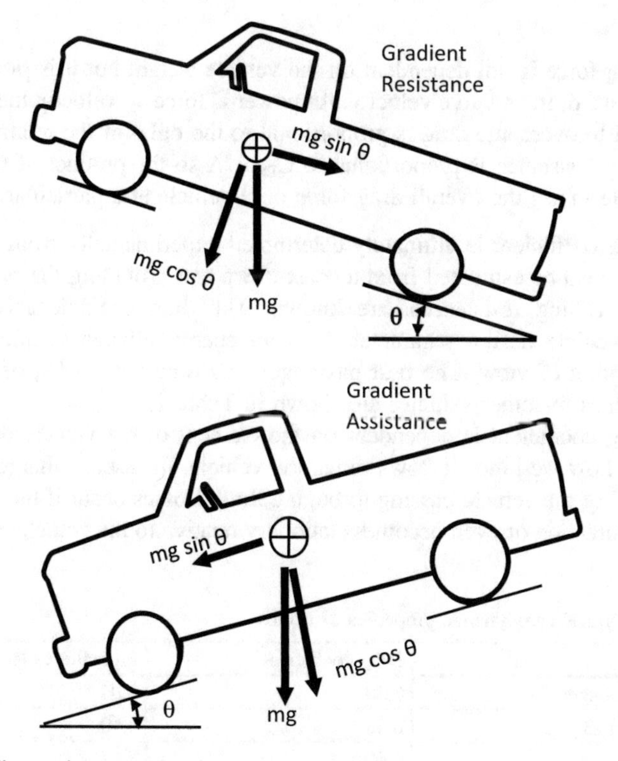

Fig. 1.3 Gradient resistance and assistance

1.3.2.2 Aerodynamic Resistance or Drag Force (D)

The aerodynamic drag force is a measure of the effectiveness of vehicle to progress through air. It can be represented as a single force acting at the centre of pressure at some distance above road level. This distance would normally be determined, initially, by Computational Fluid Dynamics (CFD) analysis and confirmed by wind tunnel testing.

The Drag Force (D) is given by the following equation:

$$D = \frac{1}{2} \rho C_D A v^2 \tag{1.4}$$

where:

C_D drag coefficient (typically 0.3 for a modern car)
ρ air density
A vehicle frontal area
v velocity of vehicle relative to surrounding air.

Note:

- The drag force is not dependent on the vehicle weight but it is proportional to the **square** of the relative velocity. As power is force × velocity then the power required to overcome drag is proportional to the **cube** of the relative velocity.
- The drag resistance is proportional to $C_D \times A$ so the product of these parameters determines the overall drag force on a vehicle at a particular velocity.

The drag coefficient is ultimately determined experimentally from wind tunnel tests. It can also be estimated from a coast-down test providing the other resisting forces, i.e. rolling resistances, are known. The drag coefficient is clearly an important vehicle design parameter from an energy efficiency, and hence fuel economy, point of view. The best passenger cars now have a $\mathbf{C_D}$ of around 0.3. Typical values for other vehicles are shown in Table 1.1.

The drag coefficient is dependent on the element of the vehicle design which determines how well the air flows round the vehicle. In essence this represents the "efficiency" of the vehicle passing through a fluid. Losses occur if the air is caused to change direction or even becomes stationary relative to the vehicle, such that the

Table 1.1 Typical aerodynamic properties of vehicles

Vehicle	Drag coefficient	Frontal area (m^2)
Modern passenger car	0.30	2.05
Delivery van (3.5 t)	0.48	4.10
Bus	0.60	7.17
Articulated truck	0.70	9.20

static pressure increases forward of the vehicle. If air flows quickly over the vehicle then the static pressure will reduce and in some instances become negative relative to ambient.

Although for full aerodynamic analysis, the compressibility of the air must be taken into account, it is instructive to consider the incompressible flow equation, a common form of Bernoulli's equation, which is valid at any arbitrary point along a streamline:

$$\frac{v^2}{2} + gz + \frac{p}{\rho} = constant \tag{1.5}$$

where:

- v air flow speed relative to the vehicle and at any point on a streamline.
- g gravitational constant.
- z elevation of the point above a reference plane.
- p pressure at the chosen point.
- ρ density of the fluid at all points in the fluid.

Note that the "gz" term can usually be ignored for a road vehicle.

Clearly, from Eq. (1.5), as the air speed falls the pressure will increase locally. Conversely as air speed increases, the local pressure will fall and may in fact become negative to ambient. This explains why papers will be "sucked" through an open sun roof at high speed.

Figure 1.4 shows streamlines flowing over a Mercedes car in a wind tunnel. It should be noted that the earlier the air separates from the vehicle, the greater the

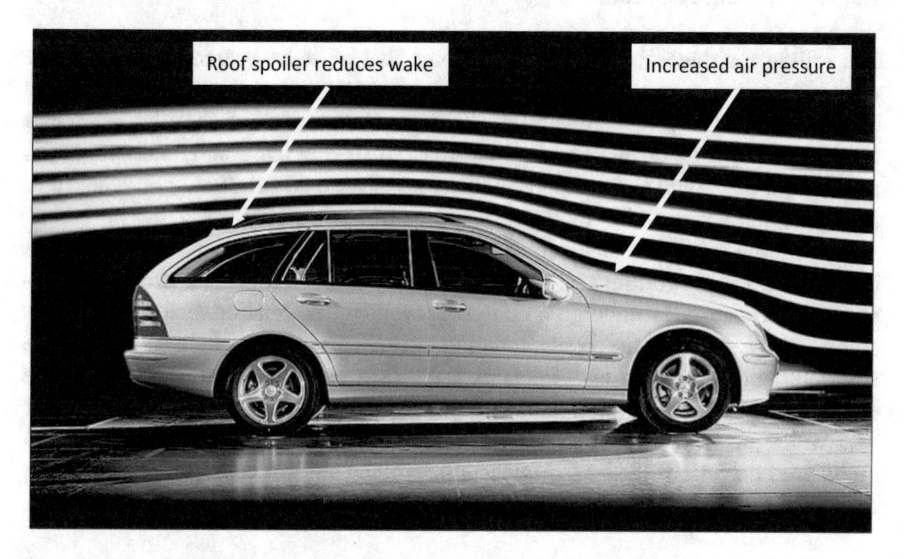

Fig. 1.4 Visualisation of streamlines in a wind tunnel test.
Ref: http://images.gerrelt.nl/roofspoiler/mercedes_windtunnel_test.jpg

"wake" (the area directly behind the vehicle), resulting in an increase in negative pressure at the rear of the vehicle and an increase in aerodynamic drag. Other losses occur if the air forms vortices in the wake of the vehicle or if the air is caused to travel a tortuous path such as that in the engine compartment or the wheel arches. Such airflows are often controlled and may be referred to as "air management", the air flow being used to cool the engine and brakes. The positive pressure areas indicated in Fig. 1.4 may be used to advantage by positioning air intakes at these points.

As well as horizontal drag forces, aerodynamic flow over a vehicle will also typically generate vertically downward forces (negative lift). This down force will aid cornering but will effectively add to the vehicle tyre/road interface force and increase rolling resistance forces during straight line driving. The aim of formula racing cars fitted with wings is to balance the increased down force required for cornering with the accompanying increased wing drag developed on the high speed straight.

The question may arise about the total area to be considered for calculating aerodynamic drag if the vehicle is towing a trailer or caravan. A vehicle storage roof box makes a good deflector of air. If no other information is available then it is usual to add the vehicle and caravan frontal areas together to give an upper limit to the total overall drag. If this proves to be incorrect then the combination will perform better than expected. On the other hand, if the frontal area is underestimated, the vehicle may be fitted with too small an engine capacity and the combination will perform badly—it will be underpowered.

1.3.2.3 Rolling Resistance (R_R)

The rolling resistance is defined as the force that must be overcome to cause the vehicle to move at constant speed over a horizontal surface, assuming no vehicle body aerodynamic forces are present. It is normally assumed that the vehicle is travelling in a straight line and that the road surface is reasonably smooth. R_R is represented as a force at the road/tyre interface of each wheel. It may be reduced to a single force acting at the road/tyre interface of each axle.

The rolling resistance arises from two main sources: the continuous deformation of the tyres during rolling and frictional effects in the mechanical driveline components. Rolling tyres undergo a continual cyclical deformation as the tyre passes continuously through the contact region area. This causes deformation of the sidewalls and tread area and, because it is not a perfectly elastic process, some energy is lost through hysteresis (see Fig. 1.5). This lost elastic energy appears as heat which may be confirmed by "feeling" the tyre temperature after a period of high speed driving. If the tyre is under-inflated then sidewall deformation increases along with the temperatures. If the vehicle continues to be driven with excessive

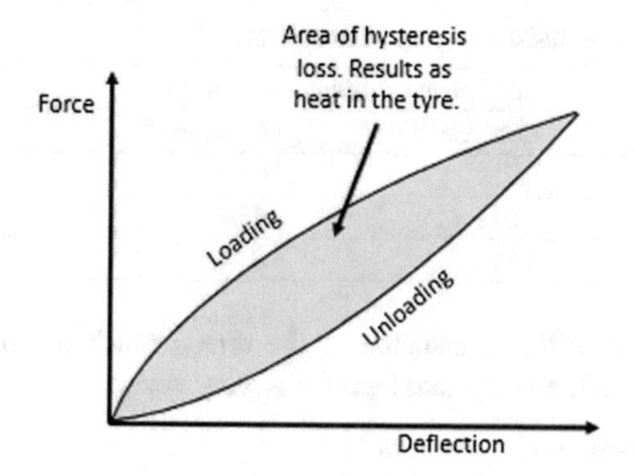

Fig. 1.5 Hysteresis loss within a tyre during loading and unloading

deflation then sidewall delamination may occur. In addition, small amounts of sliding between the tread elements and the road surface occur which add to the losses. At higher speeds, aerodynamic effects due to air drag on the rotating tyres again add to the losses.

It is common to lump all the rolling resistance losses for a vehicle together and approximate them in terms of a rolling resistance coefficient defined by the following equation:

$$R_R = N \times C_R = mg\,cos\,\theta \times C_R \tag{1.6}$$

where

R_R rolling resistance
C_R coefficient of rolling resistance
N normal force perpendicular to the surface on which the vehicle is moving.

Note that N will generally be the "mg cos θ" element of the vehicle weight (mg) acting normal to the road surface but should also include any downforce imposed by aerodynamic effects.

If the vehicle is at standstill then the initial rolling resistance is usually referred to as starting resistance, akin to static friction conditions. Starting resistance may be 50–80% more than the steady state rolling resistance: C_R typically varies from 0.012 rolling to 0.020 starting.

Rolling resistance primarily results from losses in sidewall and tread deformation resulting in hysteresis losses within the tyre which shows up as heat. The road type also influences resistance as the tyre impresses into the surface of the road. It is generally considered constant but it is in reality speed dependent, rising slightly with speed. This information would be normally supplied by tyre manufacturers. Typical values of C_R for different vehicles and road surfaces are given in Table 1.2.

The rolling resistance equation is a useful approximation which may be used for simple "first order" calculations to evaluate vehicle transmission loads, performance and fuel economy. However, rolling resistance, whether expressed as a force or a non-dimensional coefficient, is not constant in practice.

Table 1.2 Typical rolling resistance properties of vehicles

Vehicle type	Coefficient of rolling resistance (C_R)		
	Concrete	Good track	Sand
Passenger car	0.012	0.08	0.30
Truck	0.012	0.06	0.25
Tractor	0.02	0.04	0.20

In particular, the rolling resistance of the tyres, which of course, is a critical design parameter for the tyre manufacturer is very sensitive to:

- Vehicle speed—rate of hysteresis loss.
- Tyre temperature—affects compound.
- Carcass design and material properties—a thinner material results in less rolling resistance.
- Road surface—soft surface results in deformation which gives rise to higher rolling resistance.
- Slip and tread deformation when producing tractive forces—formula racing cars use "slicks" (no tread) in dry weather because they wish to minimise rolling resistance by reducing/eliminating tread deformation.
- Size—increase in width of tyre results in lower rolling resistance due to lower tyre wall deflections (see Fig. 1.6).
- Load and inflation pressure. (see Fig. 1.7).

If a normal tyre is inflated to the correct pressure, the rolling resistance reduces and vehicle economy increases. If the tyre is under-inflated, the rolling resistance will increase due to excessive tyre wall deformations. If excessive over-inflation occurs then vehicle handling is affected. Current technology is tending towards

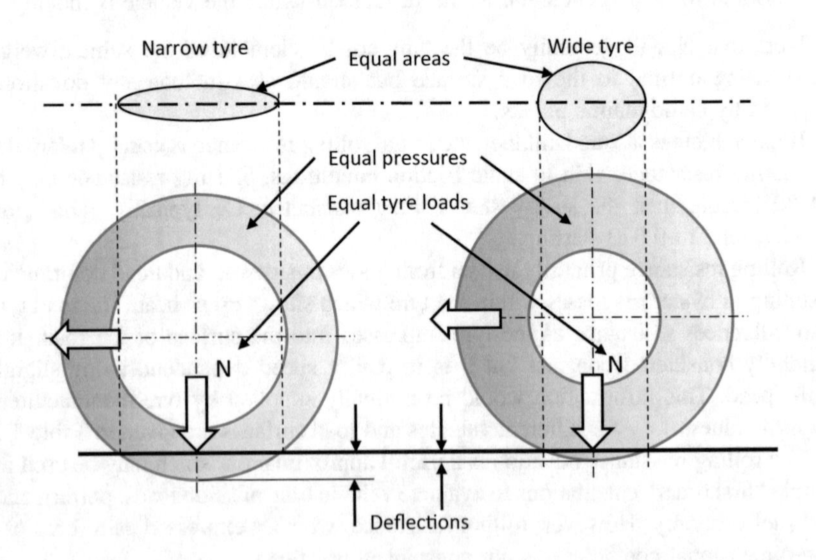

Fig. 1.6 Effect of tyre width on rolling resistance

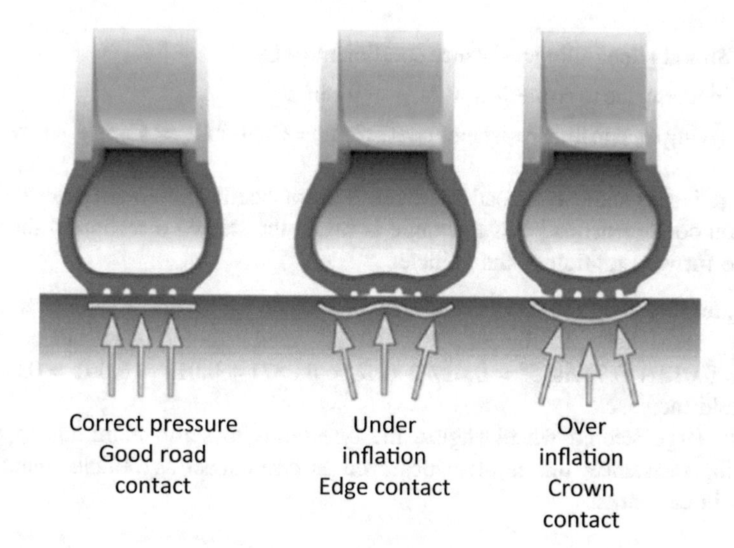

Fig. 1.7 Effect of pressure on contact area and tyre deformations.
https://i.stack.imgur.com/aJdAI.jpg

automatic pressure monitoring to take account of varying road surfaces and altitude. This is a move intended to minimise fuel consumption yet maintain optimum vehicle performance.

Rolling resistance can also change significantly during cornering. At higher speeds this tends to be less of an issue as aerodynamic drag plays a greater role in the overall resistances, and dynamic frictional coefficient falls (dynamic friction coefficient being less than the static, or "stiction", value). At low speed, such as urban driving, parking and heavy traffic conditions, aerodynamic drag is less important and resistance to manoeuvring increases as the tyre/road interface friction coefficient rises.

In general, the cornering resistance depends on the steered wheel angle (which depends on the cornering radius), steered wheel load, tyre/road interface friction level and drive configuration (front, rear or all-wheel drive). Consider the situation in Fig. 1.8:

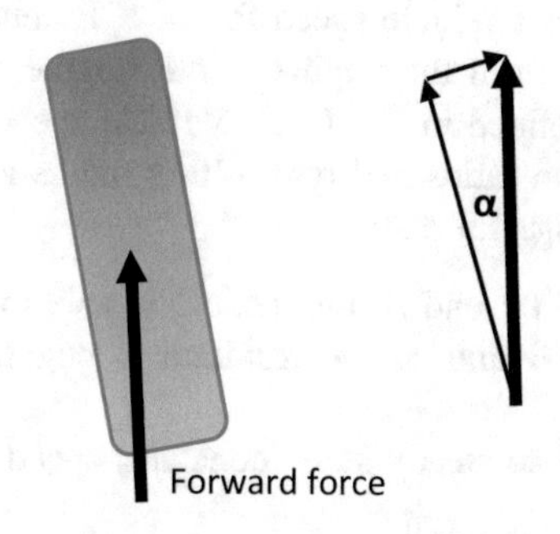

Fig. 1.8 Plan view showing cornering resistance at low speeds

Straight line rolling resistance coefficient $= C_R$
Increase due to cornering $= \Delta C_R = \mu \sin \alpha$
giving total rolling resistance coefficient $= C_R + \Delta C_R = C_R + \mu \sin \alpha$

where μ is the static tyre/road interface friction coefficient (often referred to as adhesion coefficient) and α is the angle between the steered direction of the wheel and the forward motion of the vehicle.

Example If $\alpha = 5°$ and $\mu = 0.7$ (tarmac road), then for a vehicle with straight line rolling resistance coefficient of 0.012, total rolling resistance coefficient $= 0.012 + 0.7 \sin 5° = 0.012 + (0.7 \times 0.087) = 0.012 + 0.061 = 0.073$ i.e. a six-fold increase.

With large steered wheel angles, the tyre tends to scrub more adding to the cornering resistance; this is often observed as tyre squeal as vehicles manoeuvre slowly in car parks.

1.3.3 Effect of TR and TE on Vehicle Performance

We have the total Tractive Resistance (TR) given by

$$\begin{aligned} \mathrm{TR} &= \text{Rolling Resistance} + \text{Aerodynamic Drag} \pm \text{Gradient Resistance} \\ \mathrm{TR} &= mg \cos\theta\, C_R + \frac{1}{2}\rho C_D A v^2 \pm mg \sin\theta \end{aligned} \tag{1.7}$$

where C_R is a modified resistance coefficient to take account of cornering resistances.

The acceleration of the vehicle depends on the difference between the TE and TR and on the mass of the vehicle (including rotational inertial effects) as follows:

$$TE - TR = m\ddot{x} \tag{1.8}$$

Figure 1.9 shows typical variations of the Tractive Resistances (TR) and Tractive Effort (TE) against vehicle speed for an ICE vehicle. The Tractive Effort (TE) curve is calculated from the engine torque/engine speed characteristic at a particular gear ratio as defined in Eq. (1.1). Vehicle speed (v) is derived from the engine speed, transmission ratios and tyre rolling radius as defined in Eq. (1.2).

It can be seen from Fig. 1.9 that:

At v_1, TE is greater than TR and so the vehicle is able to accelerate;
At v_2, TE is less than TR and so the resultant is negative and the vehicle will decelerate;
At v_3, TE equals TR and so steady state (constant) speed results.

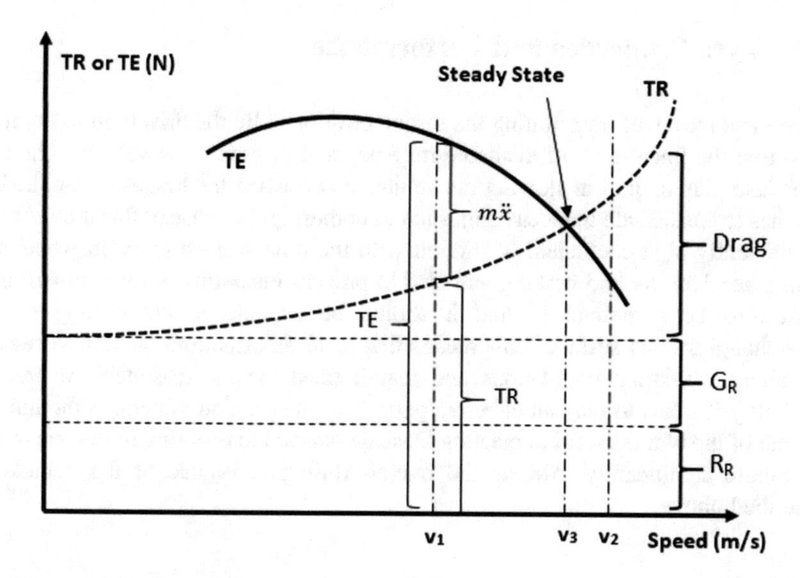

Fig. 1.9 Tractive effort/resistance against vehicle speed

Example

A vehicle has a rolling resistance coefficient 0.012. The driver allows the vehicle to free-wheel (not in gear) down a gradient when it reaches a slow steady state speed. Ignoring aerodynamic drag, calculate the slope of the gradient.

Solution:

$$TE = mg\,cos\,\theta C_R + \frac{1}{2}\rho C_D A v^2 - mg\,sin\,\theta$$

Both TE and aerodynamic drag = 0, so

$$mg\,cos\,\theta C_R = mg\,sin\,\theta$$

Giving

$$tan\,\theta = C_R = 0.012$$

So

$$\theta = 0.6875^\circ \; or\,gradient\,of\; 1\; in\; 83.333.$$

1.4 Tyre Properties and Performance

A tyre is a means of transmitting the torque developed by the drivetrain to the road such that the tractive effort available may be used to propel the vehicle. The tyre must also play its part in slowing the vehicle down when the brakes are applied. It also has to ensure safe manoeuvring, such as cornering. Because of this it must have a sufficiently high coefficient of friction with the road surface to avoid wheel slip during acceleration and braking and also to prevent instability during cornering. It must also be compliant in that it should be capable of conforming to the ever-changing road surface. This means that local deformations should be catered for along with road undulations and that it must have a reasonable degree of flexibility if it has to play an effective part of the suspension system. Although the flexure of the tyre is useful in meeting these demands, the resulting hysteresis losses contribute significantly towards the overall rolling resistance of the vehicle as described above.

1.4.1 Tyre Construction

There are two types of tyre construction—radial ply and cross (bias) ply. These are shown in Fig. 1.10. The most common tyres to be found on road cars are radial ply. The primary advantage of radial ply tyres is that the side walls are more flexible and so more tread remains in contact with the road during cornering. This is demonstrated in Fig. 1.11. Figure 1.12 shows the general construction of a radial-ply tyre in more detail.

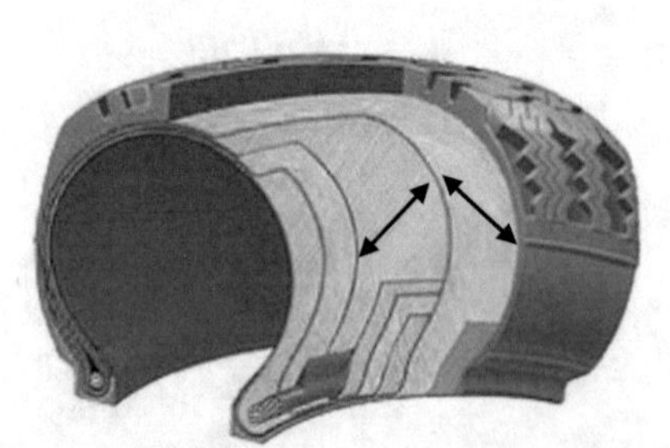

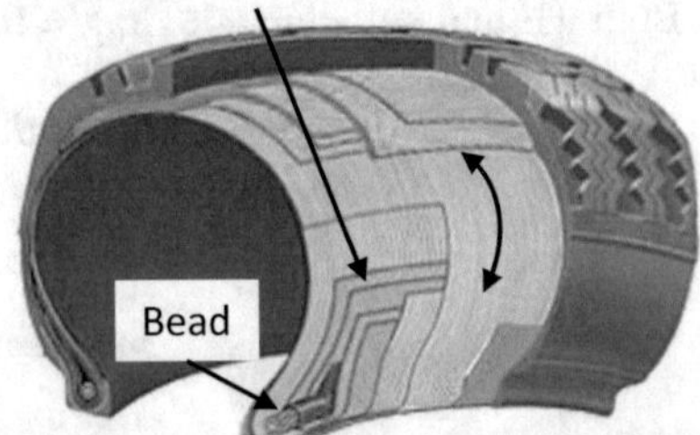

Fig. 1.10 Types of tyre construction: cross ply (left) and radial ply (right). https://upload.wikimedia.org/wikipedia/commons/1/19/Pirelli_Cinturato_Tire_cutaway.jpg

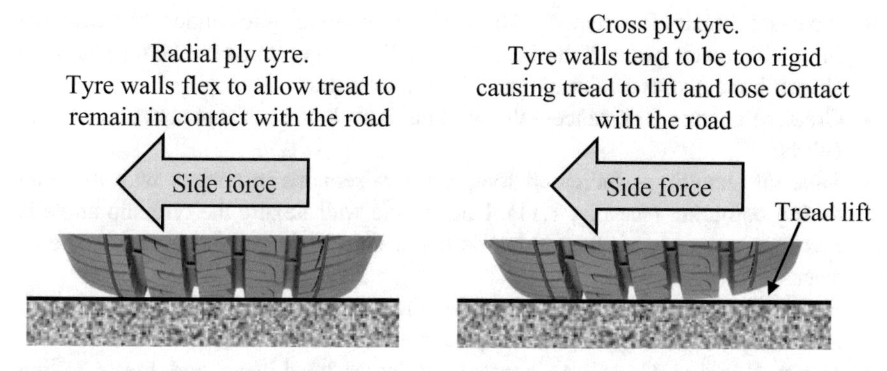

Fig. 1.11 Comparison of contact patch between radial and cross ply during cornering. https://i.stack.imgur.com/SFLKl.png

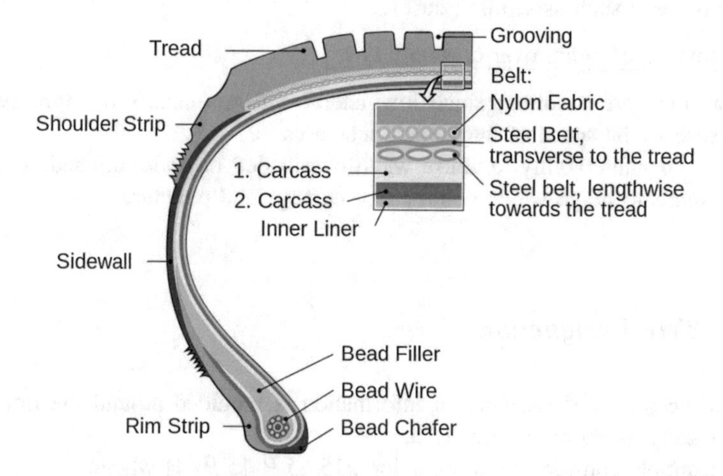

Fig. 1.12 Radial ply tyre construction. https://upload.wikimedia.org/wikipedia/commons/thumb/4/4c/Radial_Tire_%28Structure%29.svg/1024px-Radial_Tire_%28Structure%29.svg.png

Advantages of radial-ply over cross-ply tyres:

- Longer Tread Life—Strengthened bracing under tread reduces tread flexure ("squirm") in contact patch area.
- Cooler Running—Thinner side walls and less friction between plies. Runs 20–30 °C cooler than cross-ply because of lower tread squirm.
- Lower Rolling Resistance—Lower hysteresis losses due to less tread squirm as a result of flexible sidewalls.
- Enhanced Comfort—Flexible sidewalls are more forgiving with road undulations, readily absorbing uneven road surfaces. Less vibration transmitted—quieter.

- Increased Impact Resistance—The working (bracing) plies under the tread (see Fig. 1.10) better protects the inner lining. The longer cords are better placed to absorb impact stresses (strain energy) due to impacts.
- Greater Puncture Resistance—the working belts better resist penetration of road debris.
- Superior Handling—Increased footprint area remains in contact with the road during cornering (see Fig. 1.11). Due to side wall flexure the tyre slip angle is less than cross-ply so the vehicle is better able to follow the intended line of steer.
- Better Wet Traction—Steel belts stiffen the tread so it does not deform as much as cross-ply resulting in better displacement of rain water.
- Lower Running Costs—As a result of lower tread wear and lower rolling resistance.
- Reduced Sidewall Damage—Because sidewalls are more resilient (compliant) to side impacts such as scuffing curbs.

Disadvantages of radial over cross-ply tyres:

- Poor transport handling, since low lateral stiffness causes the tyre sway to increase as the speed of the vehicle increases.
- Increased vulnerability to abuse when overloaded or under-inflated. The side-wall tends to bulge which could cause damage and puncture.

1.4.2 *Tyre Designation*

The tyre designation/classification information is moulded around the rim of the tyre, typically as shown in Fig. 1.13.

For example consider the designation 215/65 R 15 95 H where:

215 indicates the width of the tyre in mm.

65 provides the aspect ratio of the tyre wall to tyre width—in this case the wall height is $0.65 \times 215 = 139.75$ mm.

R Code for radial tyre.

15 Wheel rim diameter in inches.

95 Load index—ranges from 69 to 100. Load capacity depends on load index and inflation pressure. For example a load index of 95 at 2.9 bar pressure gives 690 kg load capacity. Tables are supplied that provide such information.

H Speed symbol. Tables are provided for speed index (e.g. H represents 210 kph).

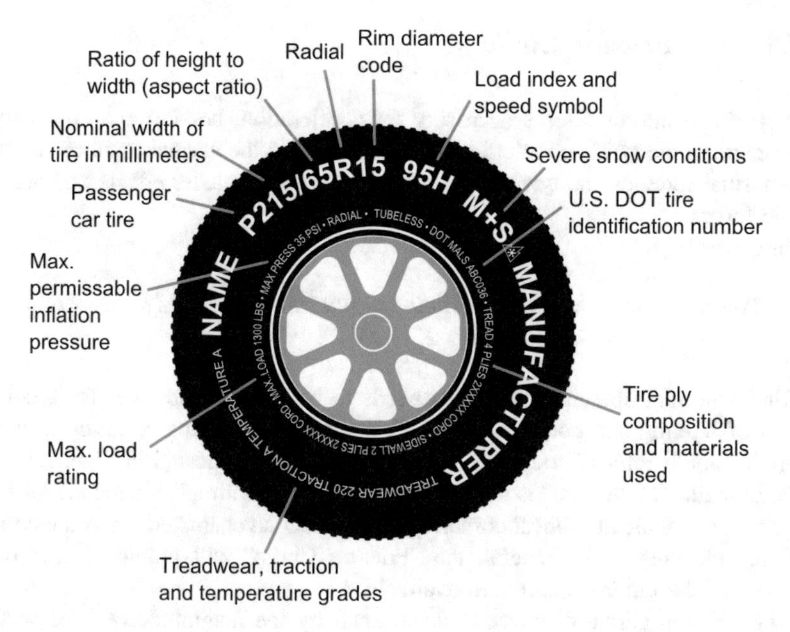

Fig. 1.13 Tyre designation. https://upload.wikimedia.org/wikipedia/commons/thumb/b/be/Tire_code_-_en.svg/288px-Tire_code_-_en.svg.png

The unladen tyre diameter D is equal to the wheel diameter + twice tyre wall height.

Thus, in the above case:

$$D = (15 \times 25.4) + 2(0.65 \times 215) = 660.5\ \text{mm}.$$

Note:

(1) The effective rolling radius of the tyre will be less than suggested by the unladen diameter due to deflections of the tyre under the normal wheel loads.
(2) The heavy weight of the tread, combined with the support belts and flexible tyre wall, may result in the tyre increasing in diameter at higher speeds. This may result in the vehicle travelling slightly faster than theoretical predictions—typically circa 6%.

1.4.3 The Friction Circle

The tyre/road interface force necessary for acceleration, braking and cornering is dependent on the tyre/road interface friction level and the normal force on the tyre. The normal force on the tyre should include any load transfer effects and aerodynamic forces.

In general:

$$Tyre/road\,force = Interface\,friction\,coefficient\ \times\ Normal\,force\,on\,tyre$$
$$F = \mu N$$

Under normal operation this force needs to be sufficient to cater for a combination of braking and cornering or acceleration and cornering resulting in a safe operating locus referred to as the "Friction Circle" as indicated in Figs. 1.14 and 1.15. In reality the "Circle" is more elliptical rather than circular as the lateral force is influenced by the non-linear cornering force/tyre load characteristic of the tyre for differing slip angles. In general, the "Friction Circle" will provide a first order estimate of the tyre/road adhesion available.

The Friction Circle diameter is determined by the magnitude of total vertical load (N) and the tyre/road interface friction coefficient (μ). This diameter represents the limit of tyre adhesion before gross sliding/spinning/lock-up of the wheel occurs. From Fig. 1.15, the resultant horizontal force acting at the interface is given by:

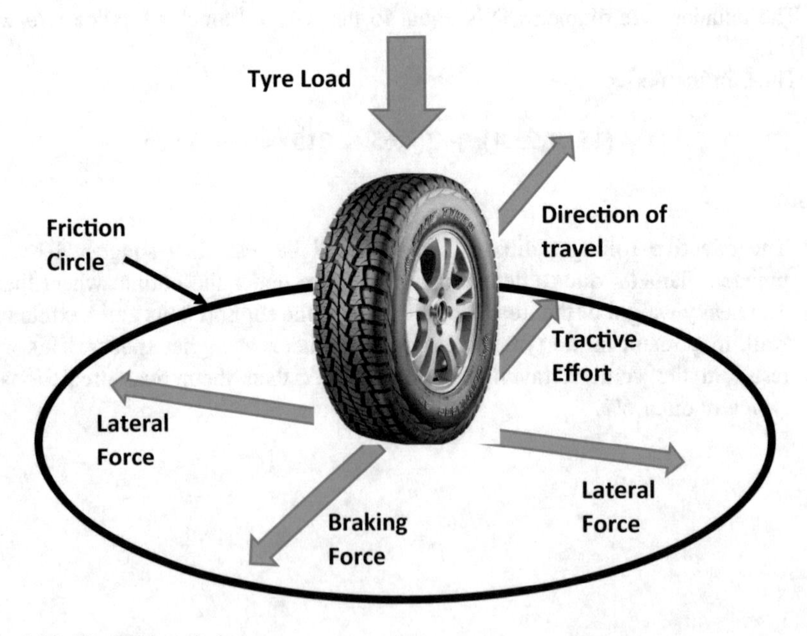

Fig. 1.14 The friction circle

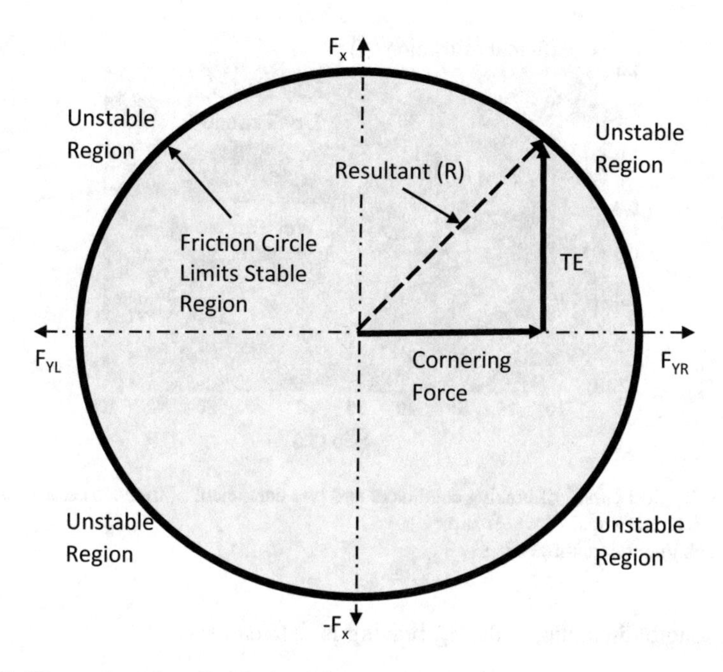

Fig. 1.15 The resultant force R calculated from the friction circle

$$R = \sqrt{F_x^2 + F_y^2} \tag{1.9}$$

where:

F_x Tractive Effort (+) or Braking Force (−)
F_y Cornering Force (which may include side wind forces)
R Resultant Force (should not exceed "μ N" for stability).

If a combination of longitudinal and lateral force (giving a resultant force R) exceeds N times μ then the tyre will slide or spin. The minimum tyre/road interface friction (μ) to avoid gross tyre slip for a particular resultant force is given by:

$$\mu = \frac{R}{N} \tag{1.10}$$

1.4.4 Limiting Frictional Force Available

Traction and braking force generated at the tyre/road contact patch are as a result of small amounts of slip occurring. For braking, this physically means that the actual forward velocity of the wheel is greater than the forward velocity of the wheel if it were free rolling at the same rotational speed.

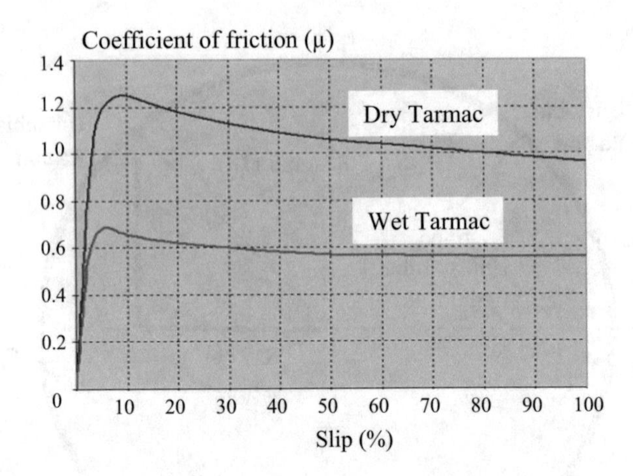

Fig. 1.16 Typical curves of braking coefficient and tyre coefficient of friction (i.e. braking force/wheel vertical load) against wheel slip.
http://i.stack.imgur.com/mbyPe.jpg

The longitudinal slip, s, during braking is defined as:

$$s = \frac{v - \Omega r}{v} \tag{1.11}$$

where:

v forward velocity of wheel
r rolling radius of tyre
Ω rotational speed of wheel.

Slip is often defined as a percentage:

$$s(\%) = \left(\frac{v - \Omega r}{v}\right) \times 100\% \tag{1.12}$$

Thus, slip ranges from zero when the wheel is free rolling to 100% when the wheel is locked but still sliding. A typical curve of the relationship between braking coefficient of friction and wheel slip is shown in Fig. 1.16.

The limiting frictional force available is normally associated with the sliding condition of the tyre when the wheel is locked up. In fact, this is not the condition under which maximum braking force is available which occurs at a slip of around 10% (see Fig. 1.16) but locked wheel sliding is the situation most commonly associated with panic braking. The public perception of this danger has diminished now that anti-lock braking systems (ABS) have become standard.

Young engineers often find confusion between rolling resistance forces, which are present whenever the vehicle is moving, as discussed in Sect. 1.3.2.3, and limiting frictional forces available during acceleration or braking, as discussed in the current Section. The two sets of coefficients however are almost two orders of magnitude different, typical values being:

$$\text{Rolling resistance coefficient} = 0.012{-}0.015$$

$$\text{Limiting tyre/road friction coefficient (on a good dry road)} = 0.9$$

The performance of a vehicle (acceleration) is generally based on the power available (power limited acceleration). As such the tyre/road normal interface force, along with the associated tyre/road friction coefficient (adhesion coefficient), determines whether that power may be used. As the tyre/road interface force depends on load transfer effects during acceleration (or braking), rather complex calculations aimed at determining the axle/wheel dynamic loads are necessary as outlined in the following sections.

1.5 Rigid Body Load Transfer Effects for Straight Line Motion

1.5.1 Vehicle Stationary or Moving at Constant Velocity on Sloping Ground

The general case of a vehicle on sloping ground is shown in Fig. 1.17. If the vehicle is stationary or moving at constant speed, then the forces on it can be analysed using the static equilibrium equations. Note that it has automatically been assumed in drawing the figure that the entire vehicle can be treated as a single lumped mass.

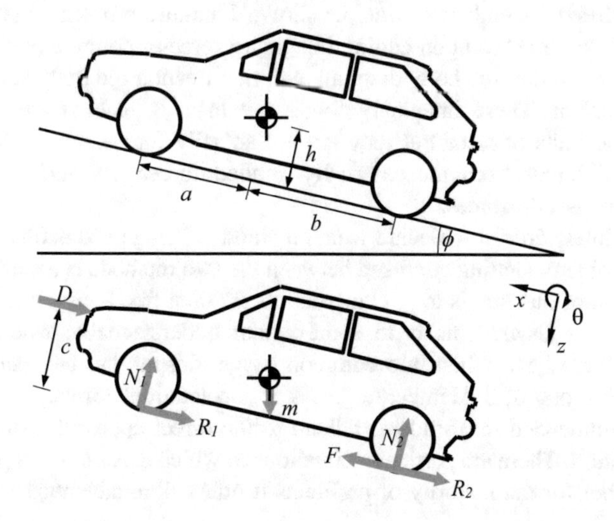

Fig. 1.17 FBD of vehicle on slope at constant speed

Note also that it is convenient to define the axis system relative to the sloping ground. The axle loads can now be found by taking moments about some point. e.g. the rear wheel contact patch:

$$N_1(a+b)+Dc+mg\,h\,\sin\,\phi-mg\,b\,\cos\,\phi=0 \tag{1.13}$$

This enables N_1 to be calculated. Then resolving perpendicular to the slope gives N_2:

$$mg\,\cos\,\phi-N_1-N_2=0 \tag{1.14}$$

If the vehicle is stationary, then the aerodynamic drag and rolling resistance forces would not be present. The force at the rear wheel contact point resulting from application of the parking brake would then equal the down slope component of gravity, $mg\,sin\,\phi$, in order to maintain static equilibrium.

1.5.2 Vehicle Accelerating/Decelerating on Level Ground

The analysis of the forces acting on the vehicle during acceleration or deceleration is slightly subtler than it first appears. In fact, it can be tackled in two ways using either Newtonian dynamics or the d'Alembert approach.

The Newtonian approach was outlined in Sect. 1.1. This is the preferred method of tackling dynamics problems. Note that if the body being analysed is not accelerating or decelerating then the $m\ddot{x}$ or $I\ddot{\theta}$ terms are zero and the equations automatically reduce to the equilibrium equations used for static analyses.

An alternative technique is d'Alembert's approach. There are occasions on which this has advantages but it is vital to understand the differences between the two approaches. D'Alembert's principle allows dynamics problems to be treated as statics problems. It relies upon adding imaginary forces—commonly referred to as *inertia forces*—to the free body diagram and then treating the body as if it were in static equilibrium. These imaginary forces are mass × acceleration ($m\ddot{x}$) terms; they have the units of force but they are not actually forces and so they are fundamentally different from the externally applied forces. As such they may be referred to as pseudo-forces.

It is absolutely crucial to decide which method is being used before starting any dynamics problem. Getting confused between the two methods is a common source of errors; a notorious trap is to put "inertia forces" on a free body diagram and then apply Newton's Laws. This is to some extent understandable when terms like *centrifugal* force have fallen into common usage, despite the fact that it is not a force at all but one of d'Alembert's "mass × acceleration" terms.

It is recommended that students adhere to the direct approach using Newton's Laws of Motion. There are particular situations in which d'Alembert's principle can be helpful, but for the majority of problems it offers little advantages.

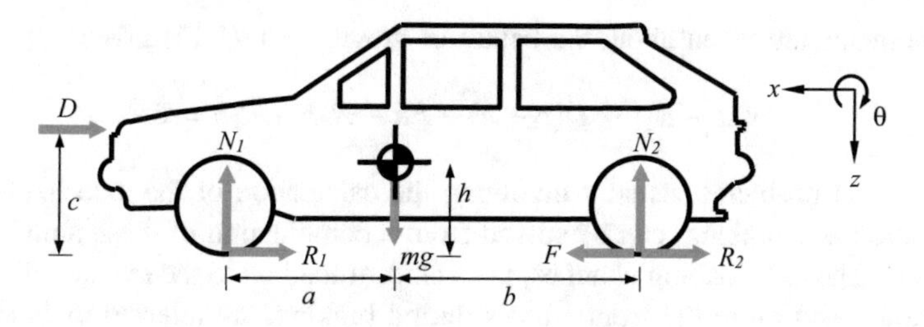

Fig. 1.18 Free body diagram of a rear wheel drive vehicle in motion

1.5.2.1 Newtonian Approach

The free body diagram for a vehicle accelerating on level ground is shown in Fig. 1.18 where:

- N_1 and N_2 are the interface loads between tyre and road
- F is the tractive effort (TE) available at the tyre road interface
- R_1 and R_2 are the rolling resistances at each axle
- D is the aerodynamic drag
- mg is the vehicle weight.

Applying the Rigid Body Laws for the vehicle centre of gravity:

$$\sum F_x = m\ddot{x} \tag{1.15}$$

$$\sum F_z = m\ddot{z} = 0 \text{ since } \ddot{z} \text{ is zero} \tag{1.16}$$

$$\sum M_G = I_G\,\ddot{\theta} = 0 \text{ since } \ddot{\theta} \text{ is zero} \tag{1.17}$$

where I_G = moment of inertia about C.G.

Equation (1.15) gives:

$$F - D - R_R = m\ddot{x} \tag{1.18}$$

Thus the tractive effort less any resistance to motion gives the force available to accelerate the vehicle.

Equation (1.16) as before gives:

$$mg - N_1 - N_2 = 0 \tag{1.19}$$

That is on a level road the total tyre/road interface forces must equal the vehicle weight.

If moments are taken about the centre of gravity, Eq. (1.17) gives:

$$N_1 a - R_1 h + D(c - h) + Fh - N_2 b - R_2 h = 0 \tag{1.20}$$

Numerical problems, usually involving the calculation of the axle loads under acceleration and braking, can be solved from a combination of these simultaneous equations. The axle loading changes, i.e. a shift of load on to the rear wheels during acceleration and on to the front wheels during braking, are referred to as *load (or weight) transfer effects*. The use of the term "*weight transfer*" is not encouraged since the actual weight (i.e. *mg*) of the vehicle does not change, nor does its centre of gravity, but the axle loads do vary dependent on the level of acceleration/deceleration.

Substituting for N_2 and F in Eq. (1.20) gives:

$$N_1 a - R_1 h + D(c - h) + (m\ddot{x} + D + R_R)h - (mg - N_1)b - R_2 h = 0 \tag{1.21}$$

But $R_1 + R_2 = R_R$ and Eq. (1.21) simplifies to:

$$m\ddot{x}h + N_1(a + b) - mgb + Dc = 0 \tag{1.22}$$

or:

$$N_1 = \frac{mgb - Dc - m\ddot{x}h}{(a + b)} \tag{1.23}$$

And therefore:

$$N_2 = \frac{mga + Dc + m\ddot{x}h}{(a + b)} \tag{1.24}$$

1.5.2.2 D'Alembert's Approach

The pseudo-force or d'Alembert force ($m\ddot{x}$) is added to the FBD as shown in Fig. 1.19 and then the problem may be treated as if it were a static rather than a dynamic problem.

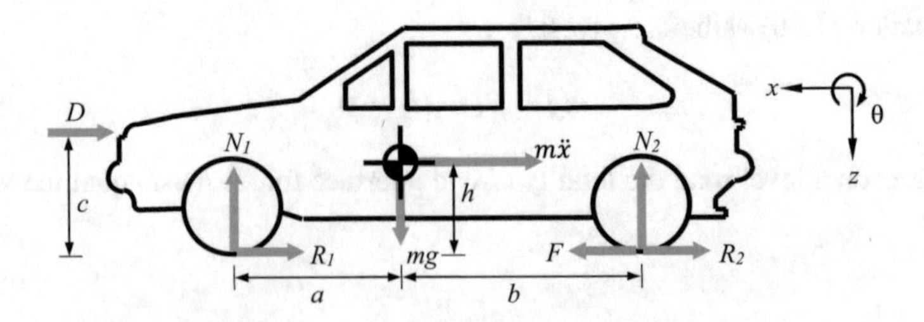

Fig. 1.19 Free body diagram for the d'Alembert approach (rear wheel drive)

The equilibrium equations can now be applied:

$$\sum F_x = 0 \tag{1.25}$$

$$\sum F_y = 0 \tag{1.26}$$

$$\sum M = 0 \tag{1.27}$$

The moment equation may be calculated around any point since the entire body is now treated as if it were in equilibrium. Note that this is not the case with the Newtonian approach where the moment equation may only be applied about the centre of gravity or a fixed point on the ground about which the body is constrained to rotate (e.g. at the pivot of a link).

This freedom to take moments about any point actually provides a slight advantage in solving this particular problem. Note that using the Newtonian approach, it is necessary to solve three simultaneous equations (Eqs. 1.22–1.24) to obtain the axle loads N_1 and N_2. Using the d'Alembert approach these can be obtained directly by, for example, taking moments about the front wheel:

$$m\ddot{x}h + mga - N_2(a+b) + Dc = 0 \tag{1.28}$$

From which:

$$N_2 = \frac{mga + Dc + m\ddot{x}h}{(a+b)} \tag{1.29}$$

Taking moments about the rear wheel gives:

$$N_1(a+b) + m\ddot{x}h - mgb + Dc = 0 \tag{1.30}$$

From which:

$$N_1 = \frac{mgb - Dc - m\ddot{x}h}{(a+b)} \tag{1.31}$$

It will be noticed again that the $m\ddot{x}h$ *and* Dc terms are positive for the rear axle and negative for the front axle. These terms represent the load transfer effect.

Hence, for this particular problem of the accelerating car, d'Alembert's approach has the benefits of being quicker and more efficient in terms of time in obtaining numerical solutions. The principal cause for concern is that the direction of the pseudo-force needs to be known. If this force is applied incorrectly the calculations would be misleading.

1.5.3 Rear Wheel, Front Wheel and Four Wheel Drive Vehicles

Although a vehicle's performance may not be limited by power it may be limited by the tractive effort which a vehicle may develop before wheel-slip occurs. That is the maximum transmission torque which the drivetrain may usefully transmit between tyre and ground and is directly related to the tyre/road interface load and the tyre/road interface friction (adhesion) coefficient. Since it has been shown that the load transfer effect causes the tyre/road interface load to vary during acceleration (or braking) then the added complication of a caravan or trailer may influence the choice of drive for the vehicle (see Sect. 1.5.4).

In general we have:

$$F = \mu_i N_i \tag{1.32}$$

where "*F*" is the tractive effort demand and N_i is the normal load on the *i*th axle.

Re-arranging gives the so-called friction demand or required adhesion coefficient μ_i to avoid wheel spin on that axle:

$$\mu_i = \frac{F}{N_i} \tag{1.33}$$

1.5.3.1 Rear Wheel Drive Vehicle (See Fig. 1.19)

From Eq. (1.31):

$$N_2 = \frac{mga + Dc + m\ddot{x}h}{(a+b)}$$

Using Eq. (1.33) this gives the friction demand at the rear wheels:

$$\mu_r = \frac{F(a+b)}{mga + Dc + m\ddot{x}h} \tag{1.34}$$

1.5.3.2 Front Wheel Drive Vehicle (See Fig. 1.20)

From Eq. (1.31):

$$N_1 = \frac{mgb - Dc - m\ddot{x}h}{(a+b)}$$

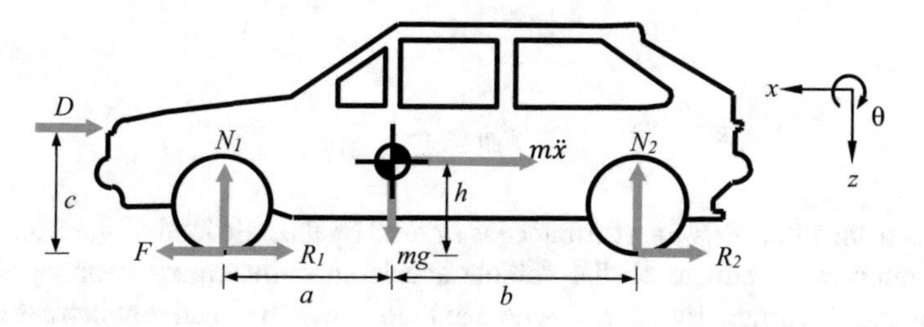

Fig. 1.20 General free body diagram for front wheel drive (d'Alembert approach)

Using Eq. (1.33) this gives the friction demand at the front wheels:

$$\mu_f = \frac{F(a+b)}{mgb - Dc - m\ddot{x}h} \tag{1.35}$$

1.5.3.3 Four Wheel Drive Vehicle (See Fig. 1.21)

Now all wheels are driven, thus:

$$F = F_f + F_r = \mu_{all}(N_1 + N_2)$$

where μ_{all} is the friction demand at all 4 wheels.

Hence:

$$F = \mu_{all}\left\{\left(\frac{mgb - Dc - m\ddot{x}h}{(a+b)}\right) + \left(\frac{mga + Dc + m\ddot{x}h}{(a+b)}\right)\right\}$$

Simplifying gives:

$$F = \mu_{all}\left(\frac{mg(a+b)}{(a+b)}\right) = \mu_{all}mg$$

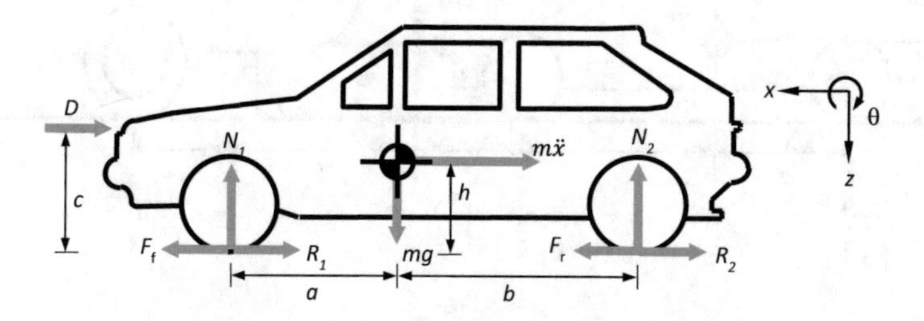

Fig. 1.21 General free body diagram for four-wheel drive (d'Alembert approach)

Or

$$\mu_{all} = \frac{F}{mg} \tag{1.36}$$

Thus front wheel drive performance is limited by the axle load of the front wheels which effectively reduces as the vehicle acceleration increases. Rear wheel drive performance is limited by the rear wheel axle load which effectively increases as the vehicle acceleration increases. With 4-wheel drive it is assumed that the full weight of the vehicle may be utilised and that the weight transfer effect plays no part in the performance of a vehicle. In practice the transfer box on a 4-wheel drive vehicle will usually be designed to split the transmission torque between front and rear axle in proportion to the anticipated axle loads of the vehicle so that the full vehicle weight may be used. If this "balance" is not maintained, possibly because of the addition of a trailer, then the full weight of the vehicle may not be employed and the traction limit will be controlled by the axle load most below the design load for the particular tyre/road adhesion coefficient. However modern 4-wheel drive vehicles employ a traction control system to distribute the torque most effectively.

1.5.4 Caravans and Trailers

From Fig. 1.22 it will be noticed that there are 2 additional forces when a caravan or trailer is coupled to a vehicle, the longitudinal force (T), known as the drawbar pull which is zero under static conditions, and the vertical hitch load (N_2). The car now has 4 unknowns and the caravan (trailer) has 3 unknowns. Under normal conditions, the vertical hitch load will be positive (downwards) on the vehicle and

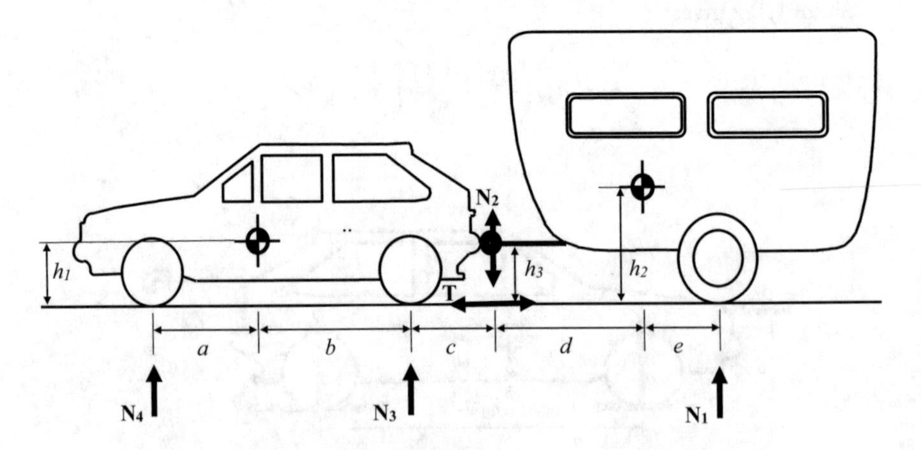

Fig. 1.22 General loads on a car/caravan combination

negative (upwards) on the trailer. Also, during acceleration, the drawbar pull will be negative (rearward) on the vehicle and positive (forward) on the trailer.

A single axle trailer will normally increase the rear axle load under static conditions. During acceleration this hitch load will reduce if sufficiently high accelerations are experienced. Combined with poor trailer load distribution, this hitch load may in fact become negative with a subsequent reduction in rear axle loading and effective increase in front axle loading. In a dynamic situation, when the combination is moving, the aerodynamic drag forces may also play an important role and then the trailer height will reflect on the hitch loads and subsequently the axle loads.

Example E1.1

With reference to Fig. 1.22, consider a car/caravan combination with the following data:

Car mass, m_1	1425 kg
Caravan mass, m_2	1191 kg
Distances	
a	1.318 m
b	1.344 m
c	1.100 m
d	2.959 m
e	0.230 m
h_1	0.6 m
h_2	0.9 m
h_3	0.5 m

Calculate the following:

(a) The static axle loads on level ground
(b) The hitch loads when the combination is accelerating at 2 m/s^2 at low forward speed (such that aerodynamic forces may be ignored) and for a rolling resistance coefficient of 0.012.

(a) **Static loads on level ground**

Under static conditions, both acceleration ($\ddot{x}$) and T = 0.
As before we have 3 equations of motion to solve as defined in Eqs. (1.15–1.17).
First consider the caravan only (see Fig. E1.1a) as car has too many unknowns to resolve

Taking moments about the tyre contact patch for the caravan only:

$$N_2(d+e) = m_2g \times e$$

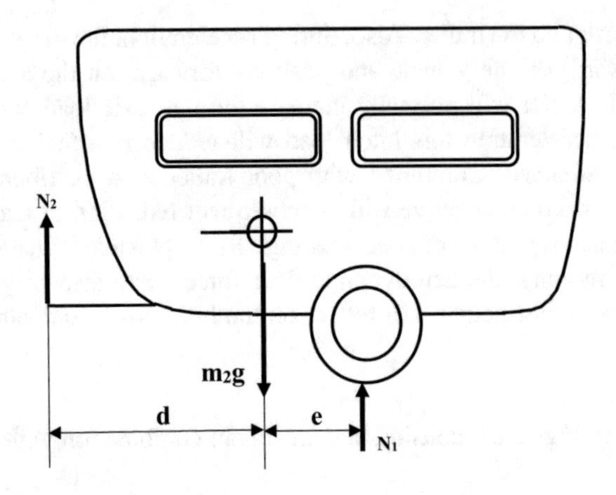

Fig. E1.1a Free body diagram for caravan only

Giving:

$$N_2 = \frac{1191 \times 9.81 \times 0.230}{(2.959 + 0.23)} = 843\,N$$

But resolving vertically:

$$m_2g = \mathrm{N_1} + \mathrm{N_2}$$

Giving

$$N_1 = (1191 \times 9.81) - 843 = 10841\,N$$

Now consider car as the vertical hitch load is now known as shown in Fig. E1.1b

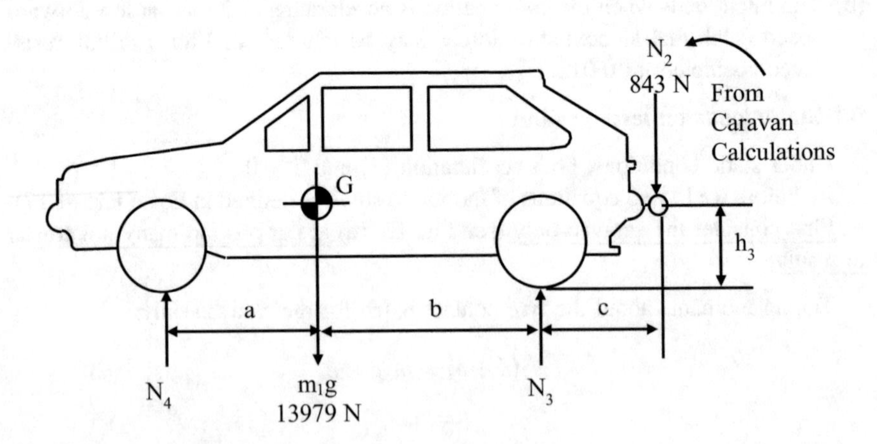

Fig. E1.1b Car only with load transferred from caravan

Resolving vertically:

$$\sum F_z = m\ddot{z} = 0$$

$$m_1 g + N_2 = (N_3 + N_4)$$

$$1425 \times 9.81 + 843 = 14822 = N_3 + N_4$$

Taking moments about front wheel patch:

$$N_3(a+b) = m_1 g a + N_2(a+b+c)$$

So

$$N_3(1.318 + 1.344) = (13979 \times 1.318) + 843(1.318 + 1.344 + 1.100)$$

Giving:

$$N_3 = 8111\,N$$

And therefore:

$$N_4 = 14822 - 8111\,N = 6711\,N$$

(b) **Dynamic loads under following conditions**:

Acceleration = 2 m/s^2
Rolling Resistance Coefficient = 0.012
Low speed therefore Aerodynamic Drag = 0
Consider caravan first—as car has 4 unknowns whilst caravan has only 3 (see Fig. E1.1c)

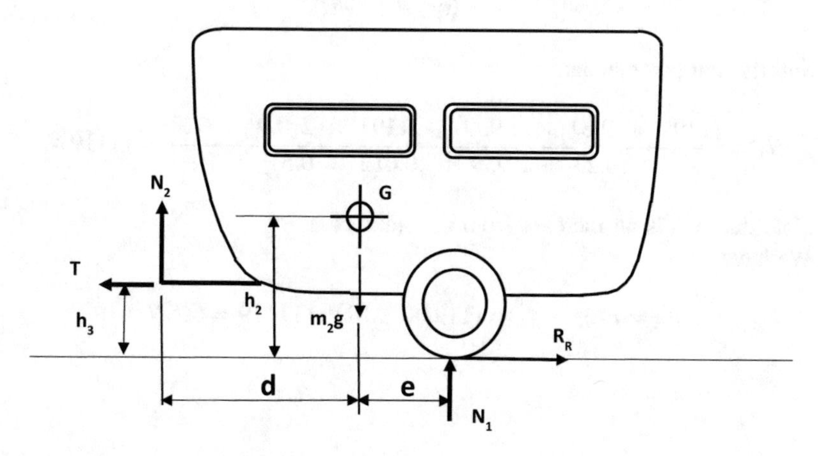

Fig. E1.1c Consideration of dynamic loads on caravan

Resolving horizontally:

$$\sum F_x = m_2\ddot{x} = T - R_R = T - C_R N_1$$

Therefore:

$$T = m_2\ddot{x} + C_R N_1$$

Resolving vertically:

$$m_2 g = N_1 + N_2$$

Therefore:

$$N_2 = m_2 g - N_1$$

Moments about C.G.:

$$N_2 d + T(h_2 - h_3) = N_1 e + R_R h_2 = N_1 e + C_R N_1 h_2$$

Substituting for T & N_2 gives:

$$(m_2 g - N_1)d + (m_2\ddot{x} + C_R N_1)(h_2 - h_3) = N_1 e + C_R N_1 h_2$$

which reduces to:

$$N_1(e + d + c_R h_3) = m_2 g d + m_2\ddot{x}(h_2 - h_3)$$

Therefore:

$$N_1 = \frac{m_2 g d + m_2\ddot{x}(h_2 - h_3)}{(e + d + c_R h_3)}$$

Substituting parameters:

$$N_1 = \frac{(1191 \times 9.81 \times 2.959) + 1191 \times 2(0.9 - 0.5)}{(0.23 + 2.959 + (0.012 \times 0.5))} = 11119N$$

Note that this is an increase from the static case.
We have:

$$N_2 = m_2 g - N_1 = (1191 \times 9.81) - 11119 = 565N$$

Note that this is a reduction from the static case.
Finally:

$$T = m_2\ddot{x} + c_R N_1 = (1191 \times 2) + (0.012 \times 11119) = 2525\,N$$

Now consider car only (see Fig. E1.1d)

Resolving horizontally:

$$\sum F_x = m_1\ddot{x} = TE - T - R_R$$

Giving:

$$TE = m_1\ddot{x} + T + R_R$$

Resolving vertically:

$$\sum F_z = 0 \quad \therefore m_1 g + N_2 = N_3 + N_4$$

Giving:

$$N_4 = m_1 g + N_2 - N_3$$

Moments about C.G.:

$$N_4 a + N_2(b+c) + TEh_1 = N_3 b + R_R h_1 + T(h_1 - h_3)$$

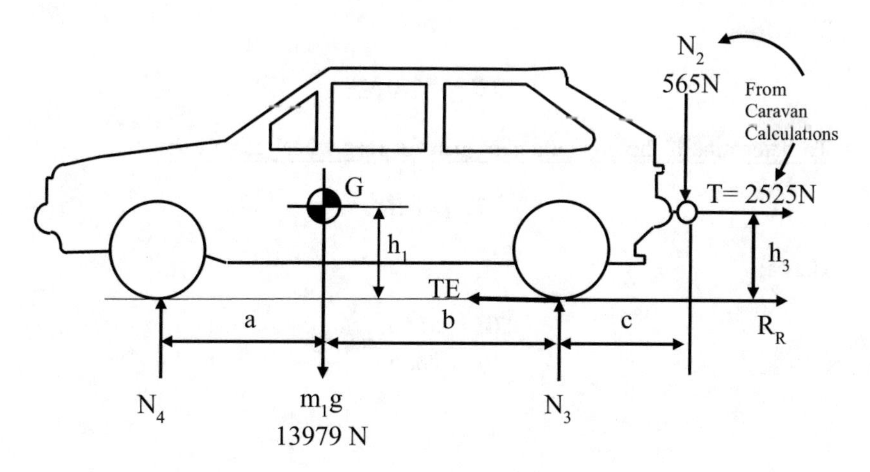

Fig. E1.1d Dynamic loads on car with forces from caravan included

Substituting for *TE* & N_4 and simplifying gives:

$$(m_1 g + N_2 - N_3)a + N_2(b + c) + (m_1\ddot{x} + T + R_R)h_1 = N_3 b + R_R h_1 + T(h_1 - h_3)$$

Rearranging gives:

$$N_3(a + b) = (m_1 g + N_2)a + N_2(b + c) + m_1\ddot{x}h_1 + Th_3$$

Substituting parameters:

$$N_3 = \frac{(1425 \times 9.81 + 565)\,1.318 + 565\,(1.344 + 1.100) + (1425 \times 2 \times 0.6) + (2525 \times 0.5)}{(1.318 + 1.344)}$$

Giving:

$$N_3 = 8836\,N$$

Note—Greater than static case
Hence:

$$N_4 = (1425 \times 9.81) + 565 - 8836$$
$$N_4 = 5707\,N$$

And

$$TE = m_1\ddot{x} + T + R_R = m_1\ddot{x} + T + c_R(N_3 + N_4)$$
$$TE = 1425 \times 2 + 2525 + 0.012\,(8836 + 5707)$$

Therefore:

$$TE = 5550\,N$$

To determine if there is sufficient grip for rear wheel drive car:

$$TE = \mu N_3$$

Hence:

$$\mu = \frac{TE}{N_3} = \frac{5550}{8836} = 0.63$$

This level of friction demand (adhesion coefficient) should be exceeded under most road conditions. Therefore rear wheel drive car is suitable for this application.

It is suggested to the reader to repeat the above example using d'Alembert approach.

1.6 Rigid Body Load Transfer Effects During Cornering

It has been shown that the longitudinal load transfer effect (LTE) due to acceleration/deceleration is given as:

$$LTE_{\ddot{x}} = \pm \frac{m\ddot{x}h}{(a+b)}$$

and that due to aerodynamic drag as:

$$LTE_D = \pm \frac{Dc}{(a+b)}$$

The positive sign in the above equations relates to the rear axle and negative sign to the front axle. Since the centre of gravity is generally along the central axis of the vehicle then the longitudinal load transfer effect may be considered as equal for each wheel on a single axle.

During cornering there is a lateral load transfer effect due to the cornering, i.e. centrifugal, forces. This lateral load transfer effect tends to increase the load on the outer wheels and reduce the load on the inner wheels. As the centre of gravity is generally not centrally displaced along the longitudinal axis of the vehicle, there is a need to proportion the cornering load transfer effect to the front and rear wheels—see Fig. 1.23.

We have:

$$Centrifugal\,force\,(CF) = \frac{mv^2}{R}$$

where m is vehicle mass, v is forward velocity of vehicle and R is the radius of the turn.

Thus the lateral load transfer effect is given by:

$$LTE_{CF} = CF\frac{h}{T}$$

where h is the height of the C.G. and T is the wheel track width.

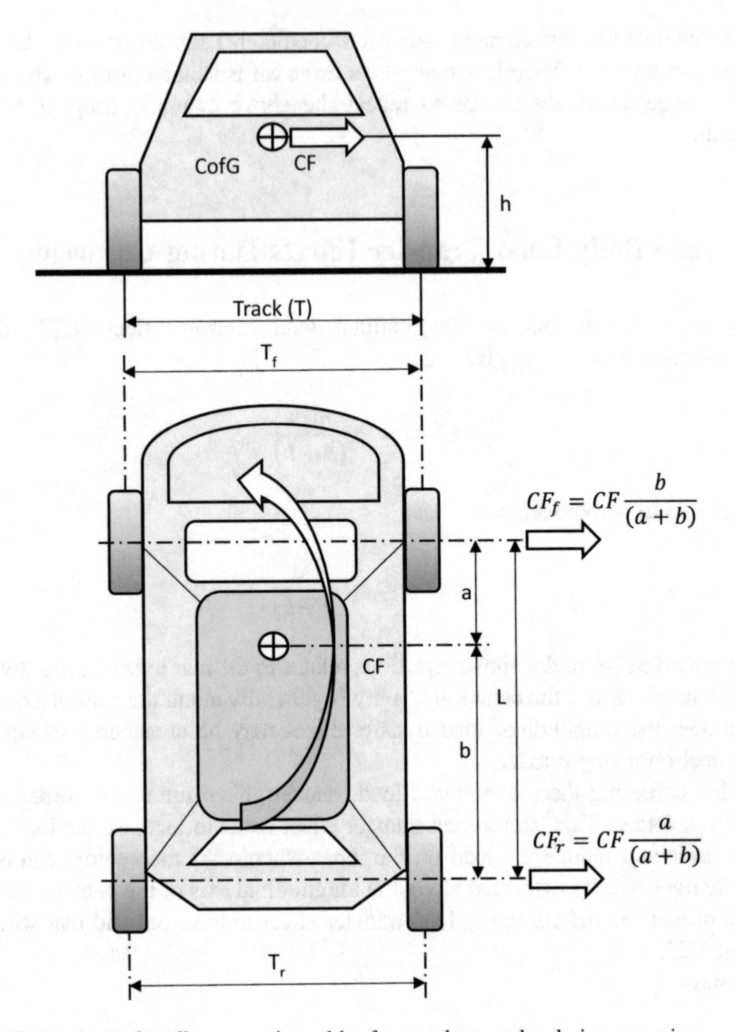

Fig. 1.23 Load transfer effects experienced by front and rear axles during cornering

More precisely, proportioned to front and rear axles:

$$LTE_f = CF_f\frac{h}{T_f} = CF\frac{b}{(a+b)} \times \frac{h}{T_f}$$

$$LTE_r = CF_r\frac{h}{T_r} = CF\frac{a}{(a+b)} \times \frac{h}{T_r}$$

Table 1.3 Dynamic tyre loads on each wheel for combined braking and cornering

Front left wheel	Load N_{1L}	Front right wheel	Load N_{1R}
Static	$+\frac{mg}{2}\left(\frac{b}{a+b}\right)$	Static	$+\frac{mg}{2}\left(\frac{b}{a+b}\right)$
Braking LTE	$+\frac{m\ddot{x}h}{2(a+b)}$	Braking LTE	$+\frac{m\ddot{x}h}{2(a+b)}$
Drag LTE	$-\frac{Dc}{2(a+b)}$	Drag LTE	$-\frac{Dc}{2(a+b)}$
Cornering LTE	$-CF\frac{b}{(a+b)}\times\frac{h}{T_f}$	Cornering LTE	$+CF\frac{b}{(a+b)}\times\frac{h}{T_f}$
Rear left wheel	**Load N_{2L}**	**Rear right wheel**	**Load N_{2RL}**
Static	$+\frac{mg}{2}\left(\frac{a}{a+b}\right)$	Static	$+\frac{mg}{2}\left(\frac{a}{a+b}\right)$
Braking LTE	$-\frac{m\ddot{x}h}{2(a+b)}$	Braking LTE	$-\frac{m\ddot{x}h}{2(a+b)}$
Drag LTE	$+\frac{Dc}{2(a+b)}$	Drag LTE	$+\frac{Dc}{2(a+b)}$
Cornering LTE	$-CF\frac{a}{(a+b)}\times\frac{h}{T_r}$	Cornering LTE	$+CF\frac{a}{(a+b)}\times\frac{h}{T_r}$

As the centre of gravity is assumed to be centrally displaced between the axles, and the longitudinal LTE has already been proportioned front/rear, then the front/rear lateral LTEs as calculated above are added to the outer wheel and deducted from the inner wheel. It is now possible to determine the load on each wheel.

By example, the individual wheel LTE terms for a vehicle braking during a left-hand turn are as shown in Table 1.3. It will be noted from Table 1.3 that summing all the individual loads for each wheel leads to the vehicle weight "mg".

To determine the lateral adhesion (friction) demands on each wheel is difficult but generally follows the usual calculation:

$$\textit{Lateral tyre/road interface force demand}\ (F) = \mu N \quad or \quad \mu = \frac{F}{N}$$

As an example, the tyre/road interface demand at each wheel is determined as follows under firstly steady state cornering (no braking or acceleration) and secondly, under cornering with acceleration/braking.

1.6.1 Steady State Cornering

Proportioning the centrifugal force to front axle:

$$CF_f = CF\frac{b}{(a+b)}$$

Proportion CF_f to each wheel based on ratio of individual tyre loading to axle load:

$$CF_{fL} = CF_f \frac{N_{1L}}{(N_{1L} + N_{1R})}$$

where subscript f & 1 refers to front wheels, L to left hand wheel & R to right hand wheel.

It is now assumed that the ability of each wheel to react against the centrifugal force will be dependent on the individual loads on that wheel.

Hence:

$$\mu_{fL} = \frac{F}{N} = \frac{CF_{fL}}{N_{1L}} = \frac{CF_f}{(N_{1L} + N_{1R})} = \frac{CF_f}{Dynamic\, front\, axle\, load} = \mu_{fR}$$

Thus the friction demand on the front left wheel is equal to that on the front right wheel.

Similarly:

$$CF_r = CF \frac{a}{(a+b)}$$

Giving:

$$\mu_{rL} = \frac{CF_{rL}}{N_{2L}} = \frac{CF_r}{(N_{2L} + N_{2R})} = \frac{CF_r}{Dynamic\, rear\, axle\, load} = \mu_{rR}$$

Thus the friction demand on the rear left wheel is equal to that on the rear right wheel.

From the above it is seen that for **steady state cornering** each wheel will demand equal adhesion on the front axle and equal adhesion on the rear axle. This means that tyre slip will be axle based rather than wheel based. As such the loss of control will be when the axle demanding the highest adhesion exceeds the friction limit for the particular tyre/road conditions. Thus the tyre characteristics play an important role in this scenario with the tyre load/slip angle characteristic determining over/under steer.

1.6.2 Non-steady State Cornering

When acceleration or braking is included, the resultant tyre/road interface force needs to be determined. The centrifugal force is a body induced force (acting at the C.G.) and could therefore be proportioned to each wheel dependant on axle and wheel loading. Braking and traction cannot be so proportioned as the torque is an input to the wheel and the motive force acts at the tyre/road interface.

Unless an anti-lock braking system (ABS) is deployed, the braking force for each wheel on the front axle will be the same, and the braking force for each wheel on the rear axle will be the same. With traction, the force per wheel will vary due to the differential and possibly yaw control on the driven axle. In general, the wheel forces are not influenced by tyre loading until the required adhesion is exceeded and the wheels will then either lock-up or slide.

With reference to Figs. 1.14 and 1.15, the resultant force in the case of combined centrifugal force (CF) and braking force (BF) is as follows:

$$Resultant = \sqrt{CF^2 + BF^2}$$

For example, the resultant force on the front left-hand wheel is given by:

$$Resultant_{fL} = \sqrt{\left(CF_f \frac{N_{1L}}{(N_{1L} + N_{1R})}\right)^2 + (BF_{fL})^2}$$

Giving:

$$\mu_{fL} = \frac{F}{N} = \frac{Resultant_{fL}}{N_{1L}}$$

Similar analyses will apply for the remaining three wheels.

Example E1.2

A car shown in Fig. E1.2 is negotiating a **left-hand corner** when it **decelerates** under the following conditions:

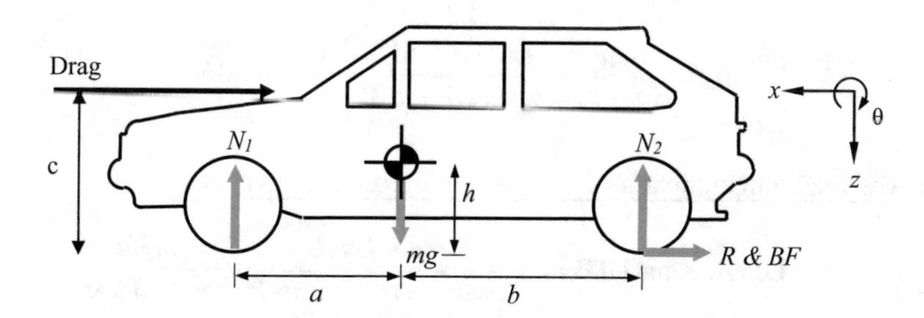

Fig. E1.2 Car undergoing cornering and braking

Speed	25 m/s
Deceleration	3.1 m/s^2
Braking force distribution front/rear	Proportioned 60% front 40% rear
Radius of corner	250 m
Rolling resistance coefficient	0.012
Drag coefficient	0.3
Density of air	1.22 kg/m^3
Frontal area	2 m^2
Mass of car	1500 kg
Equivalent mass of rotational parts	200 kg
Rolling radius of wheel	310 mm
Wheel track	1600 mm
a = 1100 mm	b = 1200 mm
c = 650 mm	h = 750 mm

Assuming both lateral cornering forces and vertical forces are proportioned according to longitudinal position of the centre of gravity, calculate:

(a) The total longitudinal load transfer during deceleration.
(b) Proportioned front and rear lateral load transfer during cornering.
(c) The road/tyre interface load of each wheel during this manoeuvre.
(d) The minimum coefficient of friction necessary for ALL wheels considering both braking and cornering.

Solution:

Firstly calculate the static axle loads:

$$N_1 = \frac{mgb}{(a+b)} = \frac{1500 \times 9.81 \times 1.2}{(1.1+1.2)} = 7677\,\mathrm{N}$$

$$N_2 = \frac{mga}{(a+b)} = \frac{1500 \times 9.81 \times 1.1}{(1.1+1.2)} = 7038\,\mathrm{N}$$

(a) The total longitudinal load transfer during deceleration.

$$\text{Longitudinal LTE} = \frac{\mathit{Force} \times \mathit{height}}{\mathit{Distance\ between\ wheels}} = \frac{m\ddot{x}h}{\mathit{wheelbase}}$$

$$= \frac{1500 \times 3.1 \times 0.75}{2.3} = 1516\,\mathrm{N}$$

(b) Proportioned front and rear lateral load transfer during cornering.

Total lateral force is given by:

$$Centrifugal\,force(CF) = \frac{mv^2}{R} = \frac{1500 \times 25^2}{250} = 3750\,\mathrm{N}$$

$$\text{Lateral LTE} = \frac{Force \times height}{Distance\ between\ wheels} = \frac{F_L h}{track} = \frac{3750 \times 0.75}{1.6} = 1758\,\mathrm{N}$$

Lateral LTE proportioned according to the location of the C.G. gives:

$$\text{Front LTE} = \frac{Force \times b}{Wheelbase} = \frac{1758 \times 1.2}{2.3} = 917\,\mathrm{N}$$

$$\text{Rear LTE} = 1758 - 917 = 841\,\mathrm{N}$$

(c) The road/tyre interface load of each wheel during this manoeuvre.

We have $Aerodynamic\,Drag = \frac{1}{2}\rho C_D A v^2 = \frac{1}{2} 1.22 \times 0.3 \times 2 \times 25^2 = 228.75\,\mathrm{N}$
Therefore

$$Drag\ LTE = \frac{Force \times height}{Distance\ between\ wheels} = \frac{D\,c}{wheelbase} = \frac{228.75 \times 0.65}{2.3} = 64.6\,\mathrm{N}$$

The above static wheel loads and LTE's are summated in Table 1.4 to give the total vertical load on each wheel.

(d) The minimum coefficient of friction necessary for ALL wheels considering both braking and cornering.

$$\text{Lateral Force (LF) at front axle} = \frac{CF \times b}{Wheelbase} = \frac{3750 \times 1.2}{2.3} = 1957\,\mathrm{N}$$

Giving LF at rear axle = 3750–1957 = 1793 N
Proportion LF to each wheel according to the wheel loads calculated in Table 1.4:
Front Left LF = 1957 $\left(\frac{3648}{9130}\right)$ = 782 N giving
Front Right LF = 1957 – 782 = 1175 N
Rear Left LF = 1793 $\left(\frac{1953}{5578}\right)$ = 628 N giving
Rear Right LF = 1793 – 628 = 1165 N
Longitudinal equation of motion for vehicle:

$$(m + m_{eq})\,\ddot{x} = (Braking\,force + Rolling\ resistance + Drag)$$

Table 1.4 Dynamic tyre loads on each wheel for worked example E1.2

	Front left (N)		**Front right (N)**
Static N_1	+7677/2 = +3839	Static N_1	3839
Long LTE	1516/2 = 758	Long LTE	758
Lat LTE	−917	Lat LTE	+917
Drag LTE	−64.6/2 = −32	Drag LTE	−32
Total	3648	Total	5482
Total front axle load	3468 + 5482 = 9130 N		
	Rear left (N)		**Rear right (N)**
Static N_2	7038/2 = 3519	Static N_2	3519
Long LTE	−758	Long LTE	−758
Lat LTE	−841	Lat LTE	+841
Drag LTE	+32	Drag LTE	+32
Total	1953		3634
Total rear axle load	1953 + 3634 = 5587 N		
Check total load	9130 + 5587 = 14717 N—OK		

Therefore, total braking force (BF) given by:

$$BF = 3.1(1500 + 200) - (1500 \times 9.81 \times 0.012) - 228.75 = 4865\ \text{N}$$

Based on the given brake force distribution (60:40), BF at each front wheel = 0.6 × 4865/2 = 1460 N and BF at each rear wheel = 0.4 × 4865/2 = 973 N

In general, resultant horizontal interface force given by:

$$RF = \sqrt{LF^2 + BF^2}$$

Therefore:

$$\text{Front Left Wheel} \quad \text{RF} = \sqrt{782^2 + 1460^2} = 1656\ \text{N}$$

$$\text{Front Right Wheel} \quad \text{RF} = \sqrt{1175^2 + 1460^2} = 2501\ \text{N}$$

$$\text{Rear Left Wheel} \quad \text{RF} = \sqrt{628^2 + 973^2} = 1158\ \text{N}$$

$$\text{Rear Right Wheel} \quad \text{RF} = \sqrt{1165^2 + 973^2} = 1518\ \text{N}$$

Remembering that at the limit F = μN, we can calculate the friction demand (adhesion coefficient) at each wheel as follows.

$$\begin{aligned}
&\text{Front Left Wheel :} && \mu = 1656/3648 = 0.454 \\
&\text{Front Right Wheel :} && \mu = 2505/5482 = 0.457 \\
&\text{Rear Left Wheel :} && \mu = 1158/1953 = 0.593 \\
&\text{Rear Right Wheel :} && \mu = 1518/3634 = 0.418
\end{aligned}$$

It is seen the friction demands are now different for each wheel. From the above, the rear left wheel demands the highest adhesion coefficient so will be the first to experience problems. However all the above friction demands are readily achievable under normal road conditions.

1.7 Concluding Remarks

This chapter has introduced basic topics surrounding the mechanics of road vehicles, knowledge of which is necessary to understand and appreciate the more detailed material contained in the remaining chapters of the book. Particular attention has been paid to the importance of the tyres as the principal means of generating sufficient traction/braking and cornering force to control the vehicle in a safe and reliable manner. The importance of being able to calculate wheel loads allowing for load transfer effects has been emphasised using rigid body assumptions. In Chap. 3 of the book we will see that more accurate calculations require the flexibilities within the vehicle body, primarily in the suspension system, to be taken into account. Nonetheless the above theory should enable the design engineer to make reliable estimations of the wheel loads and total tyre/road interface forces on all wheels during both straight line and non-steady state cornering conditions. In addition, it should be possible to predict conditions when the calculated adhesion limits may be exceeded and vehicle handling becomes a problem. Modern vehicles of course employ sophisticated control systems such as ABS to monitor wheel loads and then proportion braking or traction accordingly with the aim that the friction demands should be equal for all wheels.

Chapter 2
Steering Systems

Abstract The primary purpose of this chapter is to provide the automotive engineer with a basic understanding of steering systems. It reviews current and modern designs with accompanying theories regarding the forces within the systems, both static and dynamic. It considers power assistance and the move towards steer-by-wire as full electronic control becomes increasingly reliable. The theory is supported by worked examples and "case studies" that more readily demonstrate their application. As such the engineer should be in a position to develop appropriate in-house computer-aided design packages that meet a specific need without unnecessary peripheral features. The chapter begins by introducing requirements and regulations that govern the basic design strategies. It continues with consideration of steering geometry and current common designs. The forces imposed on the steering system, both for a stationary and moving vehicle, are introduced along with the move towards electric steering systems. The chapter concludes by considering four wheel steer and additional steering assistance.

2.1 General Aims and Functions

Within the automotive industry there is no stand-alone subject area as all systems interact with each other, the design decisions in one area impacting on that of another system. As such steering and suspension systems should be developed with close co-operation between the respective designers. As always, safety plays a significant role in any design and in particular in steering systems since drivers interact with the steering whereas they do not generally have control of the suspension system. Ride comfort and road handling are therefore regarded as a more of a quality and refinement issue. Because of the safety implications and driver interface, steering design is guided by international regulation with which all OEMs must comply. These guidelines are outline at the beginning of the next section.

The basic aim of a steering system is to rotate the steered wheels in response to driver input and thus provide overall directional control of the vehicle.

© Springer International Publishing AG 2018

D. C. Barton and J. D. Fieldhouse, *Automotive Chassis Engineering*,
https://doi.org/10.1007/978-3-319-72437-9_2

In order to do this the function of an effective steering system must be to:

- Provide a reliable connection system between the steered wheels and the driver's steering wheel input.
- Provide controlled kinematic relationships to achieve the correct steer angles of both inner and outer wheels.
- Minimise steering effort yet maintain driver "feel".

In all cases the steering system employed must be safe and must comply with the current regulations as outlined below.

2.2 Steering Requirements/Regulations

Reference should be made to:

E/ECE/324 E/ECE/TRANS/505} Rev.1/Add.78/Rev.2-21 April 2005

"CONCERNING THE ADOPTION OF UNIFORM TECHNICAL PRESCRIPTIONS FOR WHEELED VEHICLES, EQUIPMENT AND PARTS WHICH CAN BE FITTED AND/OR BE USED ON WHEELED VEHICLES AND THE CONDITIONS FOR RECIPROCAL RECOGNITION OF APPROVALS GRANTED ON THE BASIS OF THESE PRESCRIPTIONS." Addendum 78: Regulation No. 79—Revision 2

2.2.1 General Requirements

- The actuating force must be harmonious to the centre as far as the end stop and must not decrease. The reason is to ensure the driver is always in control of the input.
- It must be possible to drive the vehicle accurately, i.e. without any unusual steering corrections.
- Play in the mechanical parts is impermissible. This means that if the driver reverses direction there is no "backlash" or "take-up" in the system.
- The entirety of the mechanical transmission devices must be able to cope with all loads and stresses occurring in operation.
- Unusual driving manoeuvres, such as driving over obstacles and accident-like occurrences must not lead to any cracks or breakages. A crack is possibly the worst scenario as this may propagate and fail at some later time.
- If the actuating force at the steering wheel for a normal passenger car exceeds 150 N, power assistance is necessary. If such power assistance fails, an actuating driver force of 300 N must not be exceeded.
- For a normal passenger car, it must be possible to turn the front wheels into the position corresponding to a turning radius of 12 m in 4 s with a force of 150 N

Table 2.1 Maximum allowable steering control effort with and without power assistance

Vehicle category	Intact power assistance			With failure of power assistance		
	Maximum effort (N)	Time (s)	Turning radius (m)	Maximum effort (N)	Time (s)	Turning radius (m)
M1	150	4	12	300	4	20
M2	150	4	12	300	4	20
M3	200	4	12[a]	450[b]	6	20
N1	200	4	12	300	4	20
N2	250	4	12[a]	400	4	20
N3	200	4	12	450[b]	6	20

[a]Or full lock if 12 m not attainable
[b]500 N for rigid vehicles with 2 or more steered axles

(or full lock if 12 m radius is not attainable) at a speed of 10 km/h and in the case of failure of power assistance a turning radius of 20 m in 4 s (Table 2.1).

- Limits for other classes of vehicle are shown in Table 2.1 where:
  - Category M1: Vehicles designed and constructed for the carriage of passengers and comprising no more than eight seats in addition to the driver's seat.
  - Category M2: Vehicles designed and constructed for the carriage of passengers, comprising more than eight seats in addition to the driver's seat, and having a maximum mass not exceeding 5 tonne.
  - Category M3: Vehicles designed and constructed for the carriage of passengers, comprising more than eight seats in addition to the driver's seat, and having a maximum mass exceeding 5 tonne.
  - Category N1: Vehicles designed and constructed for the carriage of goods and having a maximum mass not exceeding 3.5 tonne.
  - Category N2: Vehicles designed and constructed for the carriage of goods and having a maximum mass exceeding 3.5 tonne but not exceeding 12 tonne.
  - Category N3: Vehicles designed and constructed for the carriage of goods and having a maximum mass exceeding 12 tonne.

2.2.2 Steering Ratio

Steering ratio is defined as the ratio of steering wheel rotation angle (lock to lock) to steer angle movement of the road wheels (lock to lock).

Normal steering wheel rotation (lock to lock):
3–3.5 turns
Typical steered wheel angles to give minimum turning circle for passenger cars:
±35°

Note:

(1) designs are now being developed where ±75° steered wheel angle is becoming realistic
(2) London taxis have a high steered wheel angle of ±60° which gives greater manoeuvrability.

Typical steering ratios (steering wheel angle to steered wheel angle):

- Passenger cars 15–20:1
  (lower value for power assisted steering and higher value for manual steering)
- Trucks 30–40:1

The actual steering ratio may vary from the design value due to steering torque reactions acting against compliance within the steering system.

2.2.3 *Steering Behaviour*

The requirements in terms of steering behaviour can be summarised as follows:

- Jolts from irregularities in the road must be damped out before they reach the steering wheel. However such damping must not cause the driver to lose contact with the road—the "feel" element.
- The basic design of the steering kinematics must satisfy the Ackermann conditions (see below). This is not necessarily always achieved, or desired, because of external influences such as tyre slip, steering geometry (camber) and suspension/geometric influences.
- By means of suitable stiffness of the steering system (particularly if rubber elements are used) the vehicle must react to minute steering corrections.
- When the steering wheel is released, the steered wheels must return automatically to the straight ahead position and must remain stable in that position.
- The steering ratio must be as low as possible in order to obtain ease of handling and give a responsive steering system.
- The resulting steering forces at the steering wheel are not only influenced by the steering ratio but also by the front axle loading, the size of the turning circle, the wheel suspension (castor, steering-axis inclination, steering roll radius), the tyre tread and the speed of vehicle (the dynamic forces).

2.3 Steering Geometry and Kinematics

2.3.1 Basic Design Needs

A fundamental requirement of all current steering systems is to connect both steered wheels on an axle to ensure they remain at some related position to each other regardless of turn angle i.e. there is a need to ensure a kinematic relationship between the two wheels. A basic way to achieve this is as shown in Fig. 2.1. Each wheel is connected to a knuckle or upright which is connected to the chassis through some swivel mechanism. Each of the knuckles is then connected to a track-rod using ball joints so that, when a lateral force is applied to this track-rod, both wheels move together, rotating/rolling about their relevant pivot and so changing the vehicle direction.

The force along the track rod may be applied by hydraulic, mechanical or electrical means or any combination of these. Such a basic system connects the wheels during steering but does not provide for the wheels to satisfy the Ackermann angle requirements. This Ackermann criteria requires each of the steered wheels to be rotated by a different amount so that they are orientated to trace a true rolling radius about a common turning point as indicated in Fig. 2.2.

True Ackermann steering is virtually impossible to achieve over the full range of steering angles using practical linkage designs. Nor is it necessarily desirable since real turning manoeuvres invariably involve some tyre sideslip as discussed below. However the trapezoidal geometry involved in the design shown in Fig. 2.3 comes close to providing Ackermann geometry which may require the inner wheel to rotate through considerably greater angles than the outer wheel as indicated in Fig. 2.4.

This most convenient geometry to approximate this need is to direct the steering arms to intersect the rear rigid axle at its centre as shown in Fig. 2.3. With such geometry the wheels rotate about the chassis pivot points to adopt the angles as indicated in Fig. 2.4. The angles of the steering arms should be noted as this will result in different torques being applied to each wheel and a possible fluctuation in steering force as felt at the steering wheel. If this does happen then there will be a need to modify the torque over the steering range to give a uniform steering wheel force.

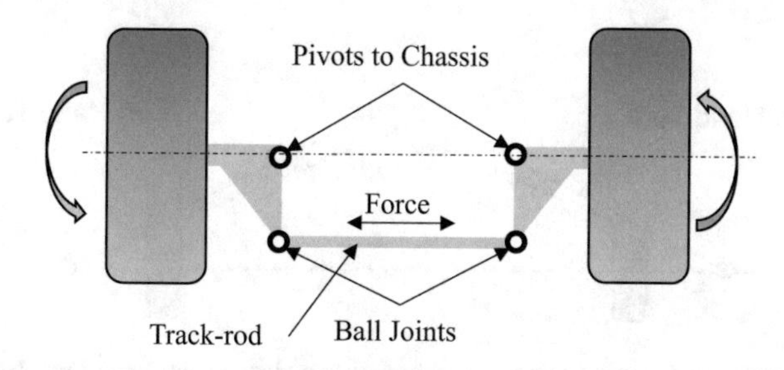

Fig. 2.1 Basic design to ensure connection of the two steered wheels

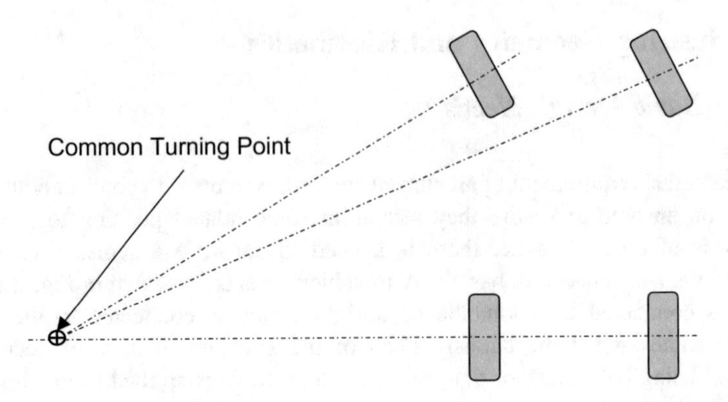

Fig. 2.2 Idealised wheel angles to allow for true rolling of the steered wheels around a constant radius curve: the Ackermann criterion

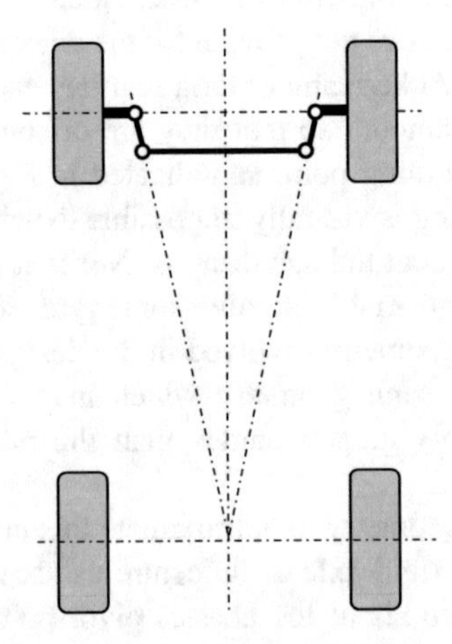

Fig. 2.3 Steering layout that approximately provides Ackermann geometry

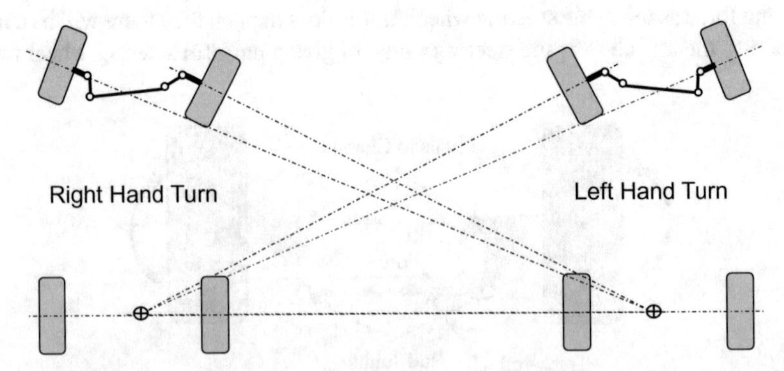

Fig. 2.4 The attitudes of the wheels when negotiating a right or left hand turn. Note movement (swing) of track arms and incorporation of tie arms to give flexibility

It must be noted that regulations state that this force at the steering wheel should not reduce as the wheel is turned away from the neutral position.

It should also be noted how the track arm moves to achieve Ackermann steering. This movement needs to be accounted for in any steering gear design that is rigidly mounted on the chassis. Tie rods are normally used to allow for such movement of the steering arms as indicated in Fig. 2.4.

Maintenance of Ackermann geometry is more important for specialised, lower speed vehicles, e.g. taxis and delivery vehicles, which spend more of their time in low speed turning manoeuvres; minimisation of tyre scrub and steering efforts are important in these conditions. The classic London taxi is designed with close to perfect Ackermann geometry up to its 60° maximum inner wheel steer angle.

2.3.2 Ideal Ackermann Steering Geometry

The nominally ideal geometry involved in a vehicle following a curved path is defined as the Ackermann geometry as shown in Fig. 2.5.

From Fig. 2.5, it can be seen that, assuming the wheels are rolling perfectly (i.e. no sideslip), the following basic expressions describe the inner and outer wheel angles:

$$\tan \delta_i = \frac{L}{[R - T/2]} \tag{2.1}$$

$$\tan \delta_{oA} = \frac{L}{[R + T/2]} \tag{2.2}$$

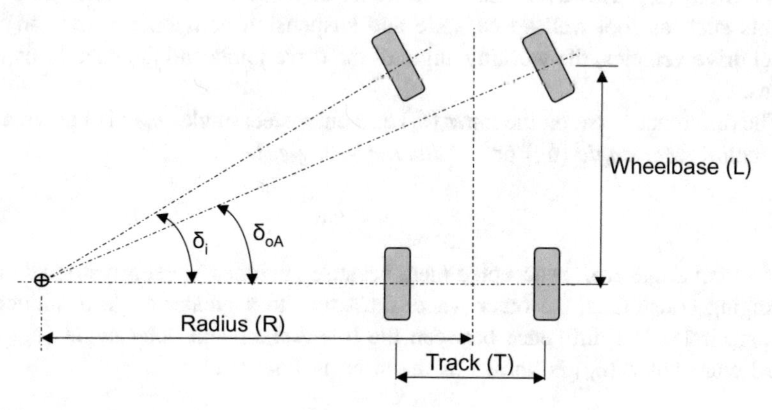

Fig. 2.5 Ackermann steering geometry

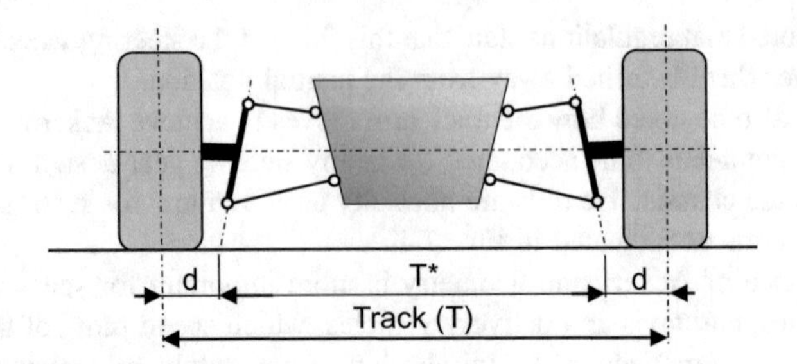

Fig. 2.6 Diagram showing intersect of steer rotation axis measured at the ground

Simplifying so that the outer angle may be calculated from the inner angle gives:

$$\cot \delta_{oA} = \cot \delta_i + R/L \tag{2.3}$$

where L is the wheelbase and R is the mean radius of turn at the rear axle.

The theoretical track turning circle diameter (D_s) as measured at the suspension intercept at the ground (Fig. 2.6) is then given by the following equation:

$$\sin \delta_{oA} = \frac{L}{D_s/2 - d} \tag{2.4}$$

Giving:

$$D_s = 2\left(\frac{L}{\sin \delta_{oA}} + d\right) \tag{2.5}$$

Clearly, to minimise D_S, the wheelbase "L" needs to be as small as possible and the outer wheel angle (δ_{oA}) as large as possible. This requires an even larger inner wheel angle (δ_i). However this inner wheel angle is limited by packaging constraints such as foot-well space, axle and suspension components and, on front wheel drive vehicles, the working angle of the drive joints and the need to fit snow chains.

The difference between the inner (δ_i) and outer steer angle (δ_{oA}) is known as the *differential steer angle* (δ_A) or *toe difference angle*, i.e.:

$$\delta_A = \delta_i - \delta_{oA} \tag{2.6}$$

If a steer angle error is accepted then, because the inner wheel is restricted due to packaging constraints, the outer wheel must turn to a greater angle to reduce the turning circle. The difference between the true Ackermann outer angle (δ_{oA}) and actual outer angle (δ_o) is known as the steering flaw (δ_F).

$$\delta_F = \delta_o - \delta_{oA} \tag{2.7}$$

Test measurements show that a reduction in the track turning circle of $\Delta D_s \approx 0.1\,\mathrm{m}/1^\circ$ steering deviation may be achieved; thus the revised track turning circle diameter becomes:

$$D_s = 2\left(\frac{L}{\sin \delta_{oA}} + d\right) - 0.1\Delta\delta_F \tag{2.8}$$

It is normal to restrict the "error" to within some industry standard with technical manuals being obliged to state the deviation, or flaw, at an inner wheel angle of 20°.

With some vehicles, the deviation may in reality become negative at low angles which indicates the outer wheel turns more than the inner wheel even at low turn angles where packaging is not such an issue. This may be an additional design intention as wheel camber is related to castor angle (see castor inclination angle), and camber becomes an important feature. At high speeds (low steer angles), it may be a desire to increase negative camber on the outer, most heavily loaded, wheel. In addition, during high speed manoeuvring, tyre sideslip angles are involved and the need to satisfy Ackermann steering geometry becomes less important.

In general it would be normal for a vehicle manufacturer to have some form of policy to achieve a target deviation, typically 60% of true Ackermann (i.e. 60% of required differential steer angle) at full lock.

If the curb to curb turning diameter was needed, then the theoretical turning diameter would be given by the equation:

$$D_s = 2\left(\frac{L}{\sin \delta_{oA}} + d + \frac{tyre\ tread\ width}{2}\right) \tag{2.9}$$

2.4 Review of Common Designs

2.4.1 Manual Steering

The main function of the steering system is to provide a connection between the steered wheels and the driver's steering wheel. It is the primary means by which the driver exercises directional control over the vehicle. In the past, regulations insisted on a mechanical link between the steering wheel and the steered wheels. Technological advances have reached a level where this mechanical link is no longer a requirement which allows true "steer-by-wire" systems to be employed. This technology is now widespread but it is still common to include a mechanical link between the steering wheel and steered wheels.

Because the forces and moments involved can be large, some form of power assistance is commonly provided in order to control the driver effort involved to acceptable levels (the work done element of force versus distance). If the system is totally power controlled then there must be some form of steering wheel resistance introduced to give the driver "feel" for the interaction between tyre and road during manoeuvres.

The design of a steering linkage involves controlling the kinematic relationships to achieve the correct steer angles at the inner and outer steered wheels as discussed above. However, these basic kinematic relationships are modified by several features: changes in suspension deflection, compliances from suspension bushes and forces/moments arising from the drivetrain (tractive effort). Hence, although some of the design principles involved in the steering kinematics may be viewed in isolation, a complete analysis of behaviour in practice must include the **entire suspension system** plus drivetrain if appropriate.

The discussion here concentrates on front wheel steering (FWS). Rear wheel steering (RWS) was generally restricted to specialised off-road (typically dumper trucks) or loading vehicles (fork trucks), although it is now being introduced in combination with FWS to create four wheel steering systems. Such advanced systems allow greater manoeuvrability for parking and better dynamic handling during high speed lane changes.

There are two fundamentally different steering system designs: the rack and pinion system and the steering box systems as discussed below. All linkage designs are very much dictated by space availability.

2.4.2 Rack and Pinion System

In passenger car design, the most common mechanism in current circulation is the rack and pinion (Fig. 2.7). Rotation of the steering wheel controls rotation of the pinion, often through a steering column which includes a pair of universal joints for access reasons, which then controls translational motion of the rack. The ends of the rack are connected to the steering arms of the steered wheels by tie rods as shown in Fig. 2.8. It is these tie rods that account for movement of the track arms as shown in Fig. 2.4.

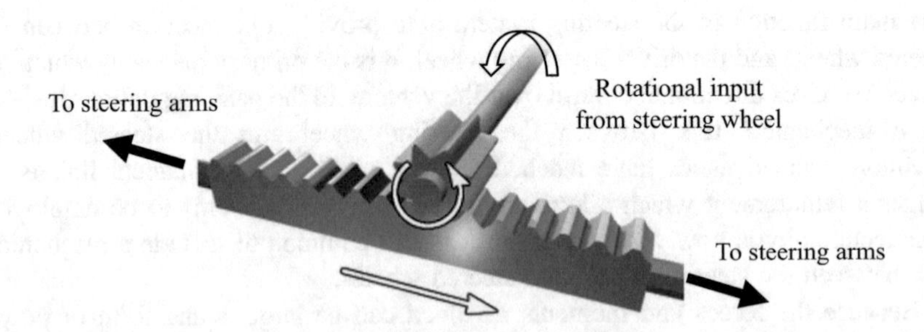

Fig. 2.7 Rack and pinion mechanism.
https://upload.wikimedia.org/wikipedia/commons/6/6b/Rack_and_pinion.png

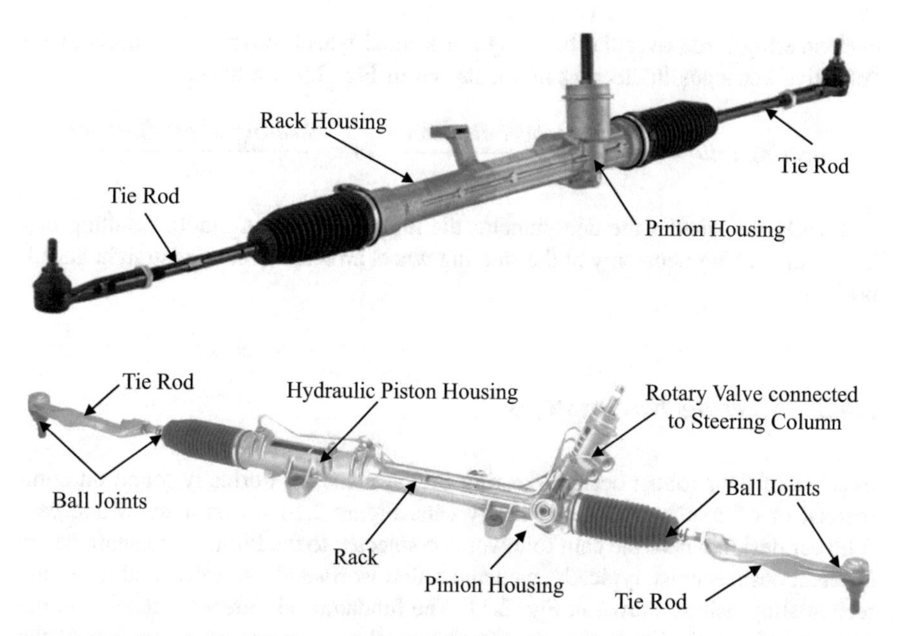

Fig. 2.8 General rack and pinion assembly upper—mechanical system (designed as a "throw away" component) lower—mechanical hydraulic power assistance.
Reproduced with kind permission of ZF © ZF Friedrichshafen AG

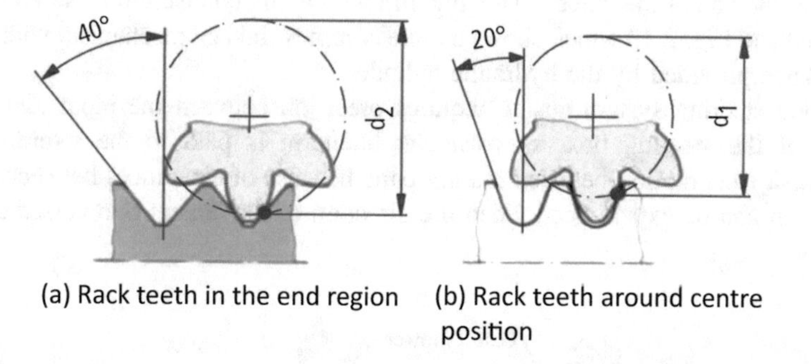

Fig. 2.9 Schematic ratio diagram for ZF rack and pinion power steering gear with variable ratio, used to give near constant steering force over full range.
Reproduced with kind permission of ZF © ZF Friedrichshafen AG

Recent developments in rack and pinion steering systems have focused on variable ratio designs, in which the effective steering gear ratio varies with steering wheel angle, and on enhancements to the power assistance systems to try and provide the best compromise between ease of control and good feedback to the driver. Figure 2.9 shows the geometry of a variable pitch rack that provides more even steering force over the full range of steered wheel movement to account for "effective" changes in steering moments. From Fig. 2.9, we have:

$$\textit{Steering ratio} = \frac{\textit{Steering wheel diameter}}{d_2} \quad or \quad \frac{\textit{Steering wheel diameter}}{d_1}$$

Clearly the smaller the denominator the higher the steering ratio, resulting in a lower force being necessary at the steering wheel around the centre (straight ahead) position.

2.4.3 Steering Box Systems

Because of their robust design, steering box systems are normally found on commercial or off-road vehicles and heavy cars. Figure 2.10 shows a screw and cam follower design where the cam follower is connected to the Pitman arm shaft. Many different designs exist, typically, cam and roller, worm and nut, worm and roller and recirculating ball as shown in Fig. 2.11. The fundamental difference relative to the rack and pinion design is that the steering gearbox converts rotary motion of the steering wheel into rotary motion of a steering arm, commonly termed a "Pitman" arm. This "Pitman" arm is connected to a steering linkage which steer the wheels. Two different installations are shown in Fig. 2.12 (simple drag link connected directly the one of the wheel's steering arm with a link connection between the two wheels) and Fig. 2.13 which shows a more complex linkage mechanism with power assistance provided by the hydraulic cylinder.

Good steering system design requires precision between the input and output sides of the steering box, so particular attention is paid to the **avoidance of backlash** (lost motion between mating parts because of clearances between them). Backlash can be experienced when the direction of movement is reversed and the

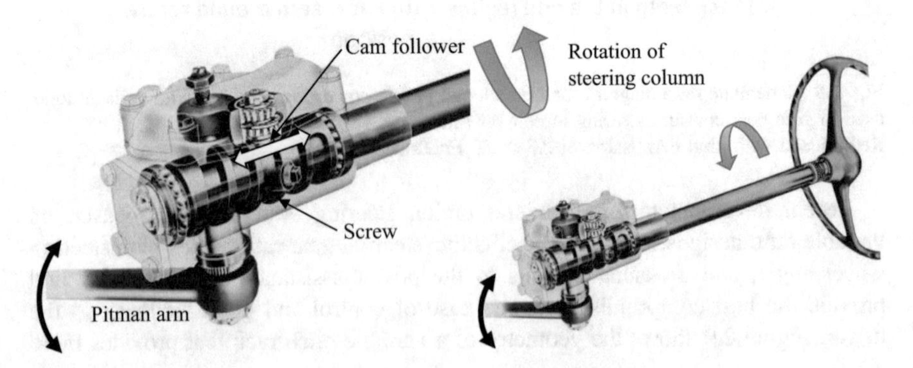

Fig. 2.10 Steering box system—screw and cam follower design. Pitman arm connected to steering linkage gear.
Reproduced with kind permission of ZF © ZF Friedrichshafen AG

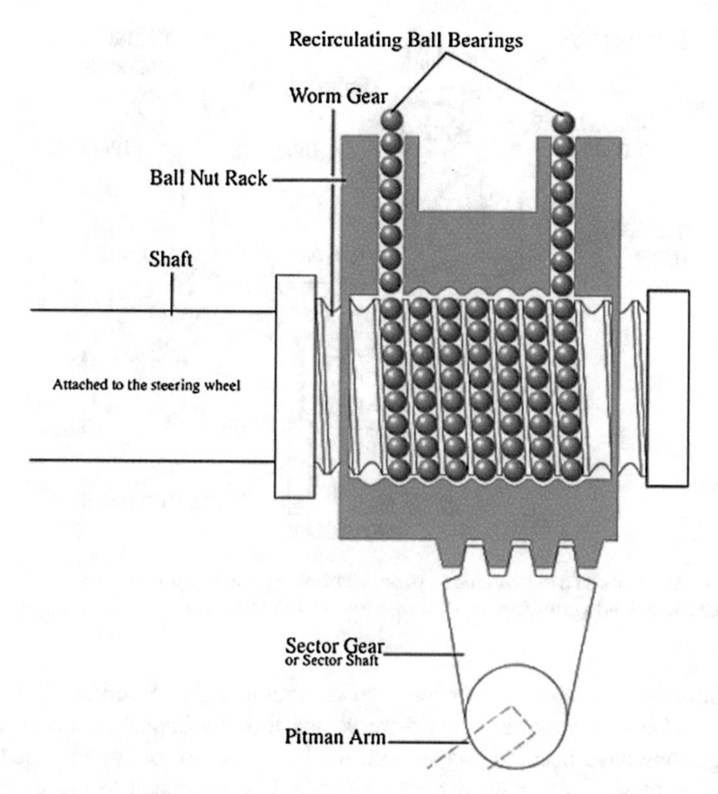

Fig. 2.11 Steering box using recirculating ball screw system. As shaft rotates the nut moves linearly but is restrained from rotating. The nut engages with a gear segment that causes the "Pitman" arm to rotate
https://upload.wikimedia.org/wikipedia/commons/0/0c/RecirculatingBall.png

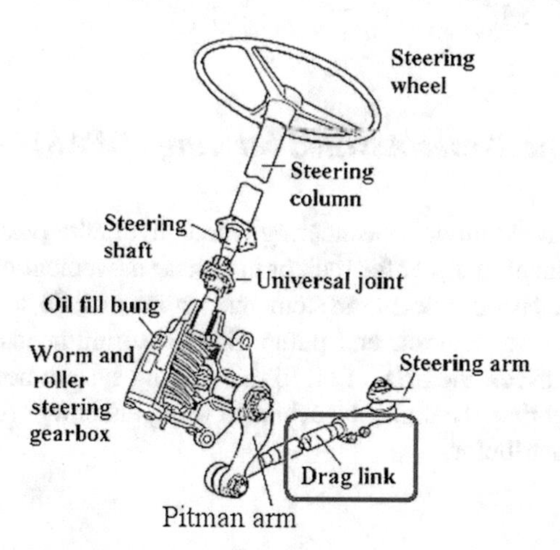

Fig. 2.12 Steering box with simple drag link to one of the steered wheels mechanical connection to the other wheel is by a link/track rod.
Reference: https://www.bing.com/images/search?&q=steering+box&qft=+filterui:license-L2_L3&FORM=R5IR45

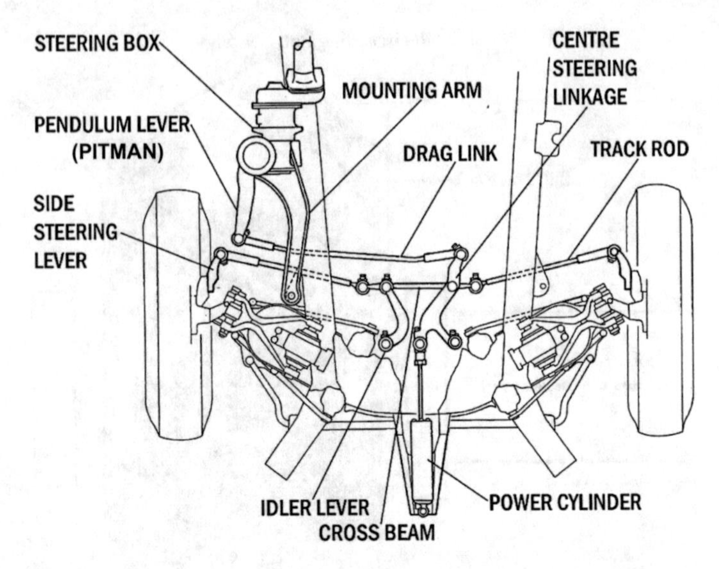

Fig. 2.13 Power assisted steering box system and linkage mechanism diagram reproduced with kind permission of Bentley Motors © Bentley Motors Ltd

lost motion is taken up before the reversal of motion begins. Steering efficiencies in the forward and reverse direction were at one time an important design criteria, although they have become less so with the increased use of sophisticated power steering systems. For a manual steering system it is beneficial to maintain a high forward efficiency in order to keep steering forces low. The reverse efficiency (back-driving), however is a compromise; a low value minimises the transmission of road roughness disturbances back to the driver but at the expense of removing some of the driver feel.

2.4.4 Hydraulic Power Assisted Steering (HPAS)

To reduce the force required at the steering wheel, hydraulic pressure may be used to assist the movement of either the rack or the linear movement of the ball screw in a steering box. In either case the system must now include a hydraulic "power pack" in the form of a reservoir and pump. Since steering needs to remain within the control of the driver, the flow of oil to the steering system needs to cease when the driver stops steering. This may be achieved using either an orbital metering unit or a rotary flow distributor.

2.4.4.1 Orbital Metering Unit

This comprises a small orbital metering pump that is supplied with fluid under pressure. Its sole function is to pass oil from the "power pack" to the steering mechanism in a controlled manner. Connection between the steering wheel and the steering mechanism is now by hydraulic hose rather than by mechanical linkages. The flow of fluid is linearly proportional to the rotation of the steering wheel. If the steering wheel is not turned then no fluid is metered through the unit and the oil bypasses it under nominal back-pressure. As such pressure is only generated as needed so energy is not wasted. This type of steering is used on heavy industrial vehicles such as earth movers, bulldozers, diggers, cranes etc.

2.4.4.2 Rotary Flow (Valve) Distributor—Fig. 2.14

In this case the flow of fluid is governed by a degree of "twist" in a torsion bar (item 6 in Fig. 2.14) within the steering column. The degree of rotary twist is limited by a loose spline at its inner end. Note that it is this mechanical connection that allows steering to be still performed if the hydraulics fail—but the manual load will be high. As the steering wheel is turned, the torsion bar is caused to twist as the road wheels remain stationary. This difference in torsional displacement causes a rotary valve to open which allows fluid to be directed to the appropriate end of the steering mechanism. If the steering wheel stops rotating the fluid continues to be transferred to the steering mechanism until the "twist" in the steering column returns to zero. The rotary valve then closes. Such a system is shown in Fig. 2.14 for a steering rack, a similar hydraulic arrangement being used for a steering box.

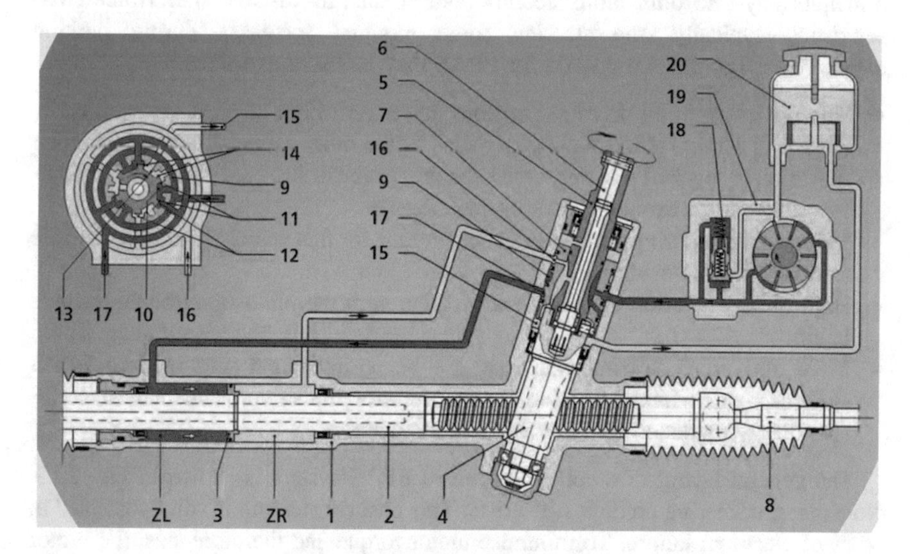

Fig. 2.14 Hydraulic power assisted steering circuit using a rotary valve. Reproduced with kind permission of ZF © ZF Friedrichshafen AG

2.4.4.3 Reaction Control Valve

The reaction control valve is a spool (shuttle) valve which is used in steering box assemblies. When the steering wheel is turned the recirculating ball screw engages with the nut so driving it in a desired direction. This is the normal function of the steering box which results in rotation of the Pitman arm. At the same time there is residual rotational drag between the screw and nut which, if allowed, causes the nut to rotate slightly in the same direction as the screw rotation. A small lever on the nut engages with the spool valve and as the nut rotates it causes the valve to move linearly. Oil under pressure is then directed to the appropriate side of the box to assist steering. If an opposite rotation is applied to the steering wheel then the nut rotates slightly in the opposite direction so causing a change in the spool position, and delivery direction. Under no input from the steering wheel, the spool valve will take up a central position due to self-centring springs and the oil is directed back to the reservoir under nominal back-pressure.

2.4.5 Electric Power Assisted Steering (EPAS)

The problems with hydraulic PAS is that there is a need to ensure the operating hydraulic pressure is always available. In virtually all cases, the pump will be a fixed deliver pump, typically a vane type pump. Even though by-pass valves may be employed (oil not "blowing-off" over a pressure valve) the engine has to provide power to overcome the back-pressures, and this requires fuel. To mitigate this problem the "assist" element may be obtained using electric motors. This operating principle may also offer more steering control such as variable speed/force characteristics—typically, the steering force required increases during parking manoeuvres. The general criteria for EPAS may be listed as follows:

- Safe operation in all driving situations and a very high level of availability.
- Highly dynamic response characteristics in the most varied driving situations.
- A sufficient level of steering assist for the driver in the case of intensive actuation forces, for example, parking manoeuvres.
- Minimal noise during all steering manoeuvres; for this vehicle function, acoustic feedback is not desirable.
- High quality steering characteristics in line with the philosophy of the vehicle brand.
- More and more steering functions are being integrated into modern EPAS systems, which improve safety or comfort for the driver, and can be correspondingly marketed by the vehicle manufacturers.

The general layout of a column-mounted EPAS system is shown in Fig. 2.15. Note that the steering column still exists. This ensures steering is still "possible" in cases of electrical failure. To minimise motor torque and therefore size, the motor

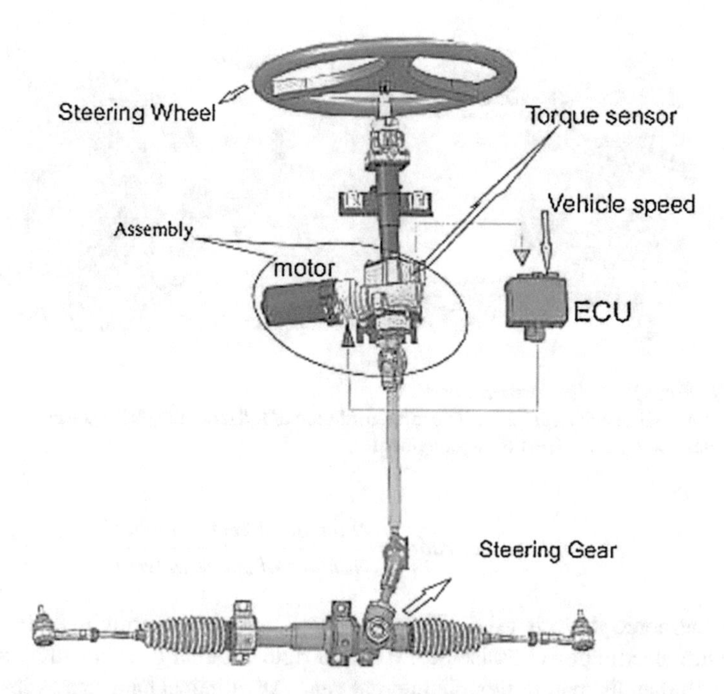

Fig. 2.15 General layout of the column mounted EPAS system.
Reference: https://i.stack.imgur.com/GLRa2.jpg

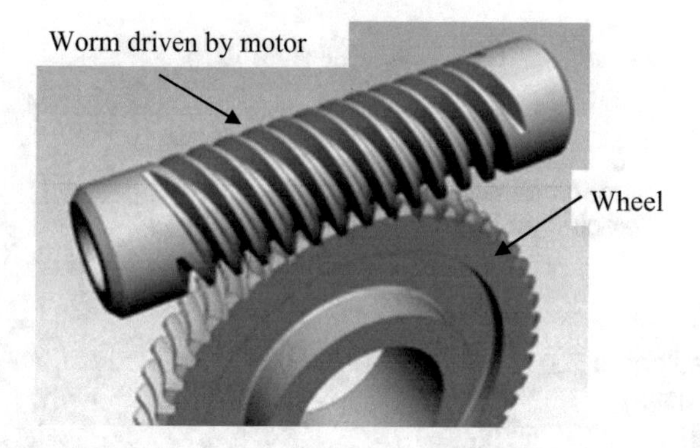

Fig. 2.16 Worm and wheel drive system.
Reference: https://upload.wikimedia.org/wikipedia/commons/thumb/c/c3/Worm_Gear.gif/220px-Worm_Gear.gif

speed will be high and as such a high gear ratio between motor and steering gear will be necessary. Two such methods of reduction are the worm and wheel as shown in Fig. 2.16 and the harmonic drive as shown in Fig. 2.17.

The "worm" is driven by an electric motor which rotates the wheel. The wheel may be fitted to a recirculating ball nut which rotates and causes a ball screw to move linearly so replacing the function of the rack. The ratio is given as:

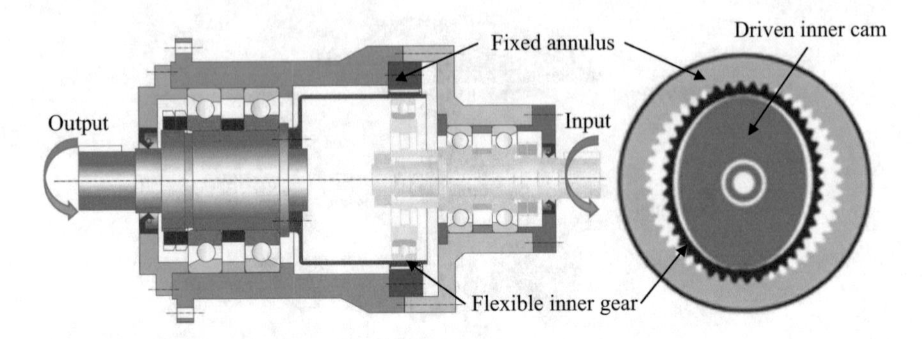

Fig. 2.17 Principle of the harmonic drive.
https://upload.wikimedia.org/wikipedia/commons/thumb/d/d6/Harmonic_drive_xsection.svg/1024px-Harmonic_drive_xsection.svg.pngmonic

$$\textit{Worm \& wheel ratio} = \frac{\textit{Number of teeth in wheel}}{\textit{Number of}\ \text{starts}\ \textit{in worm}}$$

The harmonic drive is used within a column mounted steering unit (Fig. 2.18) along with electric power assistance. It uses a rigid external gear (annulus) that has more teeth than its mating flexible internal gear. An elliptical cam rotates inside the

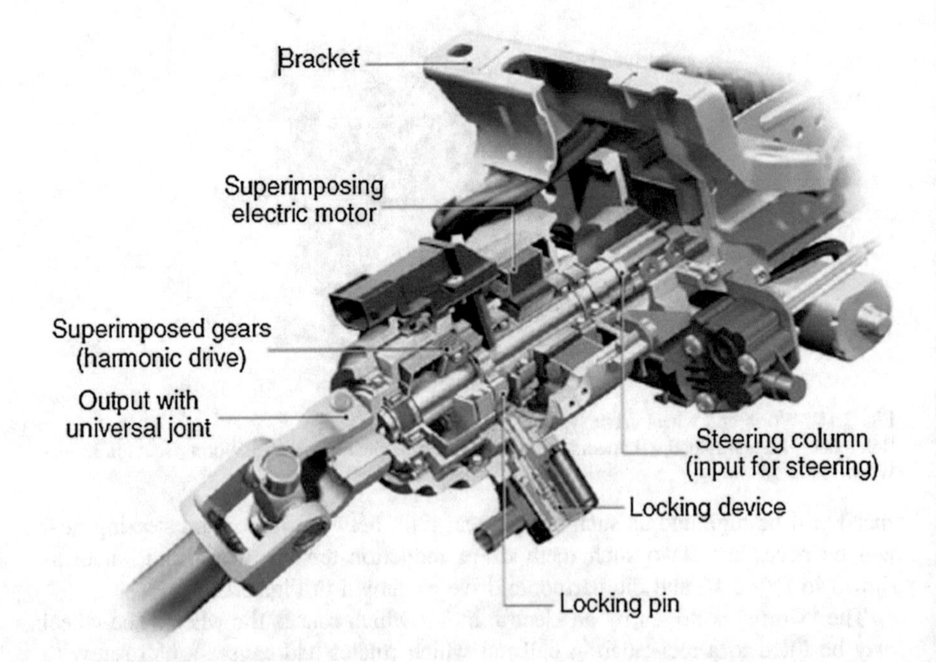

Fig. 2.18 Shows the harmonic drive within the column mounted EPAS system as used in the Audi A8.
Reproduced with kind permission of Audi © Audi AG

flexible gear causing the teeth to mesh at its extremes and to disengage at the narrow section of the cam. The difference in the number of teeth on the annulus and flexible gear causes the internal gear to rotate significantly slower (dependant on the number of teeth differential) than the input cam (motor) speed. As such the output torque will be higher by the same ratio (assuming 100% efficiency).

$$Harmonic\ drive\ ratio = \frac{Flexible\ gear\ teeth - Annulus\ teeth}{Flexible\ teeth} \tag{2.10}$$

Example If the annulus has 202 teeth and the flexible gear has 200 teeth, then:

$$Drive\ ratio = \frac{200 - 202}{200} = -0.01$$

Thus speed reduction is 100:1 and the negative sign indicates output rotation in opposite direction to input.

The worm and wheel reduction gearing is stronger than the harmonic drive and is also capable of reasonably high reduction ratios (typically 100:1). It tends to be rather bulky and heavier which is a disadvantage. Backlash is avoided by having the bearings mounted in "elastic" mounts and the worm may then be pushed into positive engagement with the wheel.

Superposition steering responds to driver input and is speed related. The torque demand is detected by induction coils within the EPAS system. The greater the rotary displacement (e.g. in parking), the greater the torque demand. There is also vehicle speed feedback that reduces torque demand as speed increases such as in high speed driving.

Dependant on the packaging constraints, the motors may be included within the steering column or on the steering mechanism itself. Drive may still be rack and pinion but rotary ball screws now become an option as shown in Figs. 2.19, 2.20 and 2.21.

With full electric steer, it may become possible to provide linear motion at the steering gear using hollow motors and recirculating ball screws. If this were adopted then each wheel may be controlled independently to account for tyre slip and individual wheel camber. In the case of the recirculating ball screw being used within the steering gear, the nut is held axially yet is caused to rotate by the electric motor. The screw is constrained from rotating so moves axially providing the axial force to steer the wheels. Such a system is shown in Fig. 2.22. The power required by the motor depends on the force demand, distance travelled and the time for the full movement.

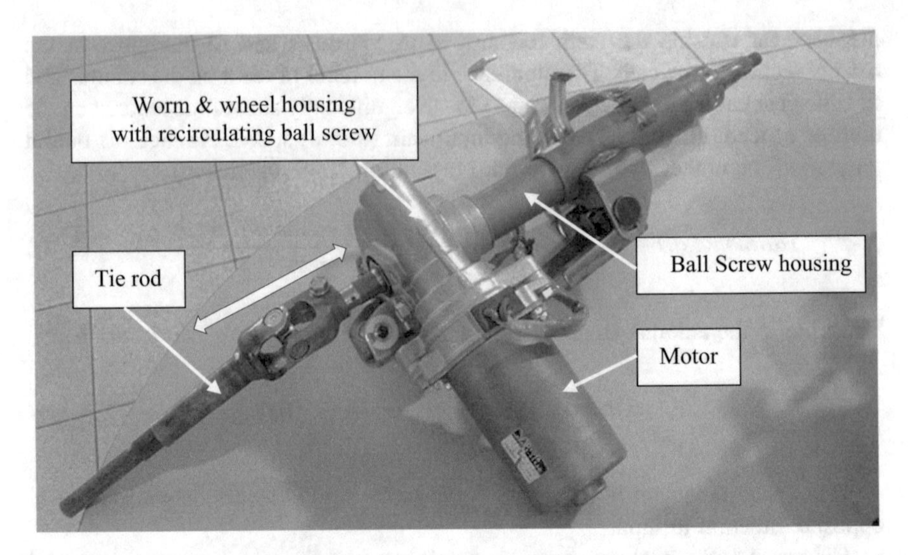

Fig. 2.19 Worm and wheel arrangement.
https://www.bing.com/images/search?&q=electric+power+ steering&qft=+filterui:license-L2_L3_L4&FORM=R5IR43

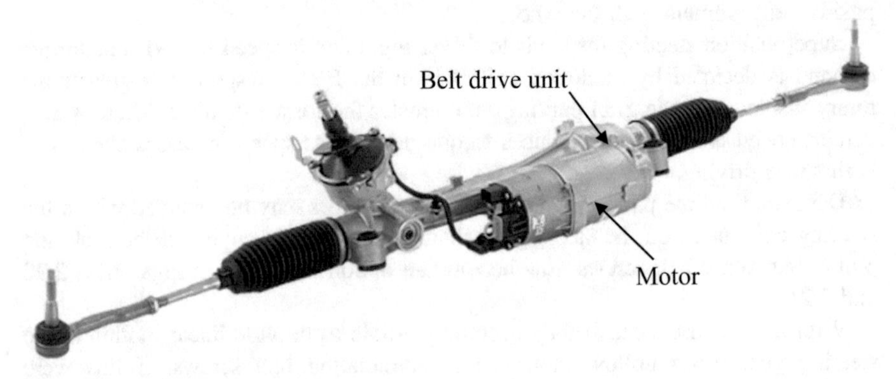

Fig. 2.20 Belt drive arrangement.
Reproduced with kind permission of ZF © ZF Friedrichshafen AG

2.4.6 Steer-by-Wire

It will be noted that within both the HPAS and EPAS systems shown above, there is still a mechanical connection between the steering wheel and the steered wheels as previously required by the regulations. As the control systems and safety protocols have advanced, the regulations have been relaxed and this is no longer a requirement. This gives the designer much more versatility in positioning the steering wheel/unit. Indeed it would be possible to have a vehicle whereby the steering could be passed across the vehicle, changing left hand to right hand drive within

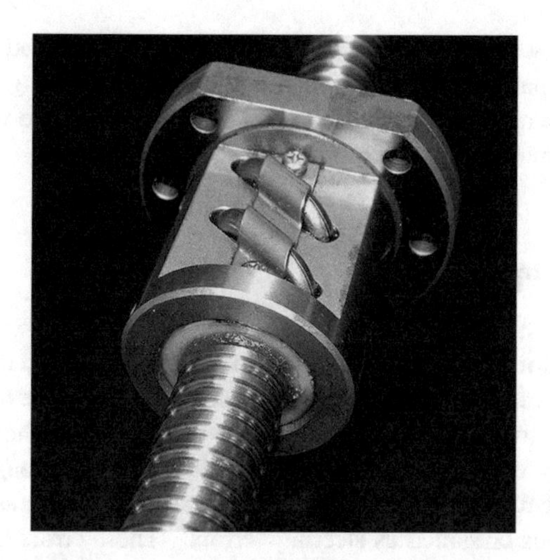

Fig. 2.21 Recirculating ball screw as used on a steering system.
Reference: https://c1.staticflickr.com/1/76/190364399_16c7137a7a_z.jpg?zz=1

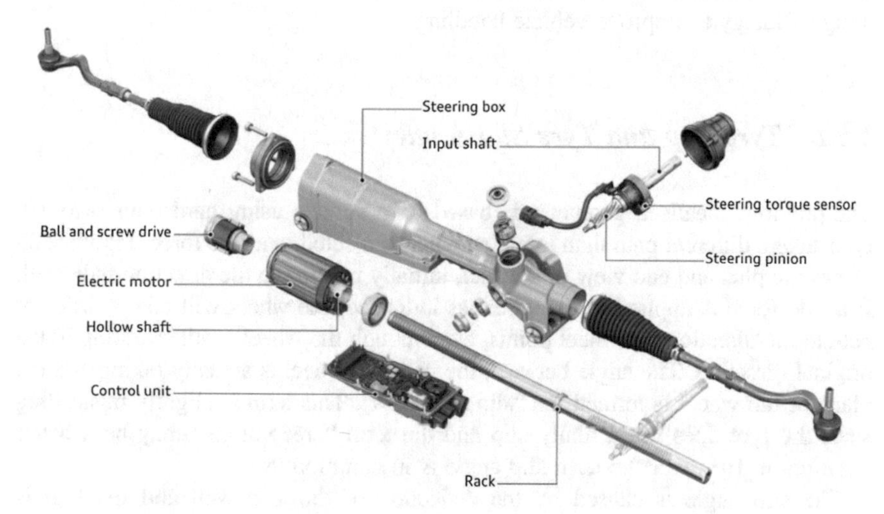

Fig. 2.22 Exploded view of hollow motor recirculating ball screw.
Reproduced with kind permission of Audi © Audi AG

minutes. The sensation of using a steering wheel still exists but is no longer restricted to conventional mechanisms—joystick, twist grips or other forms of control now being an option.

Because there is no longer a mechanical connection between the steered wheels and the steering wheel, there is a need to include a "reaction torque actuator" that gives the driver steering "feel". This may be achieved using a resistance potentiometer to provide a resistance torque that may be directly related to vehicle speed. In addition it has been mentioned above that independent steering to left and right hand wheels is now a possibility. In such cases steering may be adjusted to be speed related e.g. when parking or indeed when braking or accelerating (toe-in/out). Also acceleration/deceleration controllers can be used to integrate the wheel torques and steering to give predictive yaw control.

2.5 Steering "Errors"

The previous sections have described the various ways in which the wheels may be turned to provide vehicle direction. The ideal situation would be to have the wheels rotating around a true arc with zero tyre scrub. It has been shown that this may achieved in theory using Ackerman geometry. It is difficult enough to achieve this through geometric design alone and its realisation becomes increasingly difficult for a variety of reasons classified as steering "errors". These errors are due to general compliance and operational characteristics such as tyre slip, compliance in the mechanical system (elastomeric components) and set-up (geometric) errors. The designers' challenge is to understand these anomalies and include them in the design strategy to improve vehicle handling.

2.5.1 Tyre Slip and Tyre Slip Angle

The previous idealised geometry is based on a vehicle using hard tyres. Modern tyres take a different path than intended when subjected to a side force. Figure 2.23 shows the plan and end view of a wheel, initially moving in the direction indicated. If a side force is applied to the wheel as indicated, the wheel will take a different path to the direction the wheel points, even though the wheel is still pointing in the original direction. The angle between the path the wheel is actually taking and the plane of the wheel is termed the "slip angle" (α). This term is slightly misleading since the tyre does not actually slip and the term "creep angle" may be a better description. However the term slip angle is in common use.

The slip angle is caused by the deflection of the side wall and tread. It is proportional to the side force acting on the tyre but not in a linear manner. The greater the tyre aspect ratio (tyre wall height to width) the greater the slip. As such, low aspect ratio tyres have less slip than high aspect ratio tyres. This is true until the tyre is caused to slide. Up to this point the following applies:

$$Cornering\ Power = \frac{Side\ Force}{Slip\ Angle} \tag{2.11}$$

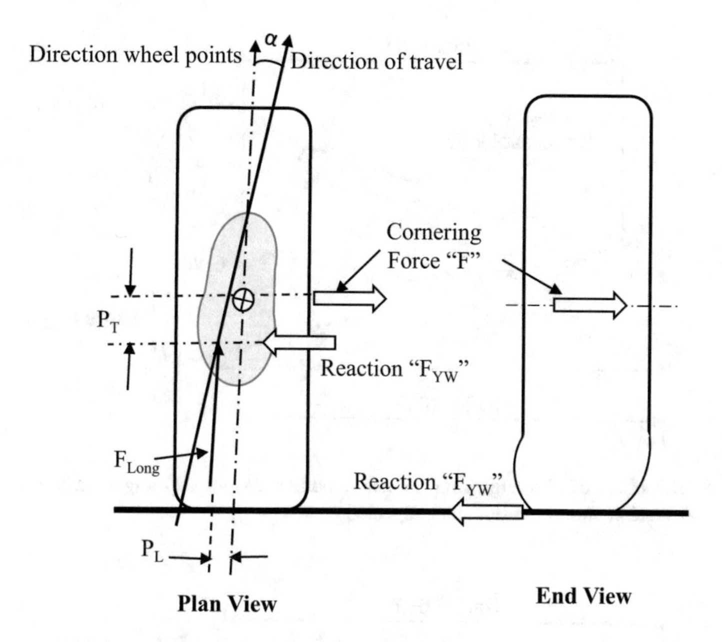

Fig. 2.23 Tyre tread contact patch distortion when subjected to a side force resulting in side-slip

The cornering power (CP) of a tyre is governed by:

- Inflation pressure—an increase raises the CP.
- Tyre construction—a radial-ply tyre has a higher CP than a diagonal-ply tyre.
- Tyre size—a low-profile tyre has a smaller wall so a higher CP is achieved.
- Camber (tilt) of wheel—tilting the wheel away from the side force increases the CP.
- Load on the wheel—if the load is increased from the normal value then CP will increase.

2.5.1.1 Effect of Slip Angles on Ackermann Geometry

As mentioned above the need to develop true Ackermann geometry is not so important in reality because of the effects of slip angle on the resulting geometry. This is shown more clearly in Figs. 2.24 and 2.25.

Understeer is preferred on normal road cars as this tends to keep the driver in control of the vehicle. Racing car drivers may prefer oversteer to enhance cornering abilities but this requires advanced driving skills that are generally absent from normal drivers. If both the front and rear slip angles are the same then, although the car still "slips", there is no over or under-steer and the term **"neutral steer"** is used.

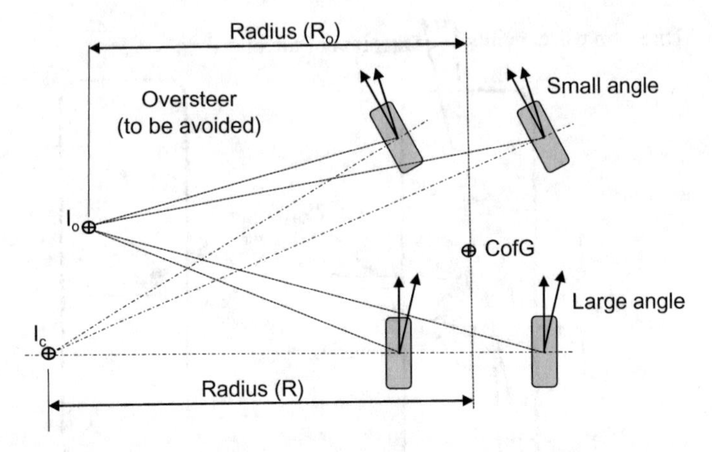

Fig. 2.24 The effect of slip angles on vehicle handling. Larger slip angles at the rear lead to oversteer. A smaller turning radius than expected results

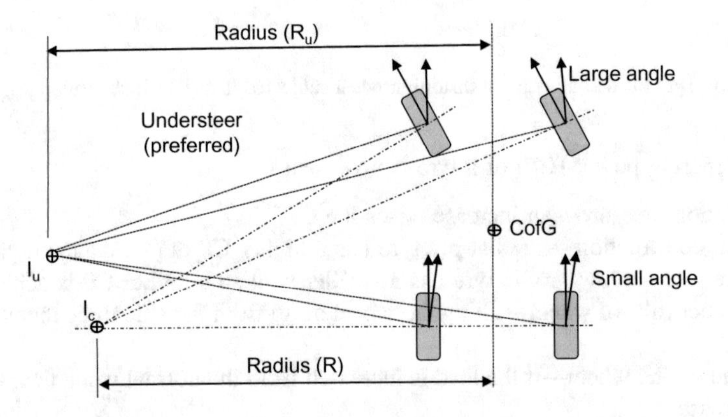

Fig. 2.25 The effect of slip angles on vehicle handling. Larger slip angles at the front lead to understeer. A larger turning radius than expected results

2.5.2 Compliance Steer—Elastokinematics

"Compliance Steer" is the change in steer angle of front or rear wheels resulting from compliance in suspension and steering linkages as produced by forces and/or moments applied at the tyre/road contact. In essence compliance steer results from deformation of elastomeric elements within the steering and suspension system as a result of wheel forces from the road/tyre interface.

There are over 70 elastomeric components (elastomer/metallic composites) on a typical car. In brief they are pseudo-bearings—that is, the two connected parts are allowed to move relative to each other but not continuously as with rolling element bearings. They are designed in a variety of forms, generally providing rotational

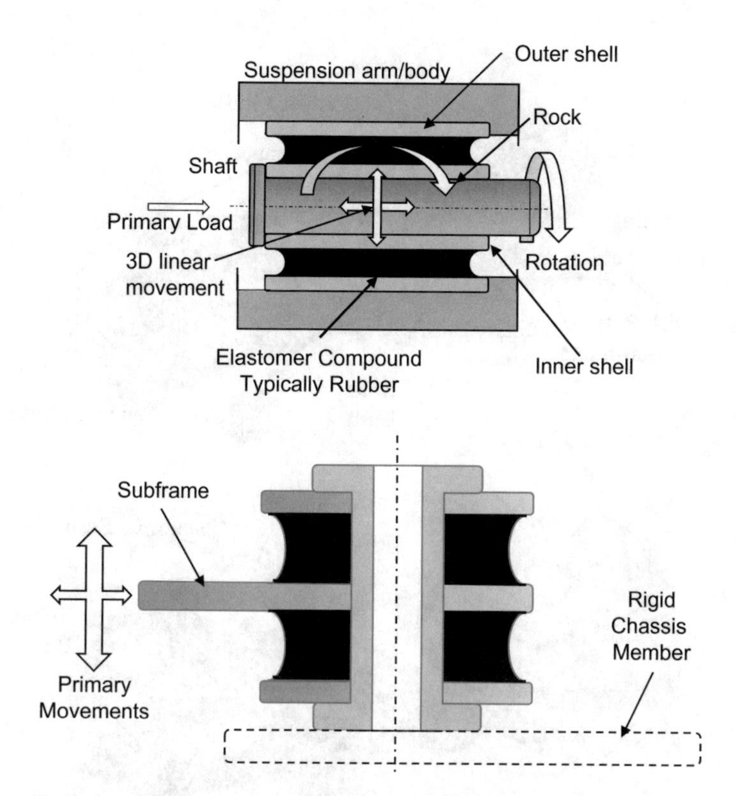

Fig. 2.26 Typical Elastomeric parts as used in a suspension systems. Upper—rotary shell type gives movement in all directions. Lower—sandwich type, used primarily to support vertical loads. See Fig. 2.27

and/or translational constraint. They are used to enhance comfort but may degrade with age and affect vehicle handling performance. The science of elastomers and elasto-kinematics is complex and may be used to provide differing characteristics in different directions of component part movement (Fig. 2.26). These figures show a cylindrical or bush type (mainly rotational) and sandwich type (mainly compression). If the elastomer compound is removed from any region then the compliance (spring stiffness) will be changed in that direction. This section is concerned only with the elastomeric parts found in suspension and steering systems such as shown in Fig. 2.27.

There are two types of Compliance Steer:

Compliance understeer—compliance steer which increases vehicle understeer or decreases vehicle oversteer.

Compliance oversteer—compliance steer which decreases vehicle understeer or increases vehicle oversteer.

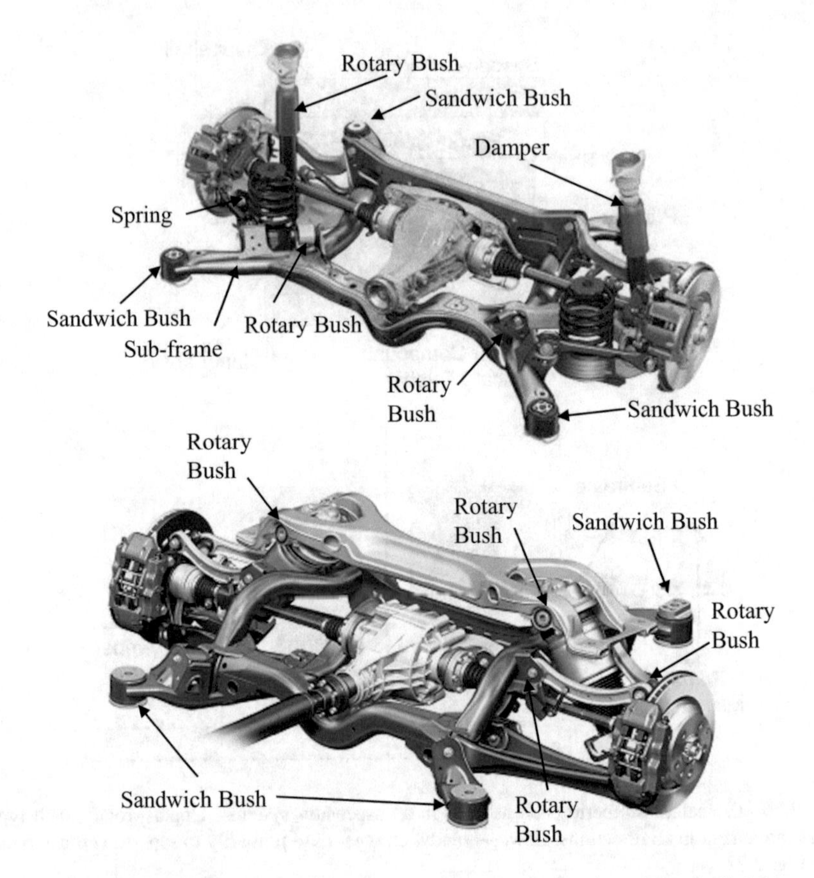

Fig. 2.27 Positions of compliant components. Upper—rear suspension of the Audi A6 lower—rear suspension of Audi Q7.
Reproduced with kind permission of Audi © Audi AG

Generally understeer is better than oversteer as the steering control remains with the driver—the driver is always turning into the corner. Oversteer may require a correcting reversal of direction by the driver. Figures 2.28 and 2.29 demonstrates the effect of such deformations on the oversteering and understeering of a vehicle for two different rear suspension arrangements.

The tendency of a vehicle to oversteer when decelerating is compounded by the compliant bushings found in most trailing arm suspensions. When the vehicle is decelerating, the trailing arm pivots towards the rear as the wheel is "pulled" backwards relative to the chassis (Fig. 2.28). This results in "toe out", which makes the vehicle unstable as it adds to the desired steering direction creating oversteer.

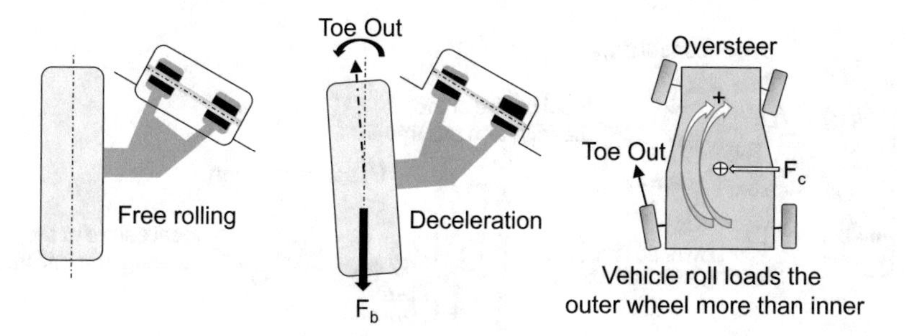

Fig. 2.28 Conventional trailing arm suspension

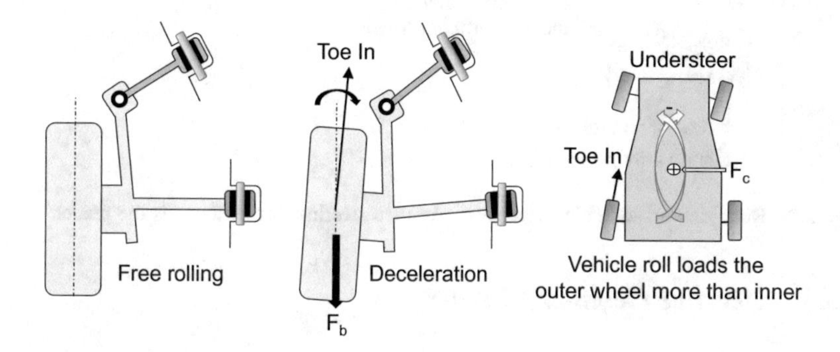

Fig. 2.29 Weissach axle semi trailing arm suspension

For the Weissach axle (Fig. 2.29), the front pivot bushing of the trailing arm is replaced by a short link. In this arrangement, when the vehicle decelerates and the wheel is "pulled" back, the result is "toe-in" at the rear. This adds to stability by reducing oversteer and is a design principle adopted by Porsche. It exemplifies how the designer has used elastomer deflections to an advantage in reducing oversteer by subtracting from the driver's chosen steering direction, leading to the desired effect of understeer.

It should be noted that any designed "toe-in" effect should take into consideration the understeer which may be experienced by a negative castor on the rear causing understeer (toe-in) as a result of side forces during cornering.

During normal operation a wheel may move linearly forward and rearward typically by the order of a total of 26 mm, that is 14 mm forward and 12 mm rearward. In general it is desired that the wheel moves linearly fore/aft and does not toe-in or out at the front wheel. As such, the positions of the ends of the track arms (connections to tie rods) and steering arms are carefully selected, along with compliant bushings, to avoid any tendency to toe-in. If these positions are selected incorrectly then geometric errors are induced resulting in bump and roll steer.

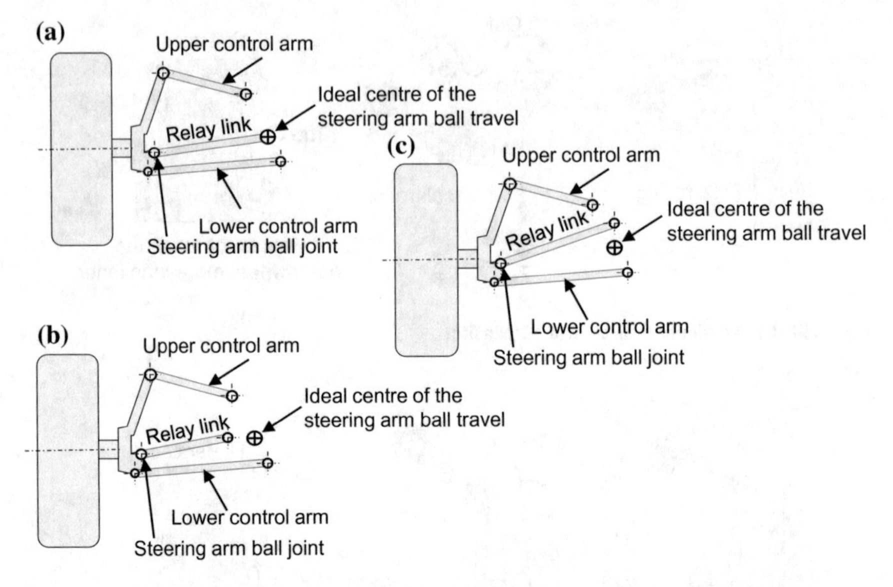

Fig. 2.30 Rear view of left side suspension. Assume steering arm is towards the reader

2.5.3 Steering Geometry Errors

The idealised treatment normally considers the suspension geometry in plan-view and in isolation. In practice this geometry is modified as the wheel moves up and down on the suspension. The additional constraint imposed on the wheel motion by the tie rod connected to the steering mechanism can cause additional steering effects during bump or roll motions. Such effects are usually small but nevertheless can have an important influence on vehicle handling, steering feel and tyre wear.

The interaction between steering and vertical suspension movement is controlled by the locus of the outer end of the tie rod. If this is coincident with the locus of the steering arm joint (attached to the hub), then no steer effects will occur. In relation to the two dimensional view shown in Fig. 2.30, the ideal centre of rotation for the tie rod can be calculated from geometry. Even this geometrical calculation may not be perfect because three dimensional effects are ignored. Hence CAD systems or purpose-design suspension kinematics packages are normally used for the suspension and steering layout.

2.5.3.1 Bump and Roll Steer

Figure 2.30b and c illustrate two possible effects in a qualitative fashion. In Fig. 2.30b the position of the tie rod, assuming it is mounted to the rear of the suspension, causes a toe out effect to occur as the wheel moves either in jounce (bounce) or rebound. However, if the inboard end of the tie rod were mounted inboard of the ideal point, then a toe-in effect would occur. If the error was symmetrical then no resulting steer effect would occur. On the other hand, if one wheel tie rod was inboard and the other outboard then during a bump or rebound incident the wheels would tend to steer together either left or right, giving rise to **bump steer**.

In Fig. 2.30c, the tie rod on the left hand wheel is mounted above the ideal point. This causes a right hand steer during bounce (toe-in) and a left hand steer during rebound (toe-out). The opposite effect occurs at the right hand wheel. Thus during body roll, both wheels steer in the same direction. For example, a right hand turn (positive steer) will cause the body to roll outwards to the left, giving rise to bounce motion at the left wheel and rebound motion at the right. Both wheels will steer to the right, (i.e. into the turn) thus giving an oversteer effect, giving rise to **roll steer**. It is clear that this arrangement should be avoided on "normal" vehicle designs.

2.5.3.2 Self-confirmation

Strike a true radius using the relay link details of Fig. 2.30a. Then, on the same drawing, strike a new radius using the outboard position of the steering link in Fig. 2.30b—note how the shorter radius blends at the steering arm ball joint but will "pull" the steering arm in at any other position (because of the shorter radius), causing the wheel to toe out (you may repeat with inboard position where the wheel will toe in as the radius is larger). Similarly strike a revised radius using the new position of the relay link as in Fig. 2.30c. Note how this curve and the idealised curve cross at the steering arm ball joint, being out-board above the ball joint and in-board below the ball joint. This causes the steering arm to "push out" (toe-in) during bump and "pull in" (toe-out) during rebound. This results in the wheels steering in the same direction during vehicle roll giving rise to roll steer.

2.6 Important Geometric Parameters in Determining Steering Forces

2.6.1 Front Wheel Geometry

The important elements of a steering system not only comprise the steering linkages as described above but also steer rotation about the steer rotation axis. The steer angle is achieved by rotation of the steered wheel about a rotation axis. This is generally not vertical and is tipped inboard at the top producing a lateral inclination angle and also tipped in a longitudinal plane to produce a castor angle. This combination of two inclined angles creates the "kingpin axis" and is this axis about which the wheel rotates. The general geometry for the lateral inclination angle and

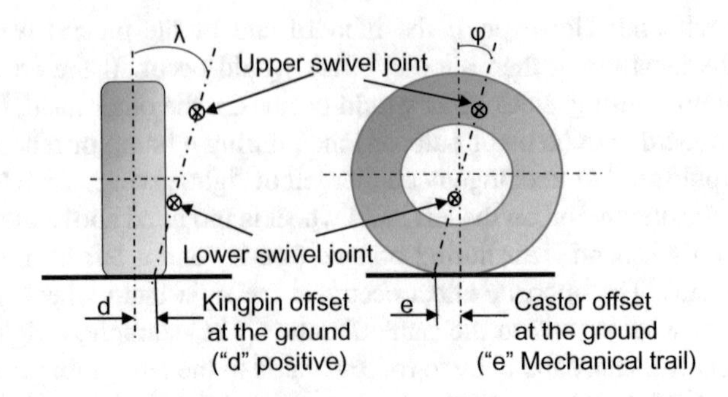

Fig. 2.31 The two axes that form the "kingpin axis"—the lateral inclination angle and the castor angle

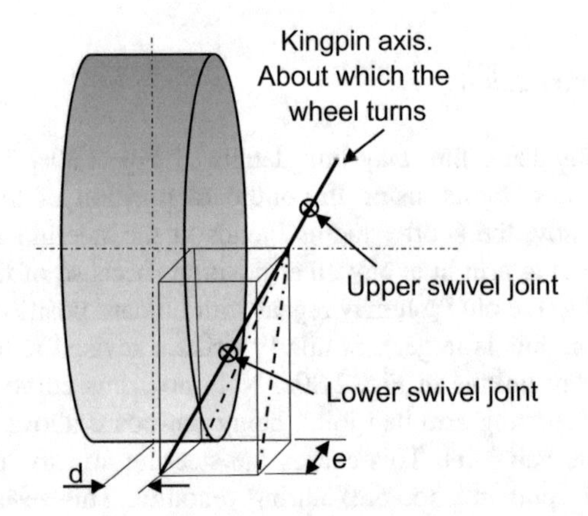

Fig. 2.32 Steer rotation geometry of a typical motor vehicle with positive lateral offset at the ground

the castor angle is as shown in Fig. 2.31. It is this geometry that determines the forces and moment reactions within the steering system.

This combined angle is the wheel axis of rotation or the swivel axis as shown in Fig. 2.32. The lateral inclination angle, commonly known as the **"kingpin"** inclination angle, may vary between 0–5° for trucks and 10–15° for passenger cars, the difference being due to vertical space availability. The intersection of the kingpin axis with the ground does not normally coincide with the centre of the tyre contact patch, there being a deliberate offset. This lateral offset (d) may be referred to as the **scrub radius** and is regarded as **negative** if it falls outboard of the tyre contact patch centre and **positive** if it falls inboard of the contact patch, as shown in Figs. 2.32 and 2.33 respectively.

The longitudinal inclination angle is referred to as **mechanical (or kinematic) castor**, positive castor being when the intercept at the ground is forward of the tyre contact patch. This distance (e) is known as the **kinematic castor trail or mechanical trail**. Castor angles normally range from 0 to 5°.

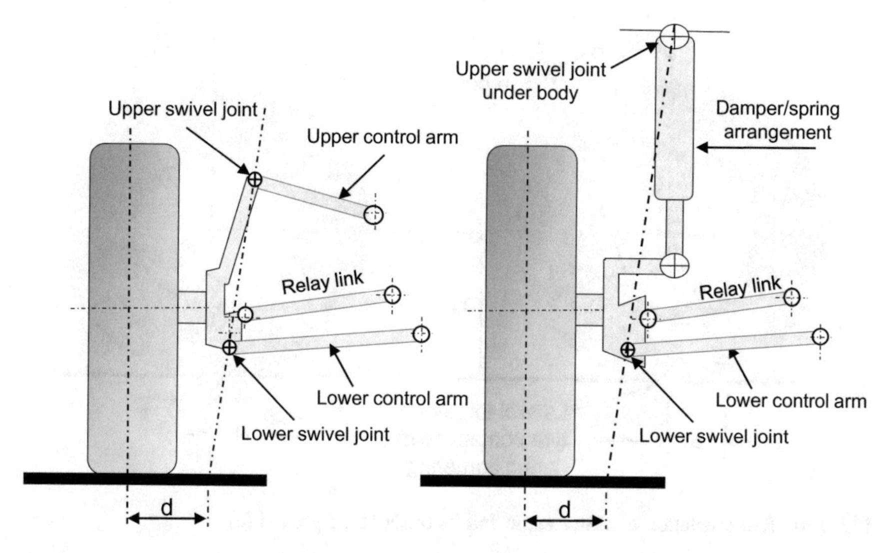

Fig. 2.33 Arrangement of double wishbone and McPherson strut with positive lateral offset at the ground

2.6.2 Kingpin Inclination Angle (Lateral Inclination Angle)

The kingpin inclination angle has the effect of causing the vehicle to rise when the wheels are turned. This creates a self-aligning moment which depends on the kingpin inclination angle, the scrub radius (offset at the ground) and the castor inclination angle.

Figure 2.33 shows arrangements for a double wishbone suspension and McPherson strut suspension both with positive lateral offset. It should be noted that the axis passed through both the upper and lower swivel joints for the double wishbone whilst for a McPherson strut the axis passes through the upper swivel joint at the body and the lower swivel joint on the lower suspension arm—it is **not** the axis through the damper/spring itself. With this arrangement the effect of the body weight is to maintain the wheel in the straight ahead position.

2.6.3 Castor Inclination Angle (Mechanical Castor)

Castor inclination axis is the inclination of the swivel axis in a longitudinal direction, shown **positive** in Fig. 2.34. The intercept of the axis with the ground provides a moment arm (**kinematic castor trail**) which in turn provides a speed dependent self-aligning moment resulting from lateral forces:

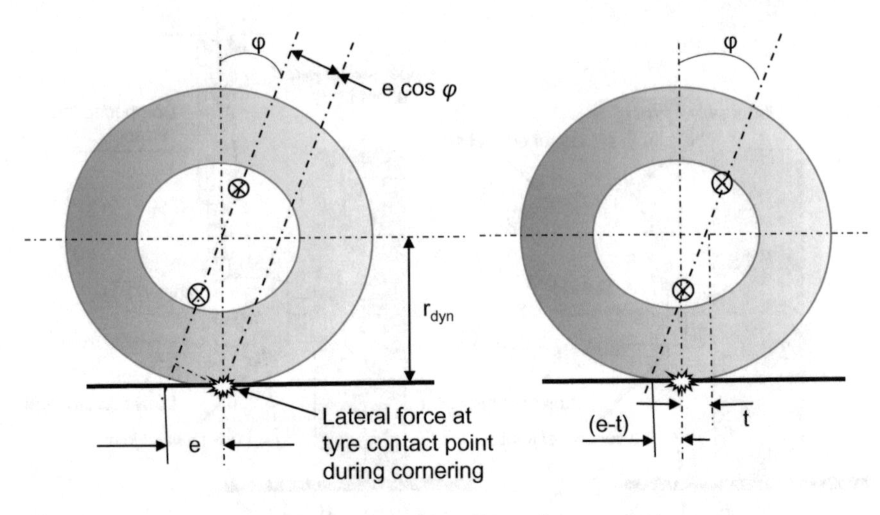

Fig. 2.34 Representation of castor angle and its offset at the ground (e)

$$Self\ aligning\ moment = \frac{mv^2}{R} \times Moment\ arm(e \cos \varphi) \tag{2.12}$$

where "v" is vehicle velocity, "m" is proportion of vehicle mass on the front tyres and "R" is turning radius.

An increased offset at the road will tend to increase self-alignment such that it may be felt at the steering wheel, thereby increasing steering effort. To take advantage of castor/camber offsets but maintain an acceptable steering effort, it may be necessary to introduce **"castor offset at the hub"**, shown as "t" in Fig. 2.34. This reduces the moment arm at the tyre contact by the distance "t" and is **negative** so giving a reduced overall offset at the ground (e − t). The benefits of negative castor offset at the hub, combined with positive castor angle, are that the kinematic castor trail is reduced so that the influence on steering from uneven road surfaces is reduced and that camber alteration is maintained when wheels are turned (Fig. 2.35).

For the steered wheels the camber angle changes as a function of the steer angle as shown in Fig. 2.35. In this example, it can be seen that, with increasing castor angle, there is a tendency to produce positive camber at the inboard wheel (circa 6.5°) and negative camber on the outside wheel (circa 3.8°) thus improving turn-in response. Since front wheel drive vehicles also create a self-righting moment due to the tractive forces, the resulting increased self-righting moment caused the designer in this case to introduce **negative castor**. This had the disadvantage of creating positive camber on the outside wheel during a turn and possibly an unstable situation—the negative castor will cause oversteer which in turn causes increased lateral forces. When negative castor is introduced to the non-steered rear wheels, the resulting moment, and system compliance, provides a toe-in effect creating understeer which assists in high speed lane changes.

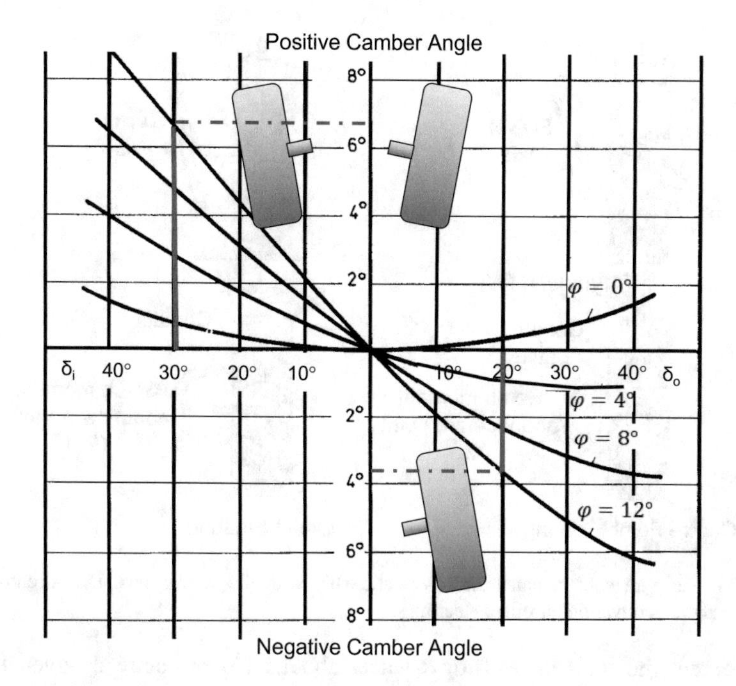

Fig. 2.35 Effects of camber angle with castor angle (where δ_i and δ_o are the inner and outer wheel steer angles)

2.7 Forces Associated with Steering a Stationary Vehicle

The maximum forces at the steering wheel occur when the vehicle is stationary; in practice, these forces also occur during slow speed parking manoeuvres. The forces arise from two effects—the frictional resistance due to the tyre scrubbing on the road and the jacking of the vehicle.

2.7.1 Tyre Scrub

The tyre contact patch and its relationship with the swivel axis is shown in Fig. 2.36. Estimating the moment acting around the scrub axis (scrub moment M_S) relies on some approximations and some empirical evidence. First, if the tyre contact patch is assumed to be circular (diameter = a) and the pressure distribution is uniform, then the moment required to turn the tyre around the centre of the contact area may be calculated by first considering the centre point steering case.

For centre-point steering the axis of swivel is about the centre of the tyre. Although this leads to light dynamic steering loads during driving it gives heavy static steering loads as the tyre "scrubs" about its contact patch. With such an

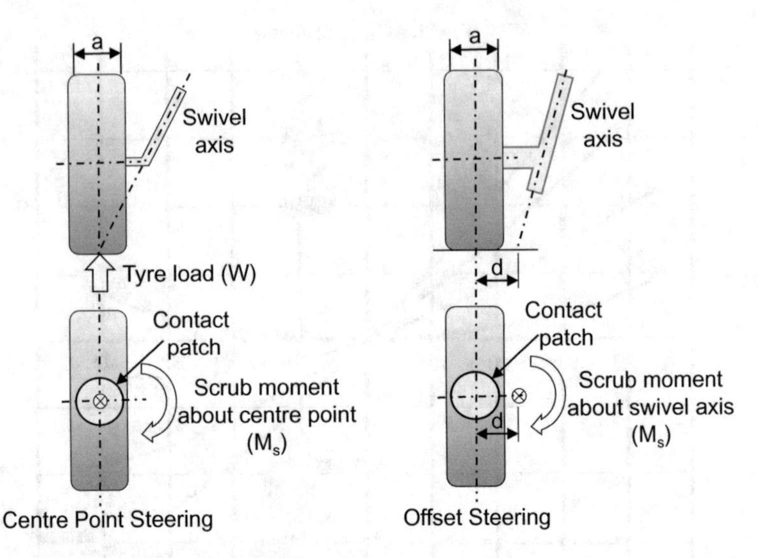

Fig. 2.36 The tyre contact patch and its relationship with the swivel axis. Left—centre point steering; right—conventional offset steering

arrangement the moment to turn a wheel about its own centre is given by the following derivation where "a" is the diameter of the tyre contact patch which is assumed equal to the contact width of the tyre.

To determine "scrub" moment or "dry parking" moment of a tyre about its centre.

Consider tyre footprint as a contact patch circle of radius R, where 2R = contact width of tyre "a", supporting a vertical load W.

Assume also:

Coefficient of friction between road and tyre μ
Uniform pressure between road and tyre p

Consider a small element within the idealised footprint as shown in Fig. 2.37. The elemental torque, dT, to rotate element about centre of contact area is given by:

$$dT = \mu \cdot p \cdot dA \cdot r = \mu \cdot \frac{W}{\pi R^2} \cdot r \cdot dr \cdot d\theta \cdot r = \mu \cdot \frac{W}{\pi R^2} \cdot r^2 \cdot dr \cdot d\theta$$

Therefore total torque is given by:

$$T = \int_0^R \int_0^{2\pi} dT = \mu \cdot \frac{W}{\pi R^2} \int_0^R \int_0^{2\pi} r^2 \cdot dr \cdot d\theta$$

$$T = \mu \cdot \frac{W}{\pi R^2} \cdot \frac{R^3}{3} \cdot 2\pi = \mu \cdot W \cdot \frac{2R}{3} = \mu \cdot W \cdot \left(\frac{a}{3}\right)$$

which is equal to the scrub turning moment i.e.

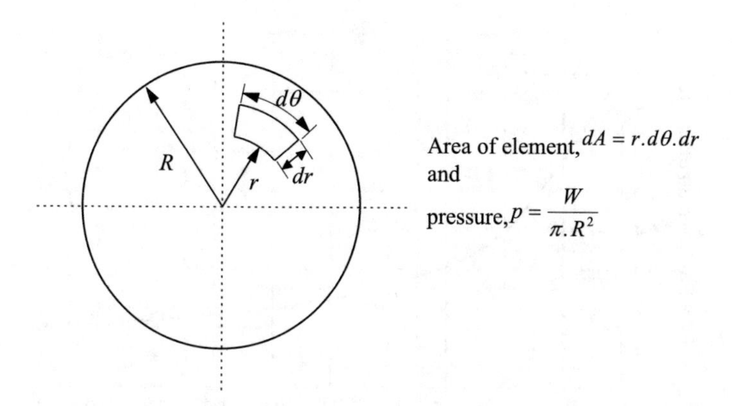

Fig. 2.37 Idealised representation of the tyre contact patch

$$\text{Turning Scrub Moment, } M_S = \mu W\left(\frac{a}{3}\right) \tag{2.13}$$

To reduce this load and provide dynamic "feel" for the driver the swivel centre is moved away from the centre of the tyre by a distance "d". With this arrangement the tyre tends to have a higher degree of roll and less "scrub".

An empirical equation is now used for dealing with this geometry which involves identifying an effective moment arm and a revised coefficient of friction.

The effective moment arm is given by the equation:

$$h = \sqrt{\left[d^2 + \left(\frac{a}{3}\right)^2\right]} \tag{2.14}$$

The equation for scrub moment now becomes:

$$\text{Turning Scrub Moment, } M_S = \mu_e W h \tag{2.15}$$

The effective coefficient of friction (μ_e) is obtained from empirical curves provided by the tyre manufacturer. A typical example is shown in Fig. 2.38 for a passenger car tyre on dry concrete. Using this curve it is seen with Centre Point Steering and d/a = 0 then the coefficient of friction is at its highest at around 0.8. The friction then rapidly drops to an approximately constant value of around 0.2 over the typical range of values of "d/a" used in practice. Such a graph may also be presented as M_S/W on the "y" axis against swivel offset "d" on the "x" axis.

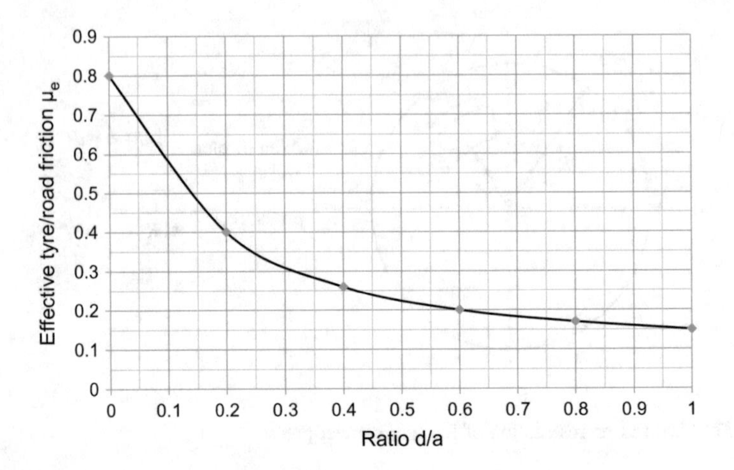

Fig. 2.38 Empirical curves for coefficient of friction

2.7.2 *Jacking of the Vehicle*

When the wheel is turned, the body will move upwards and the moment around the swivel axis to achieve this, **M_J, the jacking moment**, can be derived by first equating the rate of work involved at the swivel axis to that involved in jacking. Figure 2.39 shows the geometry involved with a view along the swivel axis and below the wheel indicating the locus of the wheel centre as the steer angle, α, increases.

Reference to Fig. 2.39 shows that the vertical displacement (x) increases with α and is given by:

$$x = d \sin \lambda (1 - \cos \alpha) \cos \lambda$$

Differentiating gives:

$$\frac{dx}{dt} = \frac{dx}{d\alpha}\frac{d\alpha}{dt}$$
$$\frac{dx}{dt} = d \sin \lambda \cos \lambda \sin \alpha \frac{d\alpha}{dt}$$

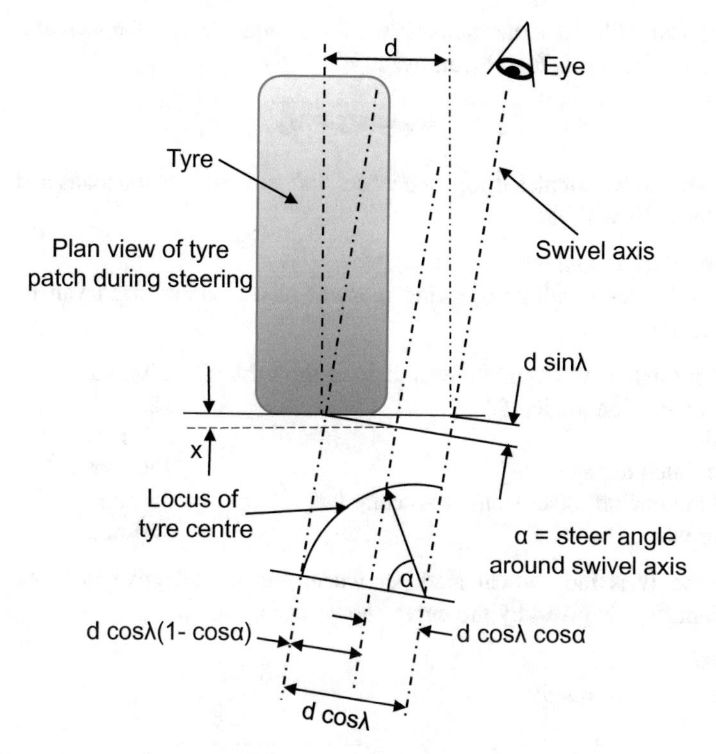

Fig. 2.39 Geometry demonstrating degree of vehicle lift as wheel is turned

The rate of doing work in jacking is therefore given by:

$$W\frac{dx}{dt} = Wd\sin\lambda\cos\lambda\sin\alpha\frac{d\alpha}{dt}$$

The rate of work around the swivel axis is:

$$M_J\frac{d\alpha}{dt}$$

Equating these rates of work leads to:

$$M_J = Wd\sin\lambda\cos\lambda\sin\alpha \tag{2.16}$$

Combined with the scrub moment, the total torque around the swivel axis (M_T) involved in steering is then given by:

$$M_T = M_S + M_J \tag{2.17}$$

This should be calculated for each wheel, left and right, as the loads and steering angles for each will vary.

Example E2.1 Part 1

To determine the scrub and jacking moment of a medium sized van having the following properties:

Total front axle load (equal on each side of the vehicle)	2400 kg
Lateral inclination angle, "λ"	12°
Tyre width	250 mm
Contact patch diameter "*a*"	250 mm
Lateral inclination offset from tyre centre "*d*"	25 mm
Steering wheel diameter	400 mm

Assume W is the vertical load per wheel and the effective interface friction coefficient "μ_e" is given by the curve shown in Fig. 2.38.

Solution

For the scrub moment:

$$h = \left[d^2 + \left(\frac{a}{3}\right)^2\right]^{1/2} = \sqrt{\left(25^2 + [250/3]^2\right)} = 87\,\text{mm}$$

and d/a = 25/250 = 0.1

From graph (Fig. 2.38), friction coefficient $\mu_e = 0.58$

Giving:

$M_S = \mu_e W h = 0.58 \times 1200 \times 9.81 \times 0.087 =$ **594 N m/wheel or 1188 N m total scrub.**

For the jacking moment:

$$M_J = Wd \sin\lambda \cos\lambda \sin\alpha$$

Left wheel = 1200 × 9.81 × 0.025 × sin 12 × cos 12 × sin 28 = **28 N m**

Right wheel = 1200 × 9.81 × 0.025 × sin 12 × cos 12 × sin 45 = **42 N m**

Summing the scrub and jacking moments gives the total moment at each wheel:

Total moment at the left hand wheel = 594 + 28 = **622 N m**

Total moment at the right hand wheel = 594 + 42 = **636 N m**

Note order of magnitude difference between scrub and jacking moments.

2.7.3 Forces at the Steering Wheel

A plan view of a steering system using a steering box is shown in Fig. E2.1a. The horizontal force in the centre of the track rod is the sum of two horizontal components arising from the moments around the swivel axles at each wheel. The force at the track rod is determined as indicated in Figs. E2.1a and E2.1b.

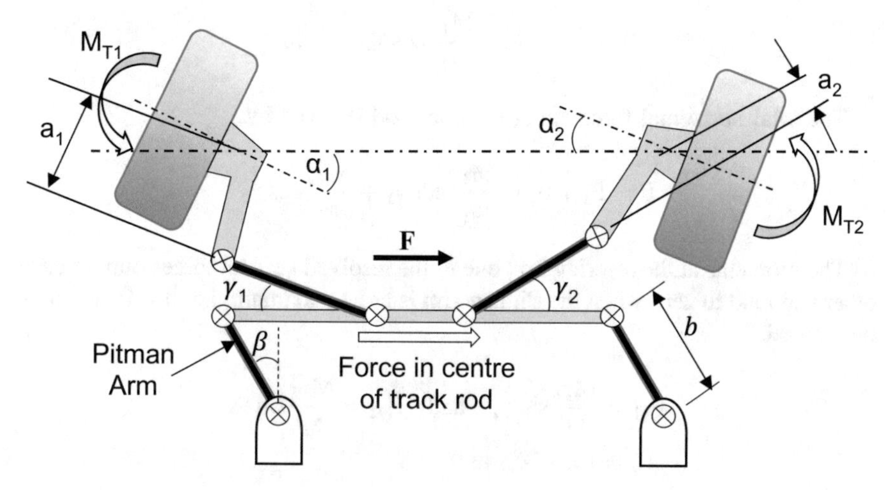

Fig. E2.1a General linkage arrangement for a steering box

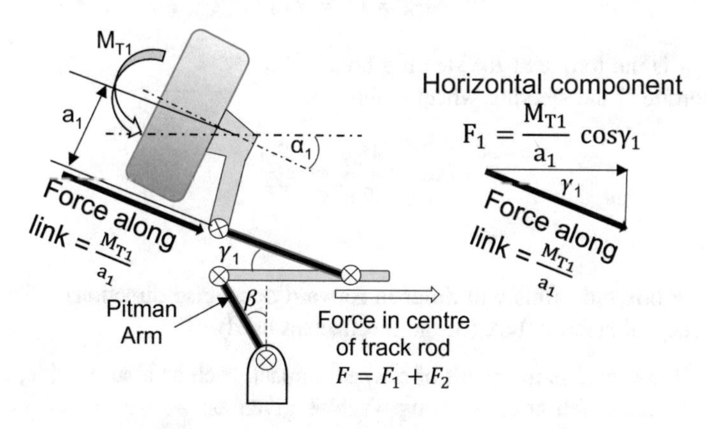

Fig. E2.1b Moments and forces associated with left hand wheel (wheel 1)

To calculate the force at the centre of the track rod.
Consider left hand wheel (wheel 1):
Resolving moment M_{T1} along link (strut) gives: Force $= M_{T1}/a_1$

This leads to horizontal force along track rod due to moment at left hand wheel (wheel 1) as

$$F_1 = \frac{M_{T1}}{a_1} \cos \gamma_1$$

Similarly for the right hand wheel (wheel 2):

$$F_2 = \frac{M_{T2}}{a_2} \cos \gamma_2$$

The total horizontal force along the track-rod is given by:

$$F = F_1 + F_2 = \frac{M_{T1}}{a_1} \cos \gamma_1 + \frac{M_{T2}}{a_2} \cos \gamma_2$$

The moments at the steering box due to the resolved vertical forces oppose each other and tend to zero when the Pitman arm is at its maximum, i.e. $\beta = 0$, and may be ignored.

$$F_{v1} = \frac{M_{T1}}{a_1} \sin \gamma_1 \quad \text{and} \quad F_{v2} = \frac{M_{T2}}{a_2} \sin \gamma_2$$

$$F_v = F_{v1} - F_{v2} \cong 0$$

The moment at the steering box is given by:

$$M_{SB} = Fb = T_{SB}$$

where T_{SB} is the torque at the steering box.

The torque at the steering wheel is then

$$T_{SW} = \frac{M_{SB}}{n\eta} \quad or \quad \frac{T_{SB}}{n\eta}$$

where

n steering box ratio (this will differ in forward or reverse direction),
η efficiency of steering box (or the mechanism used).

Note: If the force is the result of a wheel impact, such as a curb strike, then the "feedback" torque felt at the steering wheel is given as:

$$T_{SW} = \frac{M_{SB}\eta}{n} \quad or \quad \frac{T_{SB}\eta}{n}$$

i.e. the steering box efficiency now reduces the force felt at the wheel.

From Fig. E2.1c, we have:

Linear force at rack (F) × Pitch circle radius of pinion (r) = Tangential force at steering wheel (E) × Radius of steering wheel (R)

For a given steering force required at the rack, the overall steering force ratio is therefore given by:

$$\text{Steering force (gear) ratio} = F/E = R/r$$

It will be seen that the ratio is inversely proportional to the pinion radius. If the radius changes so does the ratio. This may assist in providing a constant force when the force at the rack varies due to the effective length of the steering arms changing as the road wheels turn. Such a technique is shown in Fig. 2.7 where a variable pitch rack gives rise to an effective change in the pinion radius and so a variable steering ratio.

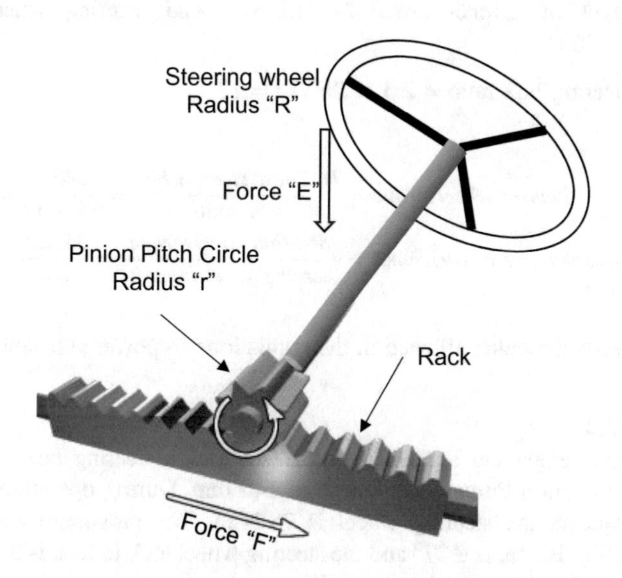

Fig. E2.1c Diagram showing mechanical advantage of steering wheel

Example E2.1 Part 2

To determine tangential hand force required at the steering wheel.

Given:

Left hand wheel (wheel 1)	Right hand wheel (wheel 2)
$a_1 = 127$ mm	$a_2 = 30$ mm
$\gamma_1 = 8°$	$\gamma_2 = 12°$
$\alpha_1 = 28°$	$\alpha_2 = 45°$

Pitman arm length (b)	120 mm
Maximum Pitman arm angle (β) (maximum)	37°
Steering wheel turns lock to lock	2.5
Efficiency	80%
Steering wheel diameter	400 mm

Horizontal force along track rod due to moment is given by general equation:

$$F = \frac{M\cos\gamma}{a}$$

$$\textit{Total horizontal force} = \frac{622 \times \cos 8}{0.127} + \frac{636.6 \times \cos 12}{0.030} = 25606.3\,\text{N}$$

$$\textit{Torque at steering box} = \textit{Force} \times \textit{moment arm} = 25606.3 \times 0.120 \times \cos 37 = 2454\,\text{N m}$$

Total travel of Pitman arm = 74° (± 37°) and steering wheel lock to lock = 2.5 × 360°

Giving steering box ratio = 2.5 × 360/74 = 12

Thus

$$\textit{Steering wheel torque} = \frac{\textit{Torque at steering box}}{\eta \times \textit{ratio}} = \frac{2454}{0.8 \times 12} = 255.63\,\text{N m}$$

$$\textit{Tangential force required by each hand} = \frac{\textit{Steering wheel torque}}{\textit{Steering wheel diameter}} = \frac{255.63}{0.4} = 640\,\text{N}$$

This exceeds the value allowed in the regulations so power assistance would be required.

Example E2.2

The steering arrangement shown in Fig. E2.1a uses a steering box with an efficiency of 80% and a Pitman arm length of 140 mm. During operation it is found that the torque at the steering wheel is 75 N m. The maximum angle "β", as indicated in Fig. E2.1a, is ±37° and the steering wheel lock to lock is 3.5 turns. It is intended to replace the steering box with a rack and pinion mechanism with an initial selection using a constant pitch rack and 7 tooth pinion with 3 module tooth form (3 mm addendum) with an expected overall efficiency of 97%.

The task is to investigate the viability of replacement of the steering box and the impact on driver perception.

Initial analysis

It is required to determine the steering wheel torque with the new arrangement and whether it will be seen by the driver as significantly different. Work done is the basis of this analysis.

Because of geometric and packaging constraints, the linear distance travelled by the rack must be the same as the travel at the end of the Pitman arm (L).

Linear travel $L = 2 \times 40 \sin 37 = 168.5\,\text{mm}$

PCD of pinion = Number of teeth × module = 7 × 3 = 21 mm

Number of revolutions of pinion to give rack movement N = 168.5/21π = 2.55 revolutions.

$$New\ torque = Initial\ torque \times \frac{Box\ efficiency}{Rack\ efficiency} \times \frac{Box\ steering\ wheel\ rotation}{Rack\ steering\ wheel\ rotation}$$

Giving

$New\ torque = 50 \times \frac{0.8}{0.97} \times \frac{3.5}{2.55} = 56.59\,\text{N m}\ i.e.\ 13\%\ increase$

This is considered too great an increase and modifications need to be considered to maintain the steering wheel torque as close as possible to the original torque of 50 N m.

Second analysis

Investigate the revised tooth module if the number of teeth on the pinion remains at 7 teeth.

Turns of steering wheel would only remain same if the efficiencies of the 2 systems were the same—but they are not. Therefore:

Number of turns $= N = 3.5 \times \frac{0.8}{0.97} = 2.89$ i.e. less turns because of increased efficiency.

But $Length\ of\ travel = 168.5 = N \times PCD \times \pi = 2.89 \times PCD \times \pi$

This gives:

$$PCD = \frac{168.5}{2.89\pi} = 18.55\,\text{mm}$$

If number of teeth remains at 7 then revised tooth module = 18.55/7 − 2.65 mm—say 2.5 mm (as a standard size). This gives $PCD = 7 \times 2.5 = 17.5\,\text{mm}$ and number of revolutions of pinion:

$$N = \frac{132.4}{\pi 17.5} = 3.06\ revolutions$$

Thus torque at steering wheel given by:

$$T = 50 \times \frac{0.8}{0.97} \times \frac{3.5}{3.06} = 47.16\,\text{N m}$$

This gives a 4% reduction in torque which is considered to be acceptable.

Third analysis

Another consideration would be to revise the number of teeth and maintain the same tooth module.

If the tooth module was to remain the same at 3 mm and the track travel was to remain the same (work done) if the torque was to be close to 50 N m then the PCD of the pinion would need to remain close to 18.55 mm. The number of teeth would then become.

$$N = PCD/m = 18.55/3 = 6.18 = 6\,\text{teeth}.$$

This is tending to being a little low from a design or manufacturing perspective but would remain as a potential solution.

In all cases the tooth **strength and wear** must be considered.

Option 1—For a PCD of 17.5 mm and module of 2.5 mm, the root diameter of the pinion would be,

$$\begin{aligned} \mathit{Root\,diameter} &= \mathit{PCD} - (2 \times \mathit{dedendum}) = 17.5 - (2 \times [2.5 \times 1.25]) \\ &- 11.25\,\text{mm} \end{aligned}$$

Note: For an uncorrected gear tooth form the dedendum = 1.25 × addendum (or module).

Option 2—A similar analysis using 6 teeth and 3 module leads to a root diameter of **10.5 mm**.

Conclusion—A modification would be best introduced if the number of teeth remained the same at 7 but reducing the module to 2.5 mm.

Example E2.3

It is intended to design the electrical drive steering system shown in Fig. E2.3a that can be described as follows:

- An DC electric motor drives a worm and wheel (w/w) gear mechanism
- This w/w gear unit rotates a recirculating ball nut that powers a screw type track rod that moves laterally to steer both wheels.

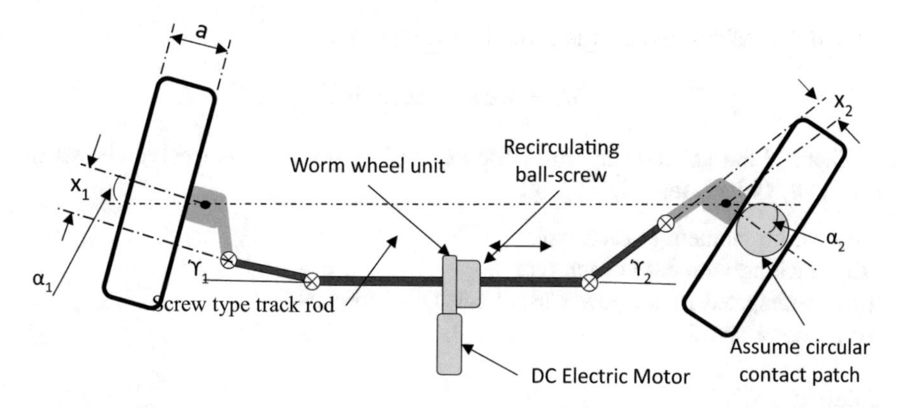

Fig. E2.3a Steering arrangement

The general detail of the arrangement is as follows:

Mass on front axle (equal per wheel)		800 kg
Lateral inclination angle	λ	12°
Contact patch diameter	a	200 mm
Lateral inclination offset from tyre centre	d	20 mm
Recirculating screw pitch	p	6 mm
Screw arrangement efficiency	η_s	90%
Worm and wheel gear ratio	R	12:1
Worm gearing running efficiency	η_w	60%
Lateral movement of screw to give lock to lock steer of wheels		180 mm
Time to move wheels lock to lock	t	6 s

The geometric parameters for each wheel at the shown point of steer are:

Wheel 1	Wheel 2
$x_1 = 127$ mm	$x_2 = 30$ mm
$\gamma_1 = 8°$	$\gamma_2 = 12°$
$\alpha_1 = 28°$	$\alpha_2 = 45°$

Given the general equation for wheel scrub (or parking torque) is as follows:

$$M_S = \mu_e W h \quad \text{where} \quad h = [d^2 + (\frac{a}{3})^2]^{\frac{1}{2}}$$

And the jacking moment is given by the equation

$$M_J = Wd \sin \lambda \cos \lambda \sin \alpha$$

Then, if the effective interface friction "μ_e" is given by the curve shown in Fig. 2.38, **Calculate:**

(i) scrub moment at each tyre
(ii) jacking moment at each tyre
(iii) anticipated motor power using calculated torques
(iv) motor speed.

Solution

(i) **To determine scrub moment at each tyre**

First determine effective coefficient of friction:
d/a = 20/200 = 0.1 giving μ_e = 0.55 from Fig. 2.38:

$$h = \sqrt{\left[d^2 + \left(\tfrac{a}{3}\right)^2\right]} = \sqrt{\left(20^2 + [200/3]^2\right)} = 69.6 \text{ mm}$$

Giving:
$M_S = \mu_e Wh = 0.55 \times 400 \times 9.81 \times 0.0696 = 150.2$ N m/wheel.

(ii) **To determine jacking moment at each tyre**

The jacking moment is given by:

$$M_J = Wd \sin \lambda \cos \lambda \sin \alpha$$

Left wheel = 400 × 9.81 × 0.02 × sin 12 × cos 12 × sin 28 = 7.5 N m
Right Wheel = 400 × 9.81 × 0.02 × sin 12 × cos 12 × sin 45 = 11.3 N m

Giving total left hand wheel moment = 150.2 + 7.5 = 157.7 N m
Giving total right hand wheel moment = 150.2 + 11.3 = 161.5 N m

(iii) **To estimate motor power using calculated torques**

Horizontal force along track rod due to moment is given by

$$F = \frac{M \cos \gamma}{a}$$

$$\textit{Total Horizontal Force} = \frac{157.7 \times \cos 8}{0.127} + \frac{161.5 \times \cos 12}{0.030} = 6495 \text{ N}$$

$$\textit{Power needed at the trackrod} = \textit{Force} \times \textit{Distance} \div \textit{Time} = \frac{6495 \times 0.180}{6} = 195 \text{ W}$$

$$\textit{Motor power} = \frac{\textit{Power at Track rod}}{\textit{Efficiencies}} = \frac{195}{0.6 \times 0.9} = 361 \text{ W}$$

(iv) **To determine the motor speed**

Worm pitch = 6 mm giving revolution of ball-screw nut = 180/6 = 30 revolutions. Therefore:

$$Motor\ Speed = \frac{Ball\ screw\ revolutions \times Worm\ ratio}{Time} = \frac{30 \times 12}{6} \times 60$$
$$= 3600\,\text{rev/min}$$

2.8 Forces Associated with Steering a Moving Vehicle

The complete sets of forces and moments acting on a rolling tyre is shown in Fig. 2.40 using the SAE J670e tyre axis system. All three forces plus three associated moments affect the moment around the swivel axis. These are:

- Vertical force (normal wheel load) and rotational aligning moment
- Lateral force (cornering) and rotational overturning moment
- Longitudinal force (traction/braking) and rotational rolling resistance moment.

The moments around the swivel axis of each wheel are summed through the steering system and eventually act through the steering gear to be felt at the steering wheel. Each of these forces are considered in turn below.

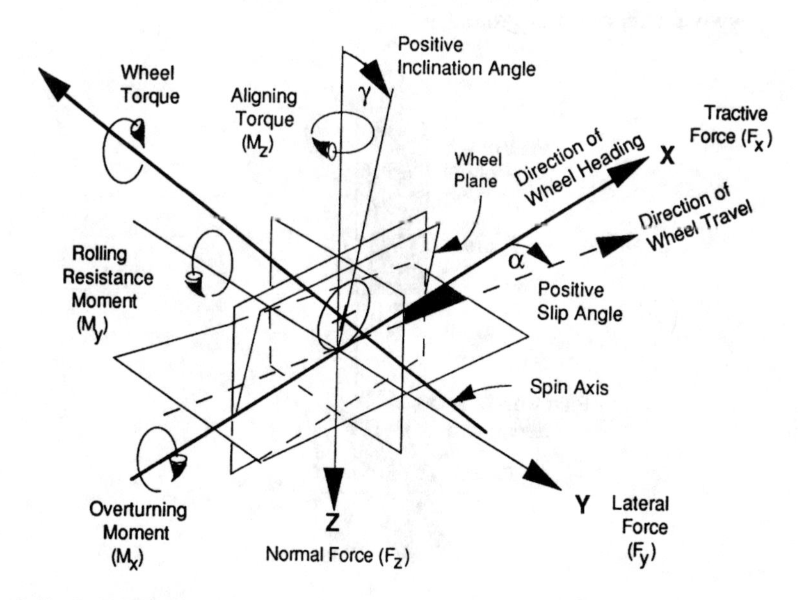

Fig. 2.40 SAE J670e tyre axis system

2.8.1 Normal Force

2.8.1.1 Effect of Lateral Inclination Angle

This has effects due to both the swivel axis inclination and the castor angle. The first of these is the same as the jacking effect discussed above. Using the SAE axis system, the swivel axis effect results approximately in a moment for the left hand wheel as shown in Fig. 2.41.

$$Moment = -F_{zl} \sin \lambda d \sin \alpha \tag{2.18}$$

The moment is negative because rotation is anticlockwise. Note that the moment is in a direction to return the wheel to a dead ahead position, creating a self-centring effect.

Similarly the moment for the right hand wheel is:

$$Moment = -F_{zr} \sin \lambda \, d \, \sin \alpha \tag{2.19}$$

Giving:

$$Total\, Moment\, about\, kingpin\, axis = -(F_{zl} + F_{zr}) \sin \lambda \, d \, \sin \alpha \tag{2.20}$$

where:

F_{zl}, F_{zr} vertical loads at left and right wheel
d swivel axis offset at ground

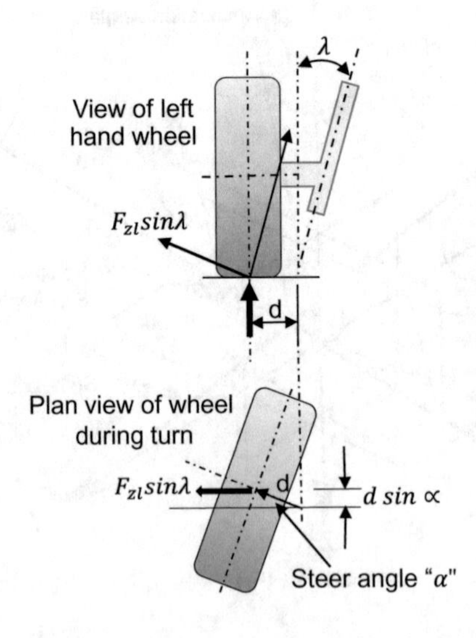

Fig. 2.41 Moment due to vertical load and lateral inclination angle

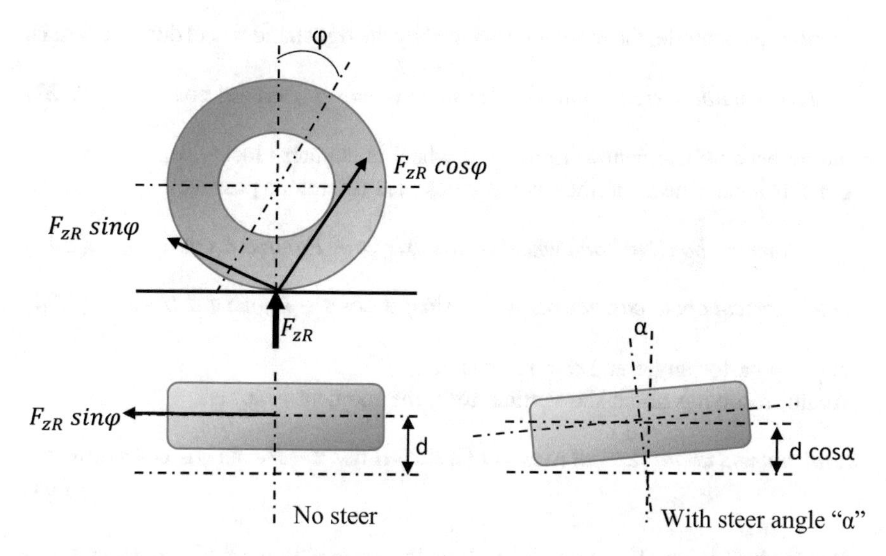

Fig. 2.42 Moment produced by vertical force due to effect of castor angle as wheel is steered

λ swivel axis inclination angle
α mean steer angle (these will differ slightly for each wheel).

During steer, the moments resulting from the vertical load and swivel axis inclination angle on both the left and right wheels act together to produce a cumulative centring moment.

It is appropriate to resolve the moments about the vertical axis giving:

$$\textit{Total Moment about vertical axis} = -(F_{zl} + F_{zr}) \sin \lambda \, d \, \sin \alpha \, \cos \lambda \qquad (2.21)$$

(Compare with Eq. 2.16 for Jacking Moment).

Note that, if "λ" is small, cos $\lambda \rightarrow 1$. Equally, by observation of the moment equations, it may be concluded (because of the double sine terms and small angles) that the moments will be small. Regardless, from a design perspective, it is best to confirm magnitudes before deciding whether the values need consideration.

2.8.1.2 Effect of Castor Inclination Angle

From Fig. 2.42, the component of F_{zr} perpendicular to the plane of the castor angle is F_{zr} sin φ and when the wheel is turned, the appropriate moment arm is approximately d cos α. Note that the moments at left and right wheel arising from this geometry oppose each other.

For the castor angle, the moment produced by the right hand wheel during a turn is:

$$\textit{Moment about right hand wheel castor axis} = -F_{zr} \sin \varphi\, d \cos \alpha \tag{2.22}$$

(negative because moment at right hand wheel is counter clockwise).
For left hand wheel the moment is clockwise (therefore positive):

$$\textit{Moment about left hand wheel castor axis} = +F_{zl} \sin \varphi\, d \cos \alpha \tag{2.23}$$

$$\textit{Total moment about castor axes} = (F_{zl} \sin \varphi\, d \cos \alpha - F_{zr} \sin \varphi\, d \cos \alpha) \tag{2.24}$$

where φ = castor angle and α = steer angle.
Again, resolving about the vertical axis, the moment becomes:

$$\textit{Total moment about vertical axes} = (F_{zl} \sin \varphi\, d \cos \alpha - F_{zr} \sin \varphi\, d \cos \alpha) \cos \varphi \tag{2.25}$$

But as "φ" is small, $\cos \varphi \to 1$. Equally, observation of the moment terms shows that the moments oppose each other so the overall effect is minimal.

Example E2-4
A car has a mass of 1200 kg with weight distribution 52% at the front and 48% at the rear. The car is right hand drive and the driver mass (m_d) is 100 kg. The proportion of driver mass distribution on the front axle is two thirds to the front offside wheel and one third to the nearside wheel, as shown in Fig. E2.4a.

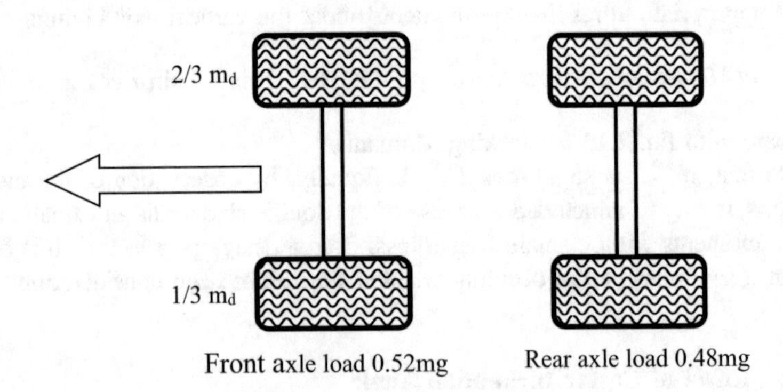

Fig. E2.4a Plan view of car and loading

On an automotive production line it is noted that the castor angles differ on each side of the car by an average of 0.1°. Other important properties of the car are:

Left hand side castor angle (φ_L) 5°
Right hand side castor angle (φ_R) 5.1°

Kingpin offset at the ground (*d*)	18 mm
Lateral inclination angle (λ)	12°
Wheelbase	2.55 m

Let:

Steer angle be given by	α
Vertical wheel load be given by	F

By first order analysis the engineer is asked to determine:

(i) The torque necessary on both road wheels to hold the wheels at zero steer angle due to the differences in castor and wheel loads.
(ii) The steer angle at which the torque from the lateral inclination angle balances out that from the castor if the steering wheel is released.
(iii) The distance the vehicle can travel before it deviates laterally by 1 m (to the edge of the lane) at this steer angle if the vehicle is initially centred and aligned in a road lane.

Solution

Total unladen front axle load = 0.52 × 1200 = 624 kg = 6121 N giving 3061 N/wheel.

Additional load due to driver = 100 kg = 981 N with distribution 2:1 right to left.

This gives 654 N right, 327 N left due to driver.

Giving right hand wheel vertical load = 3715 N and left hand wheel vertical load = 3388 N.

(i) Calculate the torque necessary on both road wheels to hold the wheels at zero steer angle due to the differences in castor and wheel loads

We have $\cos\varphi = \cos 5 = 0.996$ so let it be = 1.

Given left wheel castor to be 5° and right wheel 5.1°, the steering torque from the difference in the castor angle and wheel loads is then given by Eq. (2.25) as:

$$\begin{aligned} M_{castor} &= F_{zl}\, d \sin\varphi_l \cos\alpha - F_{zr}\, d \sin\varphi_r \cos\alpha \\ &= 3388 \times 0.018 \sin 5 \cos 0 - 3715 \times 0.018 \sin 5.1 \cos 0 \\ &= -0.63\,\mathrm{N\,m}(\textit{negative gives CCW moment and steering torque to the left}) \end{aligned}$$

(ii) Calculate the steer angle at which the torque from the lateral inclination angle balances out that from the castor if the steering wheel is released.

We have $\cos\lambda = \cos 12 = 0.978$ so let it be = 1.

The steer angle at which the torque balance occurs is then given by equating the total torque due to the lateral inclination angle from Eq. (2.21) to the torque due to the castor angle as follows:

$$
\begin{aligned}
M_{lat.incl} &= -(F_{zl} + F_{zr}) d \sin\lambda \sin\alpha \\
&= -(7103 \times 0.018 \sin 12 \sin \propto) \\
&= -26.58 \sin \propto \\
&= M_{castor} = -0.63\,\text{N m}
\end{aligned}
$$

Giving:

$$\sin \propto = 0.0237 \; or \; \propto = 1.358^\circ$$

(iii) **Calculate the distance the vehicle can travel before it deviates laterally by 1 m (to the edge of the lane) at this steer angle if the vehicle is initially centred and aligned in a road lane**

The lateral deviation may be determined from the geometry of a vehicle travelling on the arc of a circle:

$$
\begin{aligned}
R &= \frac{57.3^\circ/\text{rad}}{\propto} \times wheelbase \\
&= \frac{57.3}{1.358} \times 2.55 = 107.6\,\text{m}
\end{aligned}
$$

For a given angle of travel (ψ) on the circle, the lateral deviation (Y) is:

$Y = R(1 - \cos\psi) = 1\,\text{m}$ *and, as* $R = 107.6\,\text{m}$, *this gives* $\cos\psi = 0.9907$ *or* $\psi = 7.82^\circ$

The longitudinal distance travelled is:

$$X = R \sin\psi = 107.6 \sin 7.82 = 14.64\,\text{m}$$

Thus, without steer correction by the driver, the vehicle would leave the lane in around 14.64 m of travel down the road.

2.8.2 Lateral Force

2.8.2.1 Moment Due to General Offset

Lateral forces produce a self-centring moment because they act at the centre of the tyre at a distance of $r \tan\varphi$ behind the effective centre of rotation defined by the castor angle as shown in Fig. 2.43. The $r \tan\varphi$ term (where r = tyre radius) is often called the **castor trail** (or **mechanical trail**).

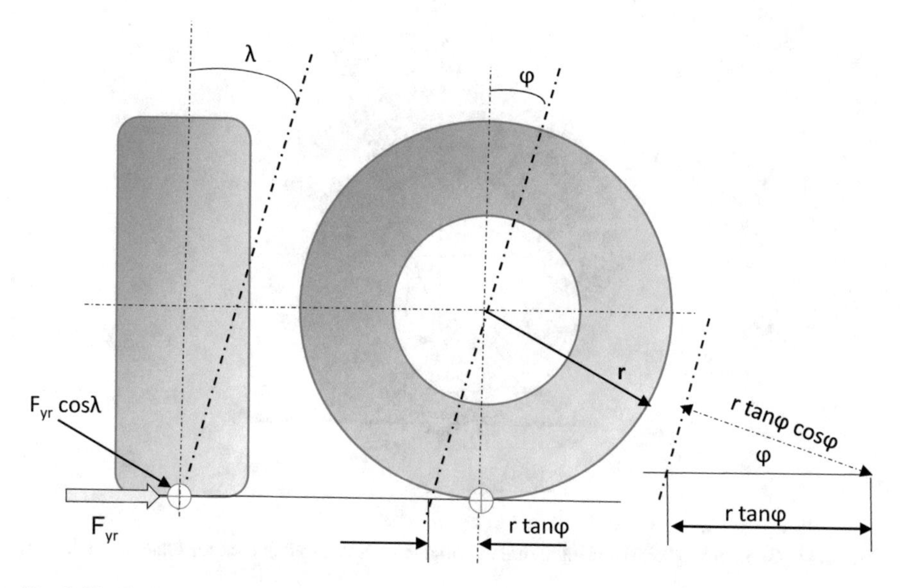

Fig. 2.43 Geometry of self-centring moment arm due to castor trail

The lateral force $\left(F_{yr}\right)$ resolved about the lateral inclination angle is $F_{yr}\cos\lambda$. This gives a self-centring moment about the right wheel castor axis:

$$M_r = F_{yr}\cos\lambda\, r\,\tan\varphi\cos\varphi \tag{2.26}$$

In fact, both wheels create a self-centring moment giving a total moment:

$$\textit{Total self}-\textit{centring moment} = \left(F_{yr}+F_{yl}\right)\cos\lambda r\tan\varphi\cos\varphi \tag{2.27}$$

where generally $F_{yr} = \mu_r \times F_{zr}$ and $F_{yl} = \mu_l \times F_{zl}$

2.8.2.2 Inclusion of Castor Offset

If **castor offset** is included then the effective mechanical castor trail is reduced by an amount equal to the castor offset (t) as shown in Fig. 2.44 to give the **kinematic trail** $(r\tan\varphi - t)$.

Equation (2.27) then becomes:

$$\textit{Total self}-\textit{centring moment} = \left(F_{yr}+F_{yl}\right)\cos\lambda(r\tan\varphi - t)\cos\varphi \tag{2.28}$$

2.8.2.3 Inclusion of Tyre (or Pneumatic) Trail

During cornering the tyre contact patch deforms into a kidney shape as shown in Fig. 2.45.

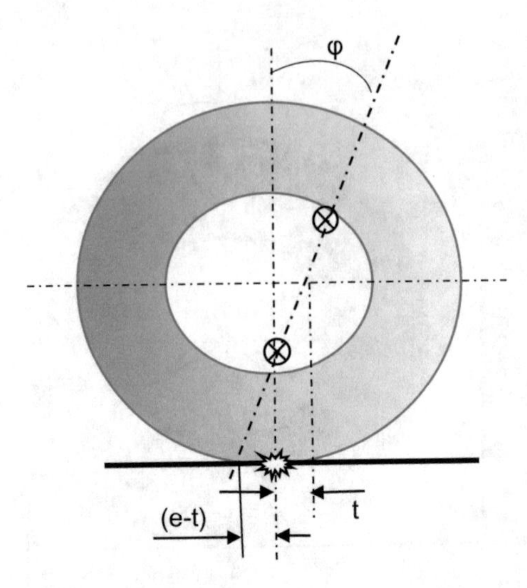

Fig. 2.44 Geometry of self-centring moment arm due to castor trail and castor offset at the hub (t)

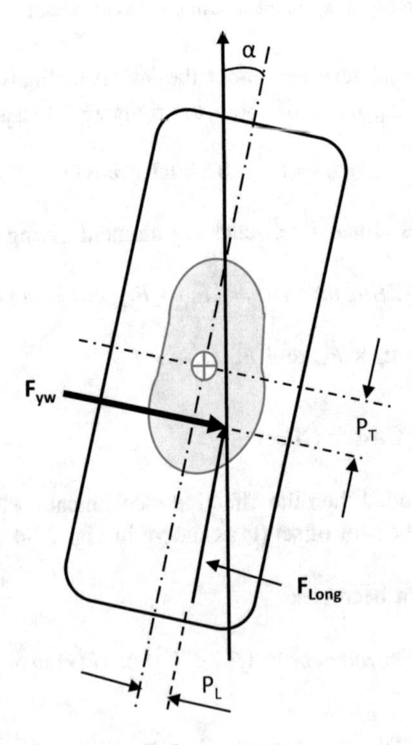

Fig. 2.45 Tyre contact patch during cornering

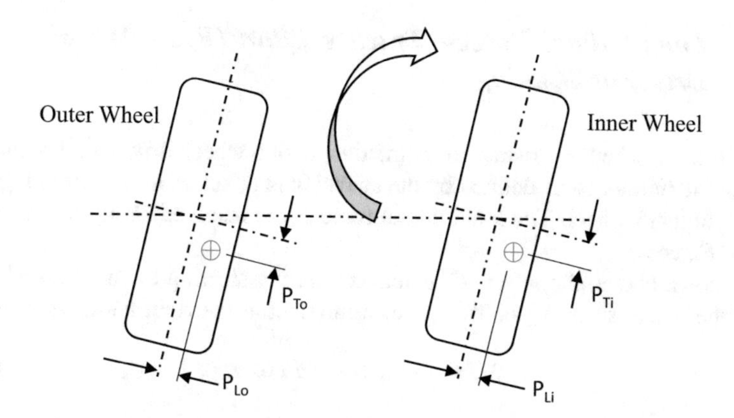

Fig. 2.46 Centre of tyre contact patch alters to rear and towards centre of turning radius. This alters centre of point of action for vertical and lateral forces

The tyre centre of pressure, where the lateral force at the wheel (F_{yw}) and longitudinal force (F_{Long}) are assumed to act, shifts rearward of the wheel centre to provide an additional trail known as **tyre trail** or **pneumatic trail** (P_T). Under braking or tractive effort during cornering, the centre of pressure also shifts away from the tyre centre by a lateral amount (P_L) to create an additional moment resulting from the tractive or braking effort.

If the pneumatic trail is included, then the equation for the outside wheel becomes:

$$\textit{Total self} - \textit{centring moment} = \left(F_{yr} + F_{yl}\right) \cos \lambda (r \tan \varphi - t + P_T) \cos \varphi \quad (2.29)$$

Note that the effective lateral tyre centre moves inboard on the outer tyre and outboard on the inner tyre. This produces the situation shown in Fig. 2.46 where the lateral tyre contact patch moment arm is different for inboard and outboard wheels.

The pneumatic trail may only be determined from test information on a particular tyre. The tyre is tested on a drum roller rig or a tyre rig with a moving flat table. The latter runs on a hydrostatic film of fluid to reduce friction. Both types of rig are able to measure the wheel load force, side force (and therefore lateral dynamic friction μ_y), slip angle and self-aligning moment in one process—the wheel load serving as a parameter. It is then possible to calculate the tyre trail for a range of wheel loads and slip angles. It should be noted that the movement is dependant of the tyre load so the pneumatic trail positions will differ for the outer and inner tyres. This will influence the overall steering "feel" (feedback) during cornering.

2.8.3 Longitudinal Force—Tractive Effort (Front Wheel Drive) or Braking

These forces, whether arising from traction (front wheel drive) or braking, act through the moment arm defined by the swivel axis offset, *d*, as shown in Fig. 2.47 for a right hand wheel. The longitudinal force F_{xr} is resolved about the castor axis to give $F_{xr}\cos\varphi$.

As shown above, the offset "d" is reduced during steering to be "$d\cos\alpha$" where "α" is the wheel steer angle. This gives a moment at the right hand wheel as:

$$M_r = +F_{xr}\cos\varphi\, d\, \cos\alpha \tag{2.30}$$

This is positive because the moment is clockwise on the right hand wheel. On the left hand wheel the moment is counter clockwise i.e. negative:

$$Total\,Moment = F_{xr}\cos\varphi\, d\cos\alpha - F_{xr}\cos\varphi\, d\cos\alpha \tag{2.31}$$

If $\alpha = 0$ this reduces to:

$$Total\,Moment = (F_{xr} - F_{xr})\cos\varphi\, d \tag{2.32}$$

Note that the effects at each side tend to cancel each other out but the resulting overall moment is sensitive to left/right wheel forces. If there is a difference between these forces then the vehicle will pull to one side during traction/braking.

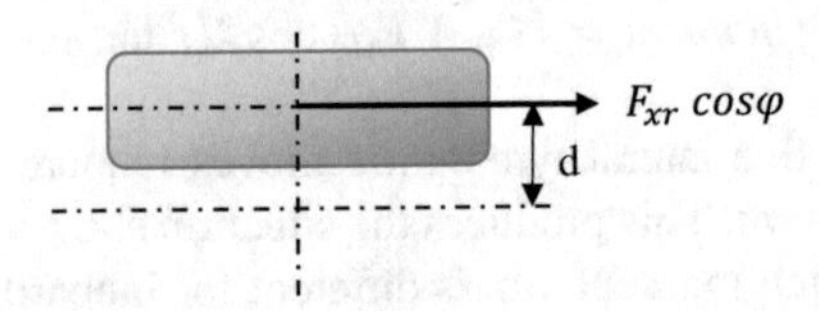

Fig. 2.47 Longitudinal force (braking) creating moment at right hand wheel due to lateral offset at the ground

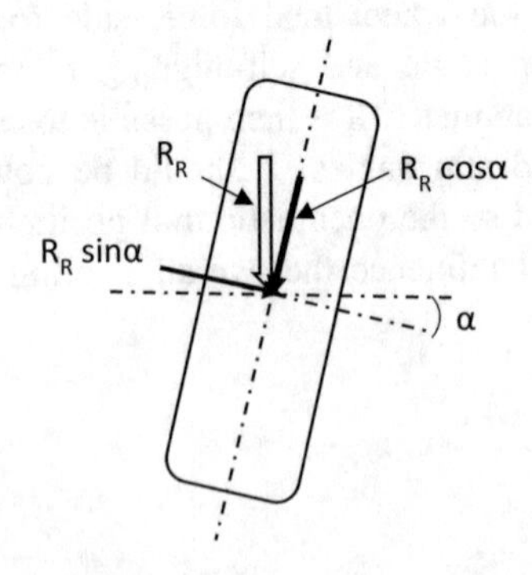

Fig. 2.48 Rolling resistance forces acting at the front wheels

If the tyre lateral offset during cornering (Fig. 2.48) is taken into account then for tractive effort due to front wheel drive:

$$\text{Moment}_{\text{(outer wheel)}} = F_{xo} \cos \varphi (d - P_{Lo}) \text{ for the outside wheel} \quad (2.33)$$

where "d" is the offset at the ground and may be negative or positive.

And for the inner wheel the moment is:

$$\text{Moment}_{\text{(inner wheel)}} = F_{xi} \cos \varphi (d + P_{Li}) \text{ for the inside wheel} \quad (2.34)$$

and if $F_{xi} = F_{xo} = F_a$ and $P_{Li} = P_{Lo} = P_L$ then:

$$\textit{Total Moment} = F_a(d - P_L) \cos \varphi - F_a(d + P_L) \cos \varphi \quad (2.35)$$

Leading to:

$$\textit{Total Moment} = -F_a 2P_L \cos \varphi \quad (2.36)$$

where F_a = axle tractive effort or braking force (when sign will change).

It must be recognised that the correct sign needs to be attributed to the offset "d" and tyre pneumatic lateral offset "P_L" as well as to the direction of the longitudinal forces.

2.8.4 Rolling Resistance and Overturning Moments

This acts as shown in Fig. 2.48 and may be resolved into two forces, one normal to the wheel ($R_R \sin \propto$) and the other along the wheel ($R_R \cos \propto$) where $\propto$ is the steer angle.

The $R_R \sin \propto$ component acts to provide a centring moment on both wheels as follows:

$$\textit{Moment} = R_R \cos \lambda (r \tan \varphi - t + P_T) \cos \varphi \quad (2.37)$$

The $R_R \cos \propto$ component provides an additional opposing moment at each wheel as follows:

$$\textit{Moment} = R_R \cos \propto \cos \varphi (d - P_{Lo}) \text{ for the outside wheel} \quad (2.38)$$

and

$$\textit{Moment} = R_R \cos \propto \cos \varphi (d + P_{Li}) \quad (2.39)$$

where

P_L pneumatic lateral offset, assumed different at inner and outer wheels
d kingpin offset at the road
RR rolling resistance, assumed equal at inner and outer wheels.

The total moment is given by:

$$Total\,Moment = R_R \cos \propto \cos \varphi (d + P_{Li}) + R_R \cos \propto \cos \varphi (d - P_{Lo})$$

or

$$Total\,Moment = 2dR_R cos \propto cos\varphi \tag{2.40}$$

Note that the lateral pneumatic lateral offset becomes irrelevant.

Compared to the other forces experienced by the wheel, these effects are so small that they are normally neglected. Typically the adhesion at the tyre/road interface for braking and tractive effort may be as high as 0.8. The rolling resistance coefficient may be as low as 0.01. As both these forces are a function of wheel load then the difference is in the order of a factor of 80.

Example E2-5
A front wheel drive car is driven round a right hand bend of 300 m radius at 60 km/h. The front axle torque applied during this manoeuvre is 1200 N m.

The front wheel arrangement is as follows:

Mass acting on front axle (equal load per wheel)	700 kg
Effective tyre rolling radius	320 mm
Castor angle (φ)	5°
Lateral inclination angle (λ)	13°
Tyre lateral offset at ground (positive)	50 mm
Castor offset behind, at wheel centre	10 mm
Pneumatic trail (behind tyre)	25 mm
Lateral pneumatic offset	±15 mm
Steering ratio	17:1

Assuming equal torque split between the two front wheels, calculate:

(i) Moment about each wheel due to lateral forces
(ii) Moment about each wheel due to tractive effort
(iii) Resultant torque at the steering wheel due to lateral and tractive forces.

Solution

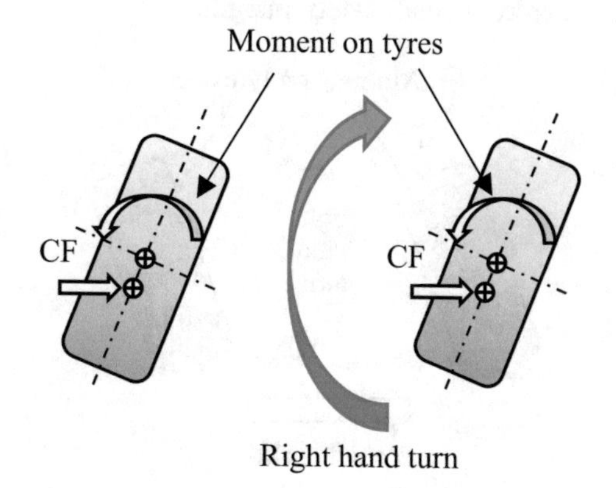

Fig. E2.5a Effect of lateral cornering forces

(i) **Calculate moment about each wheel due to lateral forces**

Consider lateral forces due to cornering (both sides) as indicated in Fig. E2.5a:

Centrifugal force (CF) = $m\omega^2R = mv^2/R = 700(60 \times 1000/3600)^2/300 = 648$ N (Total)
Castor trail = r tan φ = 320 tan 5
Pneumatic trail = 25 mm
Castor offset = 10 mm
Total trail = (320 tan 5) – 10 + 25 = 43 mm

Reference to Fig. E2.5a gives:

$$\text{Moment} = \text{Lateral force} \times \text{Trail} = 648 \times 0.043 = 27.86\,\text{N m}$$

Resolving about swivel axis gives $648 \times 0.043 \cos \nu \cos \lambda = 648 \times 0.043 \cos 5 \cos 13 = 27$ N m

Giving total moment on wheels due to lateral force, $M_{lat} = 27$ N m **counter-clockwise**.

Note: Resolving about swivel axis has minimal effect on final moment and from a designer's perspective it may be more advantageous to consider the basic value of 27.86 N m as this provides a small safety margin.

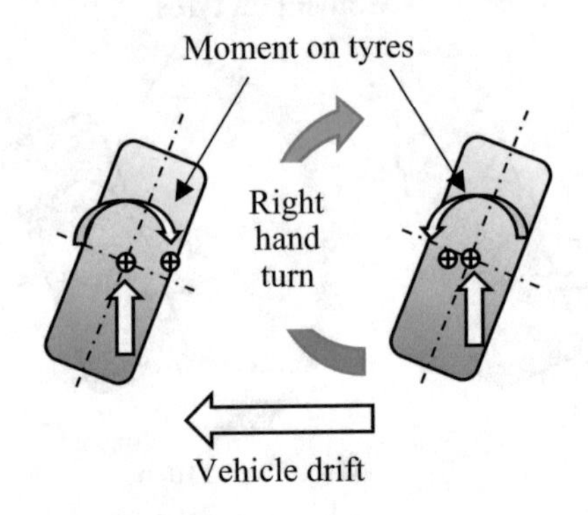

Fig. E2.5b Effect of tractive effort

(ii) **Calculate moment about each wheel due to tractive effort**

Tractive effort (TE) = Torque/Tyre rolling radius
Therefore: TE per wheel = 300/0.031 = 1935 N
From Fig. E2.5b:

$$\text{Moment} = \text{Longitudinal force} \times \text{Offset} = \text{TE} \times (50 \pm 15)$$

Consider LH wheel (Vehicle drift increases offset at the road):

$$\text{Moment} = 1935 \times (0.050 + 0.015) = 1935 \times 0.065$$

and about swivel axis = 1935 × 0.065 cos 5 = **125 N m clockwise**
Consider RH wheel (Vehicle drift reduces offset at the road):

$$\text{Moment} = 1935 \times (0.050 - 0.015) \cos 5 = \mathbf{68\ N\,m\ counter\text{-}clockwise}$$

Total moment due to longitudinal force M_{long} = 125–68 = **57 N m clockwise**

(iii) **Calculate resultant torque at the steering wheel due to lateral and tractive forces**

Total Moment M_{total} = M_{long} + M_{lat} = 57 c/w + 27 ccw = 57 − 27 = **30 N m clockwise**
Torque at steering wheel T_{sw}= M_{total}/Steering ratio = 30/17 = **1.8 N m clockwise**
Note if mechanical efficiency is considered then this would **reduce** the torque felt at the steering wheel which then becomes:

$$T_{sw} = \frac{M_{total} \times \eta}{Steering\ ratio}$$

2.9 Four Wheel Steering (4WS)

Controlling the steering of the rear wheels in addition to the front wheels can have a significant effect on the vehicle handling response. There are two distinct methods of achieving rear wheel steering: (a) by passive means, relying on the effect of the lateral forces in combination with suspension compliances, or (b) by active means in which an actuator controls the rear wheel steer angles. Passive compliance steer effects are discussed in the next chapter on Suspension Systems and Components since these effects are an intrinsic part of the suspension design. Active devices are discussed here since they are linked directly to the steering system design. Various schemes have been proposed based on mechanical, hydraulic and electrical actuation and the Japanese manufacturers have been particularly keen to market 4WS vehicles.

The 4WS systems are normally arranged so that, for low speed manoeuvres, the rear wheels steer in the opposite sense from those at the front. For high speeds, the rear wheels steer in the same sense as those at the front. At low speeds, manoeuvrability is increased by virtue of the reduced turning circle. However, a consequence of 4WS is the increased tendency of the rear of the vehicle to swing outwards (i.e. the off-tracking effect); this can cause difficulties when parking close to walls, for example. At high speeds, the benefits of in-phase rear steer are better transient response in, for example, lane change manoeuvres and improved control of the yaw damping following a transient input.

Advantages of 4WS:

- Superior cornering stability.
- Improved steering responsiveness and precision.
- High speed straight line stability.
- Notable improvement in rapid lane-changing manoeuvres.
- Smaller turning radius and tight-space manoeuvrability at low speed.

Generally, at low wheel steering angles, the rear wheels are turned in the same direction as the front wheels whilst, at high wheel steering angles, the rear wheels are turned in the opposite direction. This is demonstrated in Fig. 2.49. Clearly the steering characteristic of the rear wheels need to be linked to the characteristic of the front wheels. If the front wheels are turned through a small angle, the rear wheels turn a small angle in the same direction. As the angle of the front wheels progressively increase, the rear wheels are returned to a central position and then begin to turn in the opposite direction. This characteristic is shown in Fig. 2.50. Earlier mechanical designs achieved this using a rear steering gearbox which was mechanically connected to the front steering gear. These systems became more complex as power assisted steering was introduced but have been simplified to some degree with the introduction of electric steering.

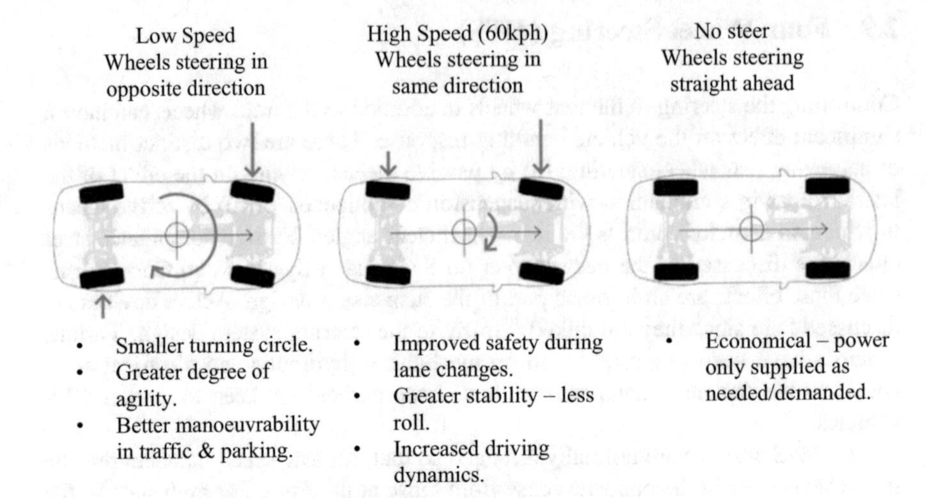

Fig. 2.49 Wheel direction for 4-wheel steer during general driving. Reproduced with kind permission of ZF © ZF Friedrichshafen AG

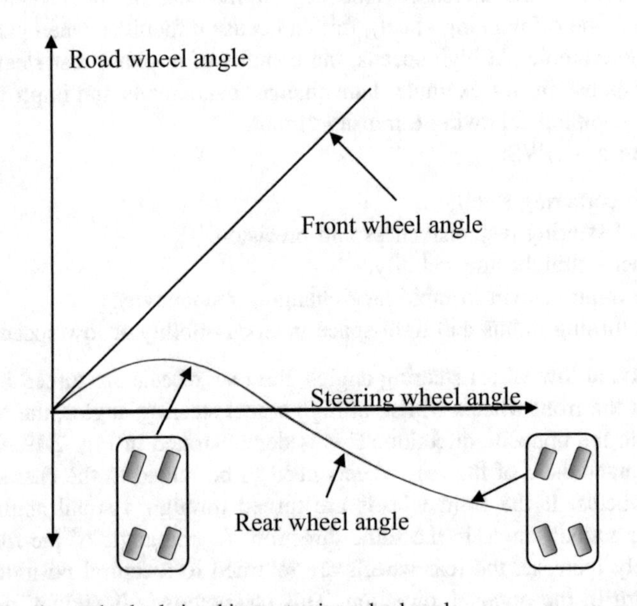

Fig. 2.50 Front/rear wheel relationship to steering wheel angle

More recent designs, termed active kinematic control, include variants for steering the rear wheels. The overall concept is shown in Fig. 2.51 that includes an active superposition steering unit and electromechanical steering at both the front and rear. The rear steering may make use of a common central module (or single actuator) or a module for each wheel, referred to as a dual actuator. The dual actuator provides steer to each of the rear wheels giving greater control over the

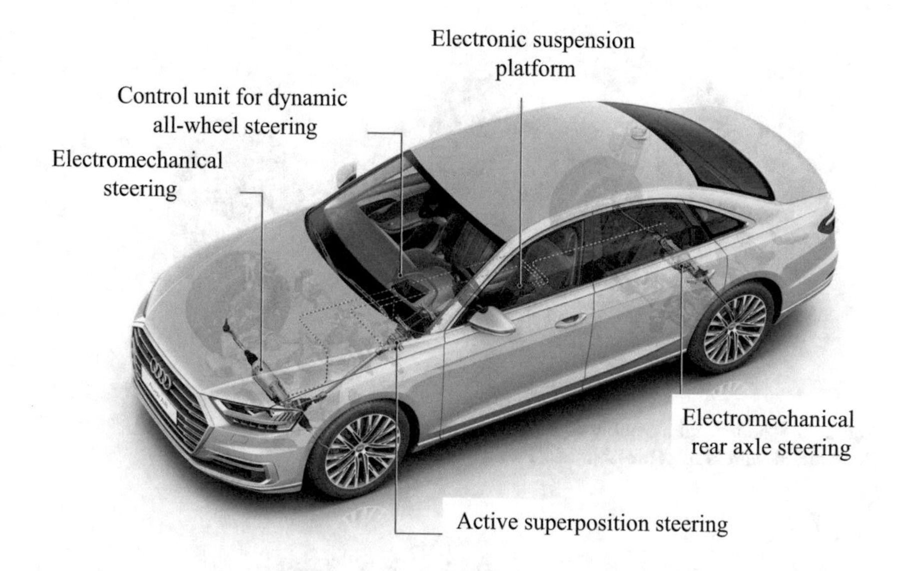

Fig. 2.51 Steer arrangement for 4-wheel steer "active kinematics control" (AKC). Reproduced with kind permission of ZF © ZF Friedrichshafen AG

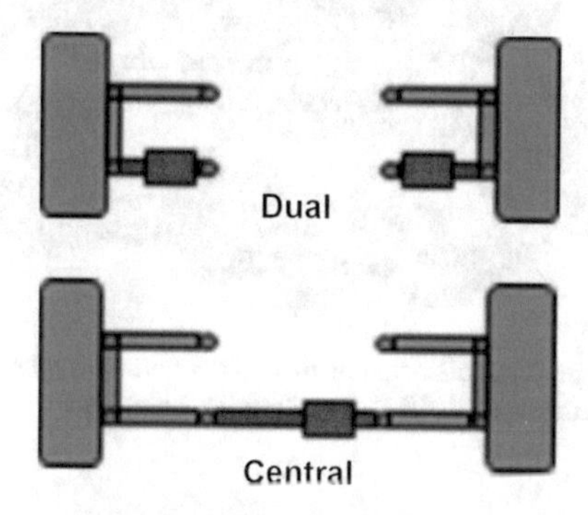

Fig. 2.52 Dual and single central actuation for 4WS. Reproduced with kind permission of ZF © ZF Friedrichshafen AG

overall geometry of the system. Both these systems are basically indicated in Fig. 2.52. It is anticipated that the front axle will also be extended to dual actuation so allowing better control of turning geometry with associated camber changes. The primary restriction will be space.

The arrangement for a single actuator is shown in Fig. 2.53 and the single (central) actuator in Fig. 2.54. A pair of dual actuators are shown in Fig. 2.55. With the dual sensors there will be a need for position sensors for each wheel to provide positioning feedback and "null" setting for straight ahead driving.

Fig. 2.53 The general assembly of a single (central) actuator. Reproduced with kind permission of ZF © ZF Friedrichshafen AG

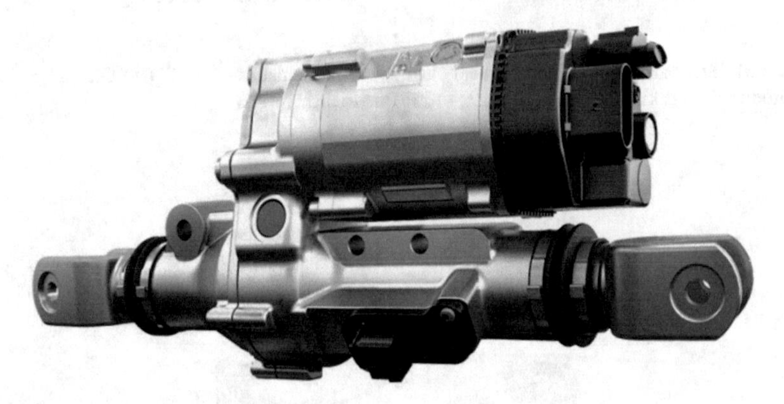

Fig. 2.54 Image of a single (central) actuator that steers both rear wheels together. Reproduced with kind permission of ZF © ZF Friedrichshafen AG

Fig. 2.55 Image of a dual actuators—each rear wheel steered independently. Reproduced with kind permission of ZF © ZF Friedrichshafen AG

2.10 Developments in Steering Assistance—Active Torque Dynamics

Consider a body mounted on four castor wheels—it is free to move in any planar direction. If it is desired to steer the body, it would be necessary to first control the direction of one axle (say lateral for the front). Direction is achieved by providing a longitudinal force at the rear but biased towards one side. This offset force provides the linear force for forward direction but also a moment to induce steer. In essence this is the method used to steer and propel a shopping trolley. The front wheels are fixed (lateral restraint) and the rear wheels are castors. The rear end is moved and the trolley is then pushed. If a similar principle is applied to a vehicle then "additional" steering assistance is possible. For four wheel drive vehicles this may be achieved in two ways as described below.

2.10.1 *Active Yaw Damping*

Principally used on four wheel drive vehicles, this system uses the front and rear differentials to modulate the torque applied to each wheel by locking and unlocking them dependent on the dynamic situation. As such the system can modulate the torque provided to each wheel in real time. If the steering system can determine that a corner is being negotiated then it can allow the vehicle to enter the corner with necessary rear wheel drive to "push" the vehicle into the corner (as there is less centrifugal force on the rear wheels). As the vehicle progresses through the corner then the wheel torque progressively increases to the front wheels so "pulling" it out of the corner (less centrifugal force on the front wheels). In general terms the torque is transferred from the rear wheels to the front wheels as it goes round the corner, the bias split being towards the front wheels as it leaves the corner. This provides maximum straightening torque as the vehicle leaves the corner so enhancing stability.

Modulation of torque side to side based on yaw and wheel speed data increases stability further, particularly on slippery surfaces and emergency situations.

2.10.2 *Active Torque Input*

This system senses a reduction in steering ability, possibly because of front end drift. In such a situation, straight ahead drive is increased to one of the rear wheels which assists cornering ability by correcting the drift. Conversely, during a combined cornering and braking manoeuvre, the braking may be biased to compensate for detected over or understeer. This method is directly related to input torque (tractive or braking) to each wheel and so may be referred to as active torque distribution.

2.11 Concluding Remarks

This chapter has introduced basic topics surrounding the steering of road vehicles. Particular attention has been paid to the general geometric and kinematic requirements in order to be able to control the vehicle in a safe and reliable manner. The importance of being able to calculate steering wheel loads allowing for both quasi-static and dynamic manoeuvres has been emphasised. In Chap. 3 we will see the close connections between the steering and suspension systems, especially for modern front wheel drive vehicles with very tight packaging constraints.

Developments in steering systems will include more active control and further enhancements in performance. For example, regulations provide information regarding turning circles but not steer angles. In general cars have a wheel steer angle of around ±35° but this may not provide the manoeuvrability required for confined parking. The development of 4-wheel steer helps improving such manoeuvres but it can lead to unnecessary expense if the vehicle is intended only for urban use. If improved parking is a requirement then increased steer angles is needed, hence ±65° is normally specified for taxis. Such needs are being addressed with concept designs for enhanced urban mobility which suggests a steering angle of up to ±75° is achievable. This will improve turning and parking manoeuvrability and will benefit all vehicles during city driving. Figure 2.56 shows such a concept design for an urban vehicle.

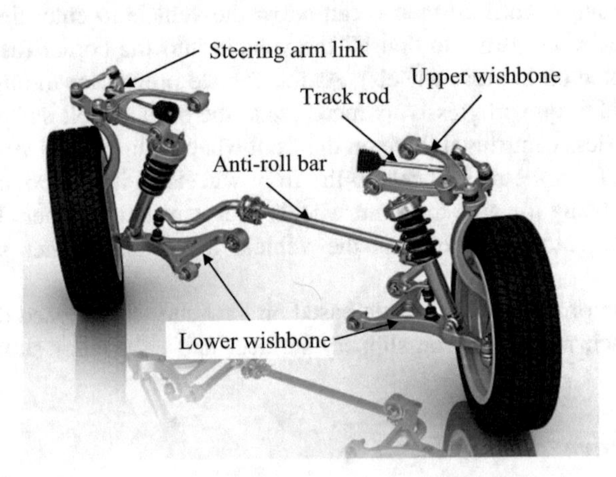

Fig. 2.56 Concept steering arrangement for enhanced manoeuvrability of an urban vehicle. Reproduced with kind permission of ZF Friedrichshafen AG © ZF Friedrichshafen AG

Chapter 3
Suspension Systems and Components

Abstract Suspension system design is as much about quality/refinement as being a handling/safety issue. This chapter begins by considering the kinematic requirements of any suspension system before discussing common suspension systems for both dependant and independent designs. Both front and rear suspension designs are considered. The chapter continues with a detailed analysis of the suspension system components—tyres, linkages, springs and d`ampers—the latter including active damping. It aims to extend the understanding of load transfer when body roll is considered along with the effects of sprung and unsprung masses. Both anti-squat and anti-dive designs are also studied. The chapter concludes with a quarter vehicle analysis whereby body bounce and wheel hop modes are identified as areas of interest. The chapter includes many numerical examples to better explain the theory and demonstrate their application.

3.1 Introduction to Suspension Design

Ride quality and handling are two of the most important issues related to vehicle refinement. Together they produce some design conflicts that have to be resolved by compromise. Also, the wide range of operating conditions experienced by a vehicle which affect both ride and handling must be taken into account by the chassis engineer. This all adds up to some very challenging tasks for the suspension designer.

While the chassis engineer has a checklist of functional requirements for a given design, there are also a number of other constraints that have to be met. These include cost, weight and packaging space limitations, together with requirements for robustness and reliability, ease of manufacture, assembly and maintenance.

Suspension design like other forms of vehicle design is affected by development times dictated by market forces. This means that for new vehicles, refined

© Springer International Publishing AG 2018
D. C. Barton and J. D. Fieldhouse, *Automotive Chassis Engineering*,
https://doi.org/10.1007/978-3-319-72437-9_3

suspensions need to be designed quickly with a minimum of rig and vehicle testing prior to launch. Consequently, considerable emphasis is placed on computer-aided design. This requires the use of sophisticated mathematical models and computer software that enables a variety of "what-if" scenarios to be tested quickly and avoids the need for a lot of prototype testing.

In order to understand the issues facing the suspension designer it is necessary to have knowledge of:

- The requirements for steering, handling and stability
- The ride requirements related to the isolation of the vehicle body from road irregularities and other sources of vibration and noise
- How tyre forces are generated as a result of braking, accelerating and cornering
- The needs for body attitude control
- Suspension loading and its influence on the size and strength of suspension members.

This chapter aims to address the above issues.

3.1.1 The Role of a Vehicle Suspension

The principal requirements for a vehicle suspension are:

- To provide good ride and handling performance—this requires the suspension to have vertical compliance to provide chassis isolation while ensuring that the wheels follow the road profile with minimum tyre load fluctuation.
- To ensure that steering control is maintained during manoeuvring—this requires the wheels to be maintained in the proper positional attitude with respect to the road surface.
- To ensure that the vehicle responds favourably to control forces produced by the tyres as a result of longitudinal braking and accelerating forces, lateral cornering forces and braking and accelerating torques—this requires the suspension geometry to be designed to resist squat, dive and roll of the vehicle body.
- To provide isolation from high frequency vibration arising from tyre excitation —this requires appropriate isolation in the suspension joints to prevent the transmission of "road noise" to the vehicle body.
- To provide the structural strength necessary to resist the loads imposed on the suspension.

It will be seen that these requirements are very difficult to achieve simultaneously, particularly when the added constraints of cost, packaging space, robustness and other factors are taken into account. This leads to some design compromises that often result in less than perfect performance for some of the desired outcomes.

3.1.2 Definitions and Terminology

There is considerable terminology associated with suspension design that may appear unfamiliar for engineers in their formative years who are meeting the subject for the first time. Most of this terminology will be described as it arises. A useful summary of the most common vehicle dynamics definitions may be found in SAE 670e Vehicle Dynamics Terminology. However care is needed as there are some differences between American and European terminology.

Figure 3.1 illustrates the SAE 670e whole vehicle reference axes together with the terms used to describe rotation about these axes. These axes have their origin at the centre of gravity of the vehicle. Note that other axis systems are sometimes used in vehicle dynamics analysis.

3.1.3 What Is a Vehicle Suspension?

The isolation of a vehicle body from the road undulations that are fed into the tyre of a vehicle at the road/wheel interface requires relative movement between the wheel and the body. This motion is, in general, controlled by some form of linkage mechanism that incorporates both stiffness and damping. It is this mechanism that is termed a *suspension*.

Suspension components, particularly springs and dampers, have a profound effect on ride and handling performance. In addition to the constraints imposed by suspension performance requirements, component designers have a range of other constraints to consider. These include weight, cost, packaging, durability and maintenance. Because of the hostile environment in which suspension components

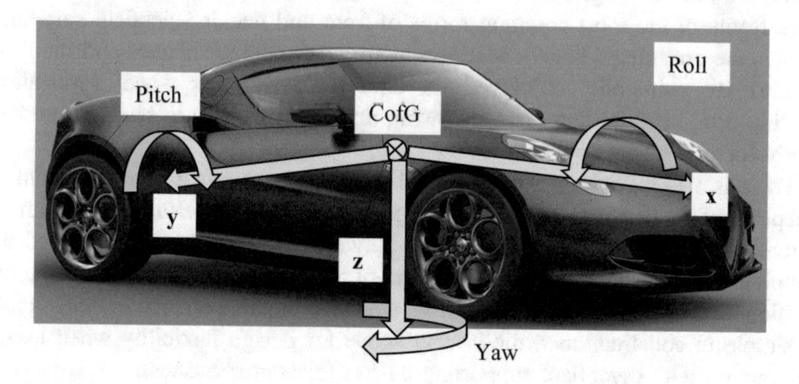

Fig. 3.1 Whole-vehicle reference axes—SAE 670e.
https://static.vecteezy.com/system/resources/previews/00k0/060/181/original/sports-car-vector.jpg

operate, and the high fluctuating loads (and hence stresses) involved, fatigue life is another of the designer's prime concerns. All suspension systems generally comprise the same components—springs, dampers and linkages. Springs are described in the Sect. 3.4, dampers in Sect. 3.5 and kinematic linkages in Sect. 3.6.

3.1.4 Suspension Classifications

In general suspensions can be broadly classified as *dependent, independent* or *semi-dependent* types.

With *dependent suspensions* the motion of a wheel on one side of the vehicle is dependent on the motion of its partner on the other side. For example, when a wheel on one side of an axle strikes a pothole, the effect of it is transmitted directly to its partner on the other side. Generally this has a detrimental effect on the ride and handling of the vehicle.

As a result of the trend to greater vehicle refinement, dependent suspensions are no longer common on passenger cars. However, they are still commonly used on commercial and off-road vehicles. Their advantages are simple construction and almost complete elimination of camber change with body roll (resulting in low tyre wear). Dependent suspensions are also commonly used at the rear of front-wheel drive light commercial vehicles and on commercial and off-highway vehicles with rear driven axles (live axles). They are occasionally used in conjunction with non-driven axles (dead axles) at the front of some commercial vehicles with rear wheel drive.

With *independent suspensions* the motion of wheel pairs is independent, so that a disturbance at one wheel is not directly transmitted to its partner. This leads to better ride and handling capabilities. This form of suspension usually has benefits in packaging and gives greater design flexibility when compared to dependent systems. Some of the most common forms of front and rear independent suspension designs are considered below. McPherson struts, double wishbones and multi-link systems are commonly employed for both front and rear wheel applications. Trailing arm, semi-trailing arm and swing axle systems tend to be used predominantly for rear wheel applications.

There is also a group of suspensions that fall some way between dependent and independent suspensions and are consequently called *semi-dependent*. With this form of suspension, the rigid connection between pairs of wheels is replaced by a compliant link. This usually takes the form of a beam that can twist providing both positional control of the wheel carrier as well as compliance. Such systems tend to be simple in construction while having scope for design flexibility when used in conjunction with compliant supporting bushes (elastomer bushes).

3.1.5 *Defining Wheel Position*

Since one of the most important functions of a suspension system is to control the position of the road wheels, it is important to understand the definitions relating to wheel location. The location of the wheels relative to both the road and the vehicle suspension is important and will in general be affected by suspension deflection and tyre loading. In the following sub-sections, parameters relating to wheel position are defined and their effect on handling behaviour is considered.

3.1.5.1 Camber Angle

This is the angle between the wheel plane and the vertical—taken to be positive when the wheel leans outwards from the vehicle (Fig. 3.2).

Camber is affected by vehicle loading and by cornering. A slight positive camber (0.1°) gives more even wear and low rolling resistance. However, on passenger cars, the setting is often negative (even when the vehicle is empty). Front axle values range from 0° to −1°20′. Negative camber improves lateral tyre grip on bends and improves handling because of camber thrust effects (these are a consequence of tyre characteristics).

A disadvantage of an independent suspension is that the wheels incline relative to the vehicle body on a bend (Fig. 3.3). This tends to produce increased negative camber on the inner wheels and increased positive camber on the outer wheels.

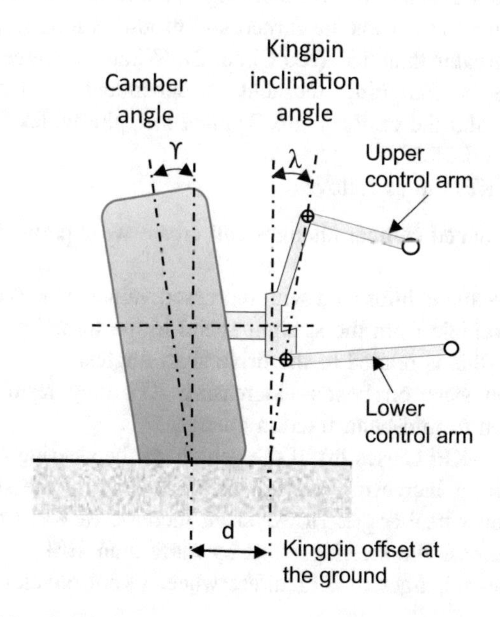

Fig. 3.2 Wheel position showing positive camber (viewed in x-direction, i.e. from the front of the vehicle)

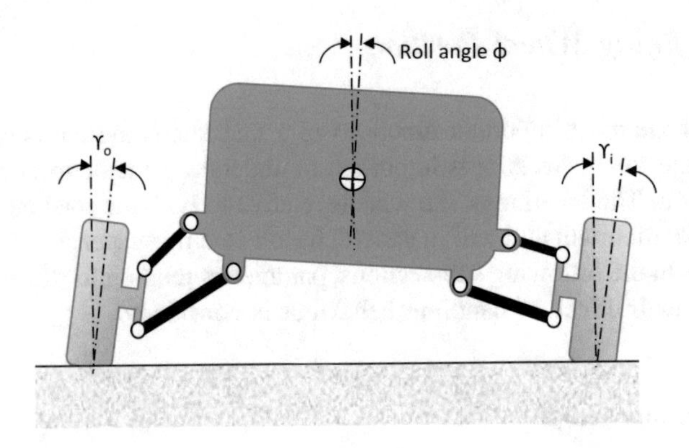

Fig. 3.3 Camber angles when left hand steer cornering with an independent suspension

The outer wheels then try to roll towards the outside of the bend tending to produce an understeer condition. Note that the change of camber (relative to the road) is a combination of changes due to suspension movement (rebound on the inside, bump on the outside) and changes due to body roll.

3.1.5.2 Kingpin Inclination (Steered Axles Only)

Sometimes called swivel pin inclination, kingpin inclination (KPI) is the angle between the kingpin axis and the vertical (Fig. 3.2). It has the effect of causing the vehicle to rise when the wheels are turned and produces a noticeable self-centring effect for KPI's greater than 15° (see Chap. 2). When the wheels are turned, the magnitude of the self-aligning moment is dependent on the kingpin angle, the kingpin offset and the castor angle. Typical kingpin angles for passenger cars are between 11° and 15.5°.

The effects of KPI are as follows:

- Tyre wear—induced camber changes can create wear particularly at high lock angles.
- Returnability—this is improved with increased values of KPI due to work done in lifting the vehicle from the straight ahead to the turned position.
- Torque steer—this is related to the drive shaft angles.
- Lift-off/traction steer on bends—increasing KPI may result in an increased vehicle reaction to mid-bend traction changes.
- Steering effort—KPI causes lift of the vehicle as the steering angle is increased. This results in an increasing amount of work done at the steering system to achieve a given wheel angle. However an increase of KPI from zero to 15° is unlikely to increase the steering effort by more than 10%.
- Tyre clearance—changes in KPI affect wheel swept envelope and hence tyre clearance.

3.1.5.3 Kingpin Offset

Kingpin offset (KPO) is the distance between the centre of the tyre contact patch and the intersection of the kingpin axis at the ground plane ("d" in Fig. 3.2), taken to be positive when the intersection point is at the inner side of the wheel. In practice the KPO varies from small positive to small negative values. For a given suspension, KPO can be changed by changing tyre width. Increasing KPO improves the returnability of the steering (see below). The disadvantage is that the offset provides a moment arm so longitudinal forces at the tyre contact patch due to braking, or striking a bump or pothole, are transmitted through the steering mechanism to the steering wheel.

The effects of kingpin offset are thus:

- Returnability—positive increases of KPO will increase returnability due to the increased lift with steering angle.
- Brake stability (outboard brakes)—with a diagonal split braking system, negative KPO will assist in counteracting the steer effect of a failed system. A similar effect occurs with "split-μ" braking where one wheel locks, or with unbalanced front brakes. For braking on a bend, negative KPO produces toe-in on the outer wheels creating a tendency to oversteer. With inboard brakes the hub offset (see below) is the critical dimension.
- Torque steer—with a conventional non-biased differential ensuring equal torque to each shaft, if one wheel loses traction, large values of KPO will cause *steering fight.*
- Steering effort—it is generally assumed that either positive or negative KPO will reduce static steering effort by allowing some rolling of the tyre on the road surface reducing tyre scrub. In practice there appears to be little effect even with large values of KPO.

3.1.5.4 Kingpin Offset at the Hub

This is defined as the horizontal distance from the kingpin axis to the intersection of the hub axis and the tyre centreline as indicated in Fig. 3.2. Hub offset is defined as positive when the tyre centre line lies outside the kingpin axis at hub centre height.

3.1.5.5 Castor Angle

Castor angle is the inclination of the kingpin pin axis projected into the fore-aft plane, initially through the wheel centre—positive in the direction shown in Fig. 3.4. The castor angle produces a self-aligning torque for non-driven wheels. It is dependent on suspension deflection and for steered wheels it influences the camber angle as a function of steering angle (see Steering Chap. 2).

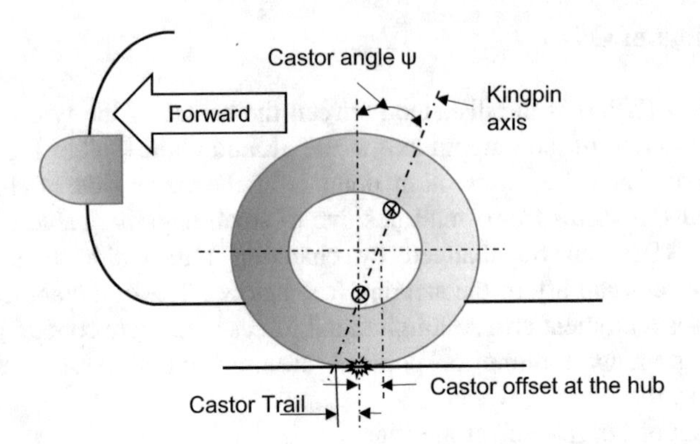

Fig. 3.4 Defining castor angle and castor trail

The effects of castor angle are as follows:

Tyre wear—induced camber changes can create shoulder wear particularly at high lock angles.

Steer response—a positive castor angle will generate negative camber on an outside wheel thus enhancing turn-in response.

3.1.5.6 Castor Trail

Castor trail, often called mechanical trail, is the longitudinal distance from the point of intersection of the kingpin axis and the ground to the centre of the tyre contact patch as shown in Fig. 3.4. It influences the magnitude of the self-righting moment. On some front wheel drive cars, there is an increased self-righting moment during cornering due to the offset of the tractive force and lateral force. This is undesirable in that the understeer effect is then too great and the steering is unduly influenced by rough road surfaces.

The effects of castor trail on vehicle behaviour are as follows:

- Straight line stability—improved with higher castor trail.
- Returnability—stronger with higher castor trail.
- Braking stability—castor angle will generally reduce with braking (due to hub wind-up and vehicle pitch) resulting in reduced castor trail leading to a degradation in braking stability, i.e. braking stability is enhanced with higher levels of castor trail.
- Steering effort—castor trail has very little effect on steering effort at stationary or at very low speeds, but otherwise steering effort increases with increasing castor trail. Steering feel at high lateral accelerations (with impending loss of grip) is also influenced by castor trail.

3.1.5.7 Castor Offset at the Hub

Castor offset at the hub is the longitudinal distance from the vertical centre line through the wheel centre to the intersection of the longitudinal axis through the wheel centre and the kingpin axis. It is positive when in front of the wheel centre as shown in Fig. 3.4. It is possible to use a negative offset to reduce mechanical trail and self-centring force felt at the steering wheel.

3.1.5.8 Toe-in/Toe-Out

This is the difference between the front and rear distances separating the centre plane of a pair of wheels (quoted at static ride height and measured to the inner rims of the wheels). Toe-in is when the wheel centre planes converge towards the front of the vehicle as shown in Fig. 3.5. Braking and rolling resistance forces tend to produce a toe-out effect, while tractive forces (front wheel drive vehicles) tend to produce the opposite effect. This leads to front wheels being set to toe-in for both non-driven and driven front axles. In the latter front wheel drive case, this is to ensure driving stability when the driver suddenly takes his foot off the accelerator. With independent front suspensions, body roll can produce changes of toe and hence *roll steer*. This is discussed in the Steering chapter.

3.1.6 Tyre Loads

Suspension loads result from the set of forces and moments that are generated at the tyre/road interface during vehicle motion. These forces depend on the static loads produced by masses of the vehicle, occupants and payload, together with the dynamic forces arising from accelerating, braking, cornering, tyre rolling resistance and aerodynamic effects. These combined tyre forces affect the handling ability of the vehicle and it is the chassis designer's task to ensure that their effects are satisfactorily controlled.

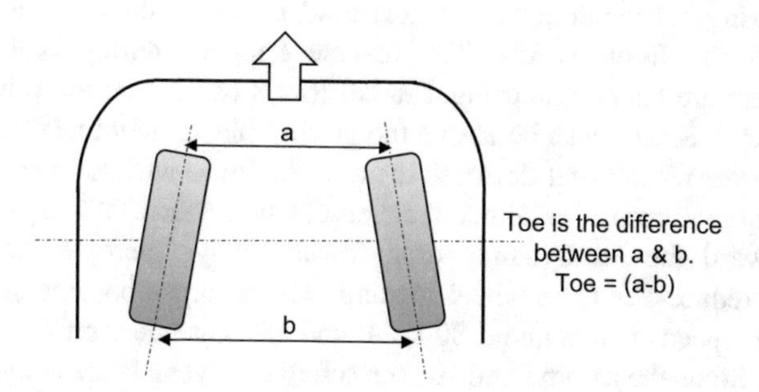

Fig. 3.5 Toe-in definition (plan view)

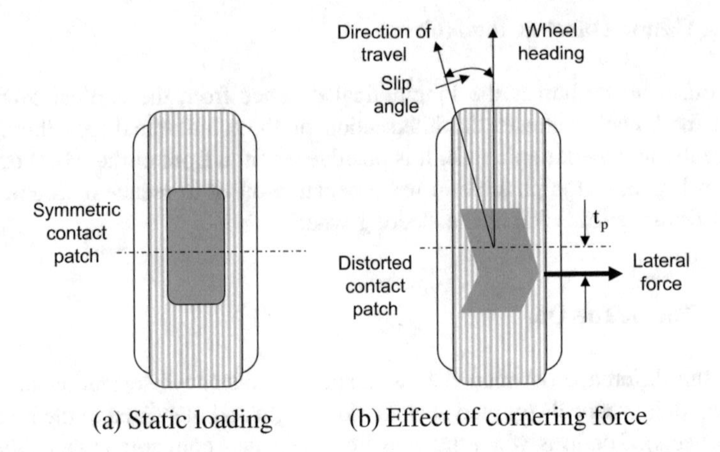

Fig. 3.6 Tyre contact patch

3.1.6.1 Tyre Contact Patch

Due to tyre compliance, the tyres deform to produce a contact area over which tyre load is distributed. This is called the *tyre contact patch*. For static loading conditions, this is a symmetrical area as shown in Fig. 3.6a. The centre of pressure in this case lies vertically below the axle centre. During motion of the vehicle, the tyre contact patch becomes distorted depending on whether the vehicle is accelerating, braking, cornering etc. These actions produce a distributed set of normal and shear forces in the tyre contact patch. It is these shear forces, and the compressive/tensile shear deflections across the interface, that cause the tyre to "creep" in the direction of the force, this "creep" generally being referred to as tyre slip.

3.1.6.2 Vertical Forces

Under static conditions, the load distribution in a vehicle determines the vertical load supported by each wheel. These forces are altered under dynamic conditions (acceleration, braking and cornering) by *load transfer effects* as described in Chap. 1. During acceleration the vertical load increases on the rear wheels and decreases on the front wheels. The reverse happens during braking. During cornering there are lateral (centrifugal force) forces exerted on the vehicle through its mass centres. Since these lie above the ground plane, they increase the vertical load at the outer wheels and decrease them at the inner wheels.

Aerodynamic forces also affect the vertical tyre loads. When the vehicle is moving forward the aerodynamic drag forces always increase the rear wheel loading and reduce the front wheel loading. These forces become significant for passenger car speeds above about 80 km/h and are dependent on vehicle type and body design. Since the aerodynamic forces acting on a vehicle are above the ground plane, they produce moments that cause pitch and roll. These in turn produce load transfer and hence affect tyre loading.

3.1.6.3 Longitudinal Tyre Forces

Longitudinal tyre forces are produced by braking and accelerating, rolling resistance and aerodynamic forces. They are affected by both wheel-slip and vertical loading.

3.1.6.4 Lateral Force and Slip Angle

As a result of the cornering force described above, there is also a shear force produced in the plane of the tyre contact patch. This produces distortion of the contact patch resulting in the direction of wheel motion differing from the wheel heading by an angle called the *slip angle.* Figure 3.6b shows this for the case of a non-steered wheel. This clearly has implications for directional control and handling stability of a vehicle. Aerodynamic forces and camber angle also affect lateral tyre forces.

3.1.6.5 Pneumatic Trail and Self-Aligning Moment

As a result of distortion of the tyre contact patch, the resultant lateral tyre force F_y acts at a distance t_p (termed the *pneumatic trail*), behind the vertical centre line through the wheel centre (Fig. 3.6b). This produces a *self-aligning moment,* M_z, of magnitude t_pF_y about an axis normal to the road surface.

3.1.6.6 Vertical Tyre Stiffness or Tyre Spring Rate

The vertical stiffness of a tyre is closely proportional to the applied vertical load and is generally dependent on the following:

- Size
- Construction
- Inflation pressure
- Camber
- Temperature
- Speed of rotation
- Lateral and longitudinal forces.

It is generally accepted that 90% of the tyre stiffness is related to inflation pressure. It is clear from the above listed variables that stiffness is a dynamic and hence variable quantity. So tyre manufacturers will not supply a given stiffness value but will provide dynamic rolling radius and circumference data at various speeds, loads, pressures and camber angles. From the given information, the chassis engineer will be in a position to calculate ride heights and gearing for a range of operation conditions.

A first order estimate of tyre vertical stiffness may be determined by measuring the rolling radius to establish the tyre deflection under load. A first order estimate for stiffness is the tyre load/deflection ratio. If the length of the contact patch is known then the rolling radius and tyre vertical deflection may be calculated to give the vertical stiffness. It should be noted that the contact patch comprises both the pressurised semi-rigid core, which supports the majority of the load, and the relatively soft rubber contact area which provides the grip or adhesion.

3.2 Selection of Vehicle Suspensions

In this section the factors that influence the selection of suspension types are examined. The kinematic requirements for suspensions and a range of suspension configurations are also discussed.

Having selected a suspension mechanism for a given vehicle a number of issues need to be considered. These include how the wheel movement changes relative to both the vehicle body and the road surface over the range of suspension travel and whether the suspension members are able to cope with the variety of loads imposed on them. Of particular concern are changes in camber and toe angles together with changes in track width (this is related to tyre scrub and hence tyre wear). The loading in some suspension mechanisms produces a need for additional support members. In general these do not affect the kinematic behaviour of the suspension mechanism.

Initially, it is assumed that suspension mechanisms undergo rigid body motion with no compliance at the joints between members. However, most modern suspensions do incorporate some compliance at certain joints to facilitate subtle wheel/body movements such as compliance steer to enhance handling performance. Also, it is assumed initially that the motion of the wheel relative to the body is 2-dimensional, i.e. it occurs in a plane perpendicular to the longitudinal axis of the vehicle. In reality, because of additional design requirements, e.g. pitch control of the vehicle body, the wheel motion needs to be 3-dimensional.

The complexity of the issues facing the chassis designer can be appreciated when one also considers how various operational factors affect chassis performance as outlined below.

3.2.1 Factors Influencing Suspension Selection

Choice of suspension is primarily determined by the engine-drive combination, which is in turn determined by the following:

- The functional requirements of the vehicle (use, performance capabilities and cargo space).
- The need to place a high proportion of the vehicle mass over the driven axles to aid traction.

The most common combinations are outlined as follows.

3.2.1.1 Front Mounted Engine and Rear Wheel Drive

In this case there is not the same restriction on engine length as there is with the front engine, front wheel drive combination. For this reason this configuration is adopted for more powerful passenger cars and estate cars. Under full load most of the vehicle mass is on the driven axle. However, when lightly loaded, e.g. only driver and passenger (the 2-up condition), this configuration leads to poor traction on wet and wintry roads, a problem that can be overcome with the use of traction control.

With this configuration, the implications for the choice of a rear suspension are that it has to accommodate the differential and drive shafts while still providing an acceptable boot space.

3.2.1.2 Front Mounted Engine and Front Wheel Drive

In this case the engine and transmission form one unit which sits in front of, over, or just behind the front axle. This leads to a very compact arrangement compared to the previous case and this is no doubt responsible for its almost universal adoption on small to medium sized saloons (up to 2 litre capacity). The main advantages include a significant load on the driven and steered wheels for normal vehicle loads providing good traction and road holding on wet and icy roads. With a fully laden vehicle this advantage tends to be lost as the centre of gravity is then further to the rear.

The implications for front suspension design are that provision must be made for drive shafts and steering gear. In general the packaging space for the suspension is limited by the size of the engine-transmission unit. This engine-drive configuration favours McPherson strut and double wishbone types of suspensions at the front. A variety of rear suspensions are possible for this drive configuration. The type of vehicle and the degree of refinement required usually influence the choice.

3.2.1.3 Four Wheel Drive

In this case it is possible to have all the wheels driven continuously or to have one of the pair of wheels always connected to the engine while the other pair are selected manually or automatically as required. The aim is of course to improve traction for all road conditions especially wet and wintry conditions. All wheel drive is particularly advantageous for off-road conditions and for improving climbing ability regardless of loading conditions.

From the point of view of suspension selection, the requirements for the front suspension are similar to that on a vehicle with front mounted engine and front wheel drive. The rear suspension requirement is similar to that for a rear suspension on a vehicle with a front mounted engine and rear wheel drive.

Suspensions have been broadly categorised in Sect. 3.1.4 as dependent, independent or semi-dependent. Solid axles have been categorised as dependent systems. However, other connections across the vehicle, such as anti-roll bars, are not classed as dependent because, although they influence suspension forces across an axle, they do not provide a geometric constraint.

3.2.1.4 Other Factors

The mass of the components associated with the wheel hub, brake components (outboard brakes), and parts of the other masses connected to the wheel hub, e.g. suspension links, suspension spring and damper, create what is known as the *unsprung mass*. This produces an additional resonant frequency when the vehicle system is subjected to road induced vibration (a 2 degree of freedom system). The effect is to increase the dynamic tyre load and cause degradation in ride comfort. The aim should therefore be to keep the unsprung mass to a minimum.

Another factor to be considered is the roll centre (and hence roll axis) location. This affects such things as the amount of body roll in cornering, the amount of lateral load transfer and the loads in the suspension links. There is a roll centre in the vertical transverse plane passing through each axle. Suspension type and geometry determine its location. Some suspension types allow much greater flexibility in the choice of roll centre location. This is discussed in detail in Sect. 3.8 below.

3.3 Kinematic Requirements for Dependent and Independent Suspensions

The main objective in suspension design is to isolate the vehicle body from road surface undulations. Ideally, this requires that the wheels of the vehicle perfectly follow the road surface undulations with no vertical movement of the vehicle body.

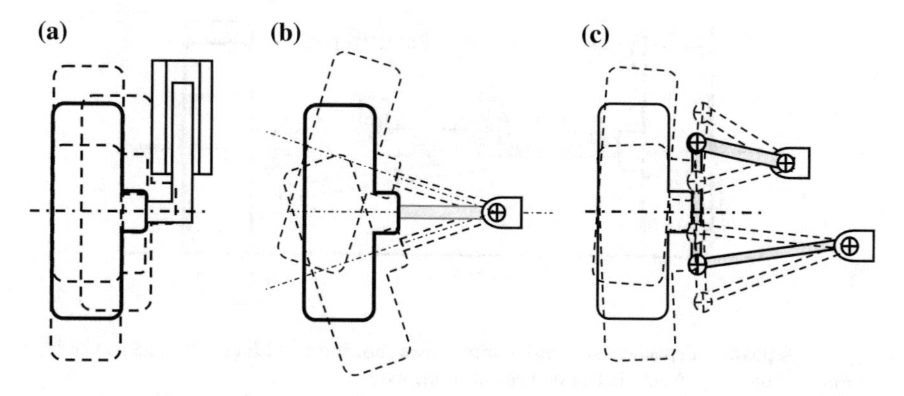

Fig. 3.7 Examples of possible independent suspension mechanisms: **a** sliding joint, **b** single pivot, **c** 4 bar linkage

In analysing this scenario, it is possible to assume that the vehicle body is stationary while we consider the motion of the wheels.

The motion requirements of the wheels relative to the vehicle body can be satisfied in a number of different ways. If each wheel can move independently of its partner on the other side of the vehicle (i.e. an independent suspension, as defined in Sect. 3.1.4), this can be achieved kinematically as shown in Fig. 3.7. In the first example (Fig. 3.7a), the wheel moves vertically relative to the vehicle body via a sliding contact system. In the second example (Fig. 3.7b), the required vertical motion is accompanied by rotation about a pivot attached to the body, while in the third example (Fig. 3.7c), the wheel and hub are attached to one of the links in a 4-bar mechanism. This provides the required vertical wheel movement accompanied by a small rotation. All of these examples have some practical disadvantages as discussed below when applications are considered.

In an independent suspension the mechanism coupling the wheel to the body of the vehicle is said to have a *single degree of freedom*, i.e. the motion of the wheel relative the body can be described by a single coordinate. If it is necessary to couple a pair of wheels together on a single axle (as in a dependent suspension), the mechanism required must have two degrees of freedom to allow vertical movement of each wheel while at the same time having a near rigid connection between the two wheels. A practical example of how this can be achieved is shown in Fig. 3.8.

3.3.1 Examples of Dependent Suspensions

Axles may be categorised into two sub-groups:

- Drive axles—sometimes called "live" axles—in which solid axles carry the driven wheels and differential.

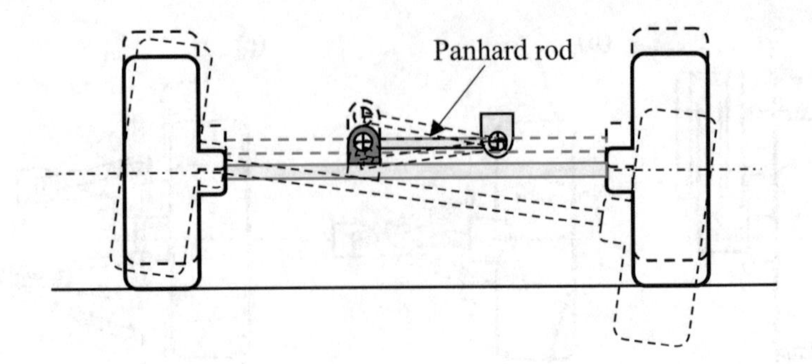

Fig. 3.8 A possible dependent suspension mechanism showing axle lift or rock. The Panhard rod allows rotation but prevents lateral movement of the axle

- Non-driven axles—usually called "dead" axles—in which a solid beam simply connects the two none driven wheels which may be steerable.

Solid axles are rarely used in modern passenger vehicle designs. They are however almost universally used for trucks and commercial vehicles and for many off-highway vehicles because of their simplicity (relative to independent systems). They virtually eliminate wheel camber (except for the small amount arising from different tyre deflections on the inner and outer wheels) and wheel alignment is easy to maintain, providing good tyre wear properties.

Many methods of mounting solid axles using dependent suspension systems have been and are still used in vehicle design. Only the main types and associated principles are covered here.

3.3.1.1 Hotchkiss Drive

The most familiar form of mounting a solid drive axle is on a pair of semi-elliptic leaf springs; this is known as the Hotchkiss arrangement (Fig. 3.9). It is a remarkably simple suspension and produces satisfactory location of the wheels with the minimum number of components. Its simplicity derives from the leaf spring properties, i.e. compliant in the vertical direction but relatively stiff laterally and longitudinally.

Developments in leaf springs over the years have overcome problems with inter-leaf friction. Single and taper leaf designs now tend to be used. However, for the low spring rates required for good ride performance, single leaf springs are generally inadequate to resist lateral loads and driving and braking torques on modern vehicles. In such cases, additional strengthening links such as Panhard rods are used as indicated in Fig. 3.8. These Panhard rods constrain movement back to the chassis and inhibits spring *wind-up* (the spring adopting an "S" shape due to wheel input torque) which results in wheel hopping.

Fig. 3.9 Hotchkiss (leaf spring) suspension system.
https://upload.wikimedia.org/wikipedia/commons/6/63/Leafs1.jpg

The detailed design of the leaf spring itself and the associated shackle mounting geometry (Fig. 3.9) which accommodates changes in effective spring length is not straightforward. Design guidelines are published in the SAE Spring Design Manual.

3.3.1.2 Four-Link Arrangements

The requirements for a solid axle to have two degrees of freedom and also withstand all types of loading can be satisfied by a number of four-link suspension mechanisms.

In the mechanism shown in Fig. 3.10, two upper and two lower links—usually trailing (but not always)—react the driving/braking torques while providing the required vertical and roll degrees of freedom. Lateral control may be provided by

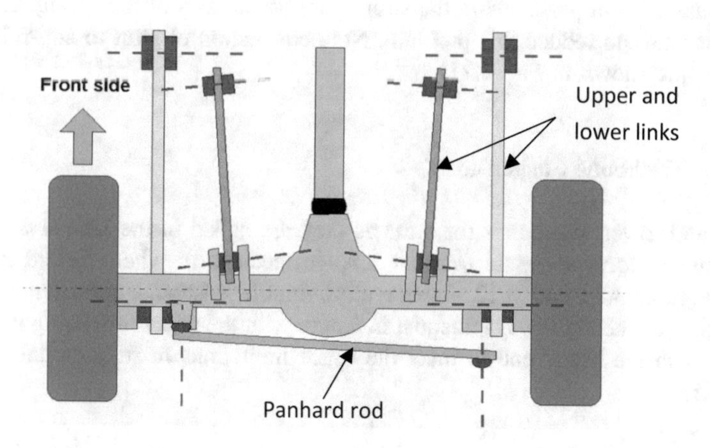

Fig. 3.10 Four link suspension.
https://upload.wikimedia.org/wikipedia/commons/6/63/Axle_-_5_Link_rigid_03.gif

angling the upper links (as shown in Fig. 3.10), triangulating the upper links, or by providing an additional Panhard rod. Any springing medium can be used, but coil or air springs are the most common. The advantages of this suspension design over the Hotchkiss arrangement are greater design flexibility in arranging the roll centre, the anti-dive/squat geometry and controlling roll steer.

3.3.2 *Examples of Independent Front Suspensions*

Almost all passenger cars, and increasing numbers of light trucks, use independent suspensions at the front. The gradual shift towards independent suspensions has resulted from their benefits in packaging, especially for front wheel drive vehicles where they are better able to fit into the limited space available around the engine/transmission system. They also permit greater suspension design flexibility and overcome problems of steering shimmy vibrations often associated with beam axle designs. Over the years a bewildering number of independent suspension designs have been proposed and the descriptions here are restricted to the generic types.

3.3.2.1 McPherson Strut Suspensions

Great simplicity is the main benefit of this design with the wheel controlled by a lower control arm in combination with the McPherson sliding strut (Fig. 3.11). The configuration shown in Fig. 3.11 is well able to withstand the longitudinal and lateral loads that will arise for various operating conditions. Advantages of the design are good packaging and simplicity, whilst its disadvantages are a high installation height (which may conflict with the bonnet line) and wheel loads that produce a moment which must be reacted by the McPherson strut. This moment may cause friction problems in the strut. Angling the axis of the spring relative to the strut axis can reduce this problem. Note connection of strut to anti-roll bar in the example shown in Fig. 3.11.

3.3.2.2 Wishbone Suspensions

Upper and lower wishbones (or arms as they are called in the United States) are combined to form a classic four-bar linkage mechanism when viewed from the front. Figures 3.7c and 3.12 show typical double wishbone arrangements. The wishbones are nearly always unequal in length with the upper arm being invariably shorter than the lower one to meet the space limitations in front-engine vehicles (Fig. 3.12).

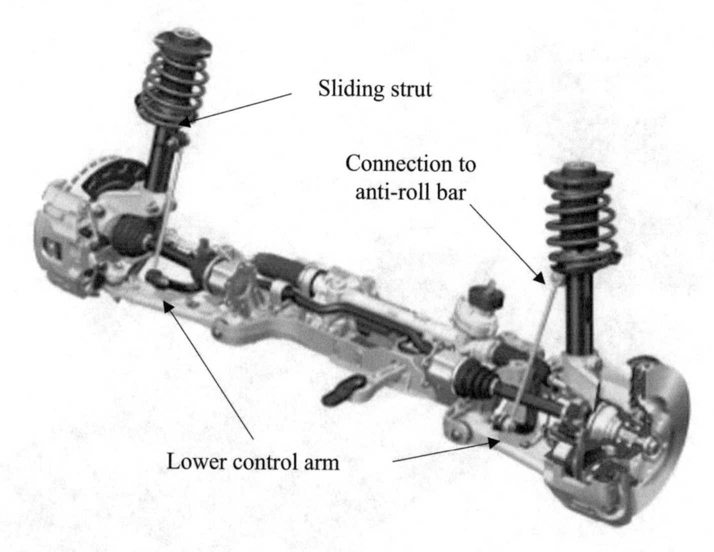

Fig. 3.11 McPherson strut suspension.
Reproduced with kind permission of Audi © Audi AG

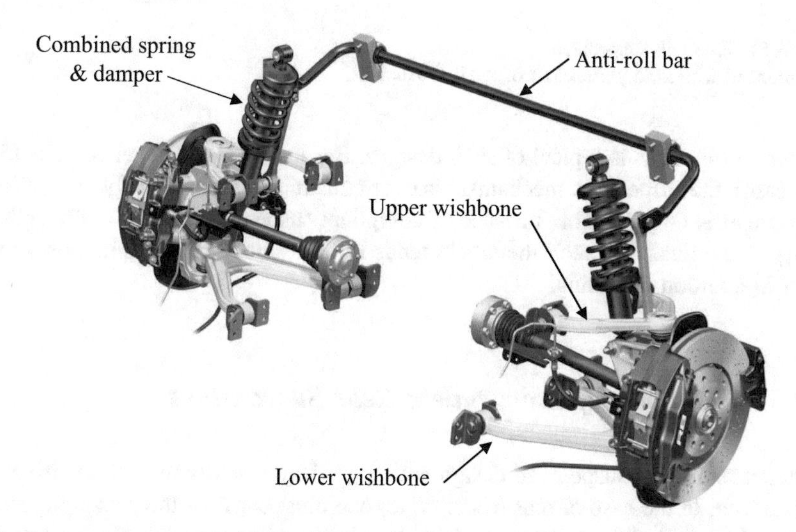

Fig. 3.12 Double wishbone suspension.
Reproduced with kind permission of Audi © Audi AG

3.3.2.3 Multilink Designs

Multi-link suspensions are now a common feature on passenger cars from mid-range models upward. They are usually based on three, four, or five links per wheel station and are designed to give more complete control over wheel positioning. This comes at a price due to the added number of components, and the extra effort required in setting them up. The front suspension of a five link design as

Fig. 3.13 Five link suspension.
Reproduced with kind permission of Audi © Audi AG

shown in Fig. 3.13 is typical of such designs. Because of the number of links (5 in this case), the suspension mechanism is over-constrained kinematically. Suspension movement is only possible because of compliant suspension bushes at the ends of some of the links. As such the wheel tends to orbit about the kingpin axis during steer in addition to rolling.

3.3.3 Examples of Independent Rear Suspensions

Differences in rear suspension design are dictated by whether the axle is driven or non-driven. In the case of rear wheel drive, one must consider the packaging space (effected by both differential and a need to minimise intrusion into the boot space) and also torque-steer effects. This latter point is particularly relevant for larger, more powerful, saloon cars. Some examples are discussed below.

3.3.3.1 Trailing Arm Suspension

A simple single trailing arm (Fig. 3.14) can be used with a torsion bar spring or with a rubber or hydro-elastic one. With this design the wheel camber is the same as the body roll angle.

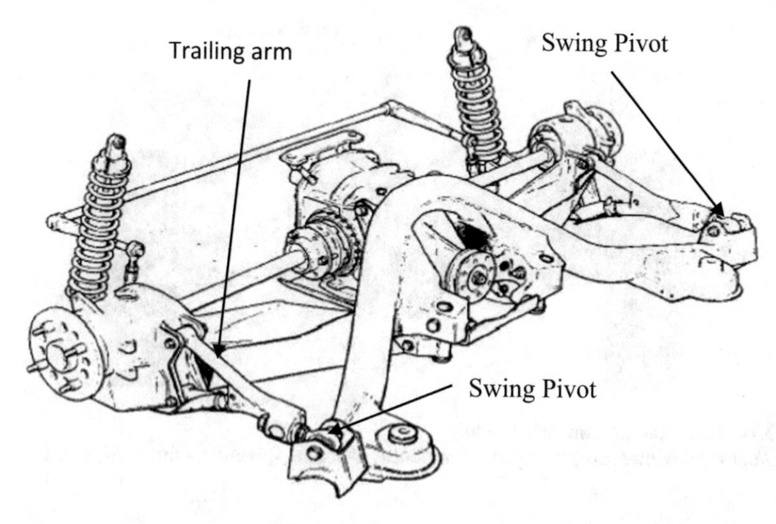

Fig. 3.14 Rear trailing arm suspension.
http://tech-racingcars.wdfiles.com/local–files/ford-escort-mk-v-rs-cosworth/WRC_rear_suspension.jpg

3.3.3.2 Swing Axle Suspension

This is also a simple way of achieving an independent rear suspension (Fig. 3.15). The length of the swing axle controls the camber behaviour but this has to be relatively short to accommodate the differential. Camber changes (and consequently scrub) are significant during normal wheel travel. Also this design is particularly prone to the problem of *jacking* (vertical movement of the vehicle) and is little used nowadays.

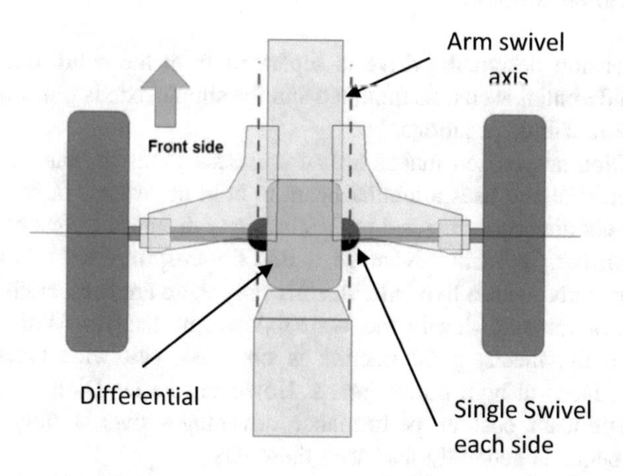

Fig. 3.15 Swing axle suspension.
https://upload.wikimedia.org/wikipedia/commons/thumb/d/d1/Axle_-_Swing_axle_01.png/640px-Axle_-_Swing_axle_01.png

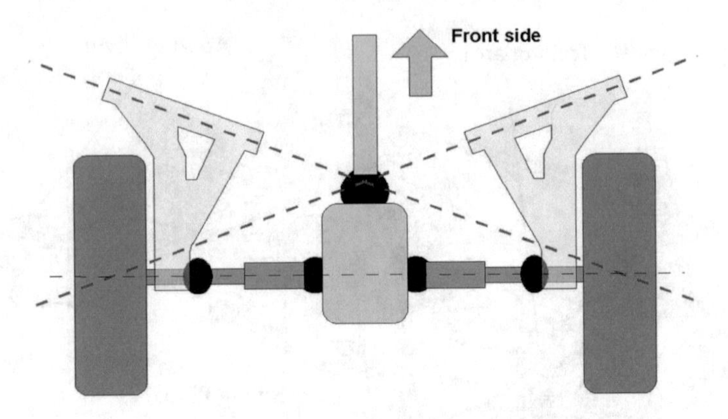

Fig. 3.16 Semi-trailing arm suspension.
https://upload.wikimedia.org/wikipedia/commons/9/94/Axle_-_Semi_trailing-arm_23.gif

3.3.3.3 Semi-trailing Arm Systems

This design is a cross between a pure trailing arm and a swing axle (Fig. 3.16). The additional flexibility in the design allows a compromise in the control of camber and jacking. However care must be taken with the geometry to control the amount of steer which results from the inclination of the pivot axis. More recent variants of this type of design try to exploit small amounts of rear steering to improve overall vehicle handling qualities.

3.3.4 Examples of Semi-independent Rear Suspensions

3.3.4.1 De Dion System

In this suspension design the drive is separated from the solid axle (Fig. 3.17). Hence, the differential is chassis mounted and the simple axle is typically located by one of many four-link variations.

The De Dion suspension makes use of universal joints at both the wheel hubs and the differential and uses a tubular beam to hold the wheels in parallel. The De Dion tube is not directly connected to the chassis or designed to flex so does not act as an anti-roll bar. Its main advantage is that the unsprung mass is reduced dramatically compared with a live axle. Additionally there are camber changes during axle bounce or rebound due to the Watt linkage mechanism. With zero camber settings on both wheels, good traction is obtained with wide tyres and wheel hopping is reduced at high power inputs. However, the De Dion suspension does not have significant cost or performance advantages over a fully independent system and hence is generally less used these days.

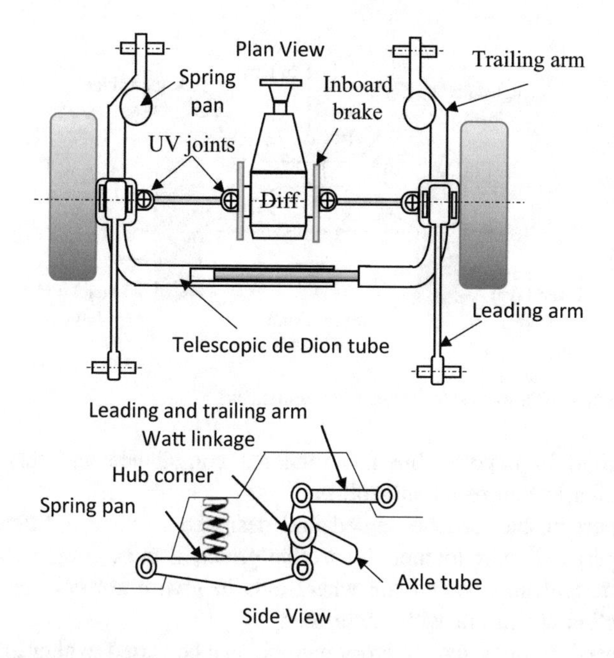

Fig. 3.17 De Dion suspension system

3.3.4.2 Trailing Twist Axle System

This arrangement has become common at the rear of small front wheel drive vehicles due to its relative simplicity and low cost. It is a cross between a fully independent trailing arm arrangement and a dependent beam axle arrangement. The axle is replaced by a crossbeam that is allowed to twist by an amount controlled by the geometry of the suspension and the structural properties of all the members.

An example is shown in Fig. 3.18. Such designs are known by various names, e.g. trailing twist-axle and semi-dependent suspensions. The cross member is

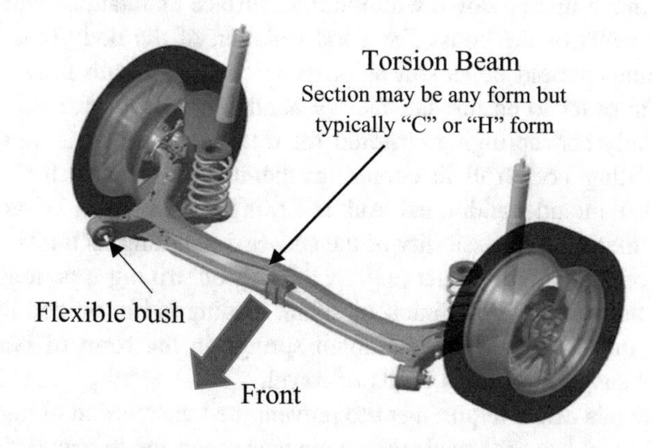

Fig. 3.18 Trailing twist axle.
http://i.stack.imgur.com/87Xnp.jpg

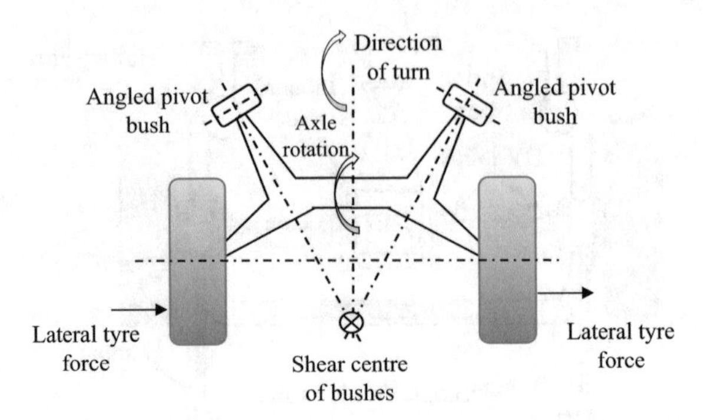

Fig. 3.19 Trailing twist axle with lateral force understeer

effectively rigid in bending but has torsional compliance and this performs a function similar to that of an anti-roll bar.

The support bushes can be angled and designed to have a different stiffness along and perpendicular to their respective principal axes (Fig. 3.19). This can produce a shear centre behind the wheel axis to give a compliance steer in the correct direction consistent with understeer.

The shape of the open section cross-member can be varied to alter the position of its shear centre (a feature of open sectioned members in torsion, typically rolled channel section). This permits some flexibility in the choice of roll centre location as discussed in Sect. 3.8 below.

3.4 Springs

Suspension systems require a variety of compliances to ensure good ride, handling and NVH performance. The need for compliance between the unsprung and sprung masses in order to provide good vibration isolation has long been recognised. In essence, a suspension spring fitted between the wheel and body of a vehicle allows the wheel to move up and down with the road surface undulations without causing similar movements of the body. For good isolation of the body (and hence good ride), the springs should be as soft as possible consistent with providing uniform tyre loading in order to ensure satisfactory handling performance.

The relatively soft springing required for ride requirements is normally inadequate for resisting body roll in cornering; therefore it is usual for a suspension system to also include additional roll stiffening in the form of anti-roll bars. Furthermore, there is the possibility of the suspension hitting its limits of travel as a result of abnormal ground inputs (e.g. as a result of striking a pothole). It is then necessary to ensure that a minimum of shock loading is transmitted to the sprung mass. This requires the use of additional springs in the form of bump stops to decelerate the suspension at its limits of travel.

Finally, there is also a requirement to prevent the transmission of high frequency vibration (>20 Hz) from the road surface via the suspension to connection points on the chassis. This is achieved by using rubber bush connections (elastomeric components) between suspension members—see Steering chapter.

The following compliant elements are thus required in suspension systems: suspension springs, anti-roll bars, bump-stops and rubber bushes. In this section attention is concentrated on suspension springs and anti-roll bars.

3.4.1 Spring Types and Characteristics

The main types of suspension spring are:

- Steel springs (leaf springs, coil springs and torsion bars)
- Hydro-pneumatic springs
- Air bag springs.

3.4.1.1 Leaf Springs

Sometimes called semi-elliptic springs, these have been used since the earliest developments in motor vehicle technology. They rely on beam bending principles to provide their compliance and are a simple and robust form of suspension spring still widely used in heavy commercial applications such as lorries and vans. In some suspensions (e.g. the Hotchkiss type) they are used to provide both vertical compliance and lateral constraint for the wheel travel. Size and weight are among their disadvantages.

Leaf springs can be of single or multi-leaf construction. In the latter case (Fig. 3.20) interleaf friction can affect their performance and this can be reduced with the use of interleaf plastic inserts. Rebound clips are used to bind the leaves together for rebound motion. The swinging shackle accommodates the change in length of the spring produced by bump loading. The main leaf of the spring is formed at each end into an eye shape and attached to the sprung mass via rubber bushes. Suspension travel is limited by a rubber bump stop attached to the central rebound clip. Structurally, leaf springs are designed to produce constant stress along their length when loaded.

Spring loading can be determined by considering the forces (see Sect. 3.10) acting on the spring and shackle as a result of wheel loading (Fig. 3.21a). The spring is a three force member with F_A, F_W and F_C acting at A, B and C respectively. The wheel load F_W, is vertical and the direction of F_C is parallel to the shackle (a two-force member). The direction of F_A must pass through the intersection of the forces F_W and F_C (point P) for the link to be in equilibrium. Knowing the magnitude of the wheel load enables the other two forces to be determined. The number, length, width and thickness of the leaves determine the stiffness (rate) of the spring. Angling of the shackle link can be used to give a variable rate. When the angle $\theta° < 90°$ (Fig. 3.21), the spring rate will increase (i.e. have a rising rate) with bump loading.

Fig. 3.20 Example of leaf spring designs.
https://upload.wikimedia.org/wikipedia/commons/a/a8/Leaf_spring_011.JPG

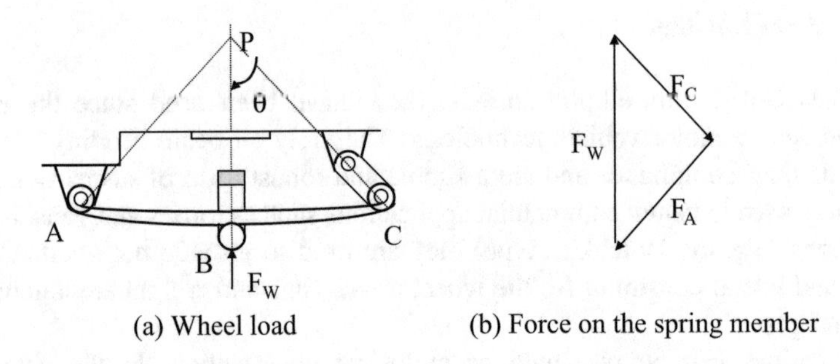

Fig. 3.21 Leaf spring loading

3.4.1.2 Helical Coil Springs

This type of spring shown in Fig. 3.22 provides a light and compact form of compliance that is an important feature in terms of weight and packaging constraints. It requires little maintenance and provides the opportunity for co-axial mounting with a damper. Its disadvantages are that because of low levels of structural damping, there is a possibility of surging (resonance along the length of coils) and the spring as a whole does not provide any lateral support for guiding the wheel motion.

Most suspension coil springs are of the open coil variety (helix angle greater than 15°). This means that the coil cross-sections are subjected to a combination of torsion, bending and shear loads as shown in Fig. 3.23. Spring rate is related to the wire and coil diameters, the number of coils and the shear modulus of the spring material. Cylindrical springs with a uniform pitch produce a linear rate. Variable rate springs are produced by varying either the coil diameter (conical types) and/or the pitch of the coils along their length. In the case of variable pitch springs, the coils are designed to progressively "bottom-out" as the spring is loaded, the shorter pitch coils making contact first. The number of working coils are progressively

Fig. 3.22 Helical coil spring.
https://www.bing.com/images/search?&q=Steel+Coil+Springs&qft=+filterui:license-L2_L3_L4&FORM=R5IR43

reduced thereby increasing stiffness with load. It would not be common for a coil spring to be fully compressed (all coils touching) as this distorts the coils, increases stresses and reduces spring life.

Calculation of spring rate
Referring to Fig. 3.24:

$$\text{Strain Energy "U"} = \text{Area under graph} = 1/2 \times \text{Force} \times \text{Distance} = \frac{1}{2}Fx \tag{3.1}$$

In the case of torsional loading $U = 1/2 \times \text{Torque} \times \text{Angle of twist} = \frac{1}{2}T\theta$

However, from theory of torsional loading, we know $\theta = \frac{TL}{JG}$

Therefore:

$$U = \frac{T^2L}{2GJ} \tag{3.2}$$

where L = length under torsion = length of wire $= \pi\, D\, n$ where n is number of active coils and G = shear modules of spring material.

For a solid circular section, we have polar second moment of area $J = \frac{\pi d^4}{32}$ where d = diameter of wire,

Also for a close coiled spring as shown in Fig. 3.25 (helix angle α is small, therefore sin α = 0 and cos α = 1), we have torque $T = F\frac{D}{2}$

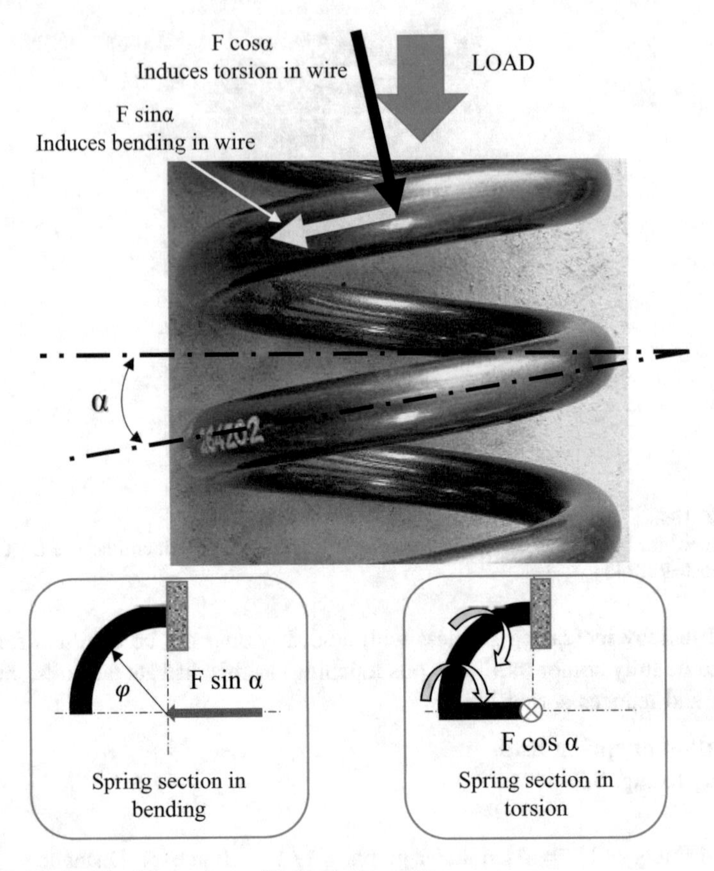

Fig. 3.23 Typical forces on open coil spring wire section

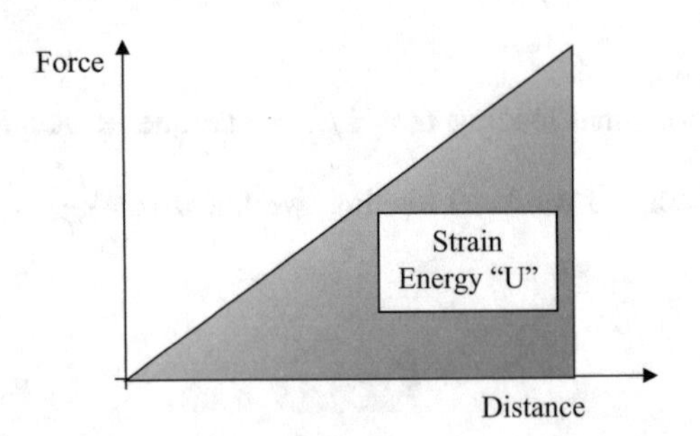

Fig. 3.24 Fundamental energy diagram of any constant rate spring

Substituting for L, J and T in Eq. (3.2) gives:

$$U = \frac{F^2 D^2 \pi D n 32}{2 \times 4 \times G\, \pi d^4} = \frac{4F^2 D^3 n}{G d^4}$$

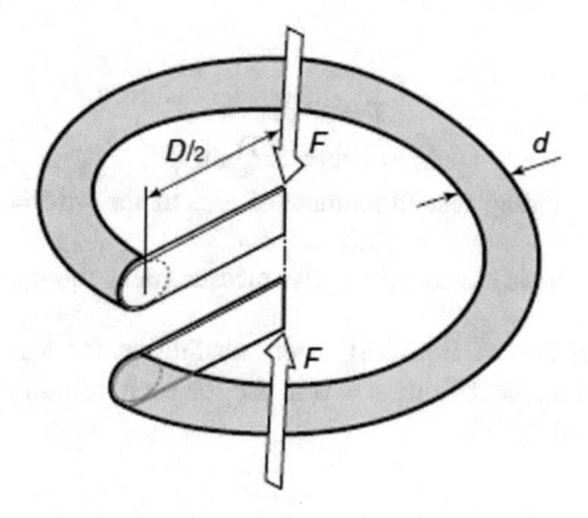

Fig. 3.25 Torsional loading on a "Close Coil" helical spring

By Castigilano's theorem, we have:

$$\frac{dU}{dF} = x$$

Thus, the differential of strain energy with respect to a load F gives deflection (x) in the direction of that load:

$$x = \frac{dU}{dF} = \frac{8FD^3n}{Gd^4}$$

Therefore spring rate:

$$k = \frac{F}{x} = \frac{Gd^4}{8D^3n} \tag{3.3}$$

This is the general equation used for helical coil spring stiffness calculations.

Also, from simple torsional theory, we have the following exprerssion for the shear stress in wire:

$$\tau = \frac{G\,d\,x}{\pi\,D^2n} = \frac{8F\,D}{\pi\,d^3} \tag{3.4}$$

Note: If the spring was open coil (α "large" so cos α ≠ 1 and sin α ≠ 0), then we have to allow for the elastic energy due to bending as well as torsion of the coils and the total strain energy becomes:

$$U = \frac{T^2L}{2GJ} + \frac{M^2dx}{2EI} \tag{3.5}$$

where:

$$T = F\frac{D}{2}\cos\alpha$$
$$\mathrm{M} = \mathrm{P}\sin\alpha \times \frac{\mathrm{D}}{2}\sin\varphi$$
$$\mathrm{I} = \text{bending second moment of area of the wire} = \frac{\pi d^4}{64}$$
$$dx = \frac{D}{2}d\varphi$$
and $d\varphi$ is elemental angular measurement about coil

For the bending part of Eq. (3.5), after substitution for M, it is necessary to integrate the $sin^2\varphi$ element from $\varphi = 0$ to 2π for each coil and then multiply by number of active coils, "n".

3.4.1.3 Hydro-Pneumatic Springs

In this case the spring effect is produced by compressing a constant mass of gas (typically nitrogen) in a variable volume enclosure. The principle of operation of a basic diaphragm accumulator spring is shown in Fig. 3.26. As the wheel deflects in bump, the piston moves upwards transmitting the motion to the fluid and compressing the gas via the flexible diaphragm. The gas pressure increases as its volume decreases to produce a hardening spring characteristic.

The principle was exploited in the Moulton-Dunlop hydro-gas suspension where damping was incorporated in the hydro-pneumatic units. Front and rear units were connected to give pitch control. A further development of the principle has been the system developed by Citroen. This incorporates a hydraulic pump to supply pressurised fluid to four hydro-pneumatic struts (one at each wheel-station). Height correction of the vehicle body is achieved with regulator valves that are adjusted by roll-bar movement or manual adjustment by the driver.

In general, hydro-pneumatic systems are complex (and expensive) and maintenance can also be a problem in the long term. Their cost can, however, be offset by good performance. Patents cover the two systems discussed above, but there is still scope for development of alternative hydro-pneumatic systems. Some of these concepts are incorporated into controllable "active" suspensions.

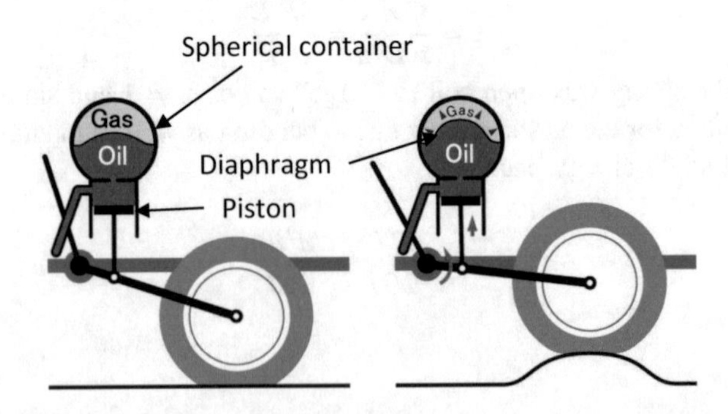

Fig. 3.26 Principles of a hydro-pneumatic suspension spring. https://upload.wikimedia.org/wikipedia/commons/b/bb/Sus_hydropneumatic002.png

3.4.1.4 Air Bag (Bellow) Suspension Springs

Figure 3.27 shows a typical suspension as used on many heavy duty vehicles, including trailers, and Fig. 3.28 shows a diagrammatic representation of a rolling seal air-cushion. The air bag is totally enclosed by a rolling diaphragm type seal and is provided by air pressure from the vehicle on-board compressor via storage reservoirs. The maximum operating pressure is in the region of 2–5 bar dependant on vehicle load during a severe bump before the stops are engaged.

The distances L_1 to L_2 in Fig. 3.28 give the wheel to suspension ratio, sometimes referred to as the transmission ratio.

Fig. 3.27 Typical heavy truck air cushion suspension.
Reproduced with kind permission of BPW Group © BPW Group

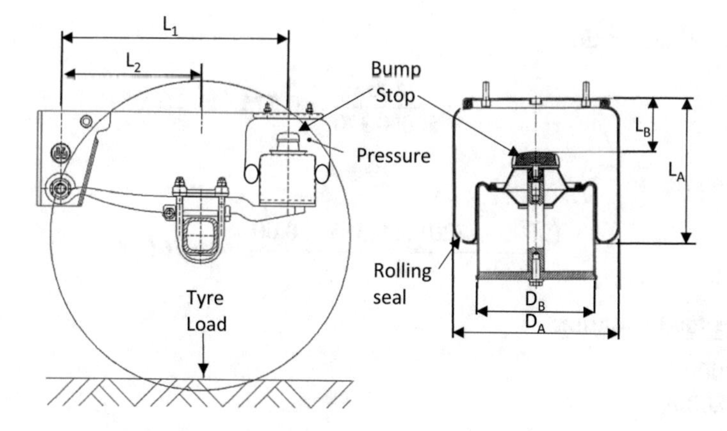

Fig. 3.28 Diagrammatic representation of a rolling seal air-cushion.
Reproduced with kind permission of BPW Group © BPW Group

The pressure (P) in the suspension bag (bellows) may be determined by the following:

$$P = \frac{(m_t - m_u)r_t p}{2} \quad (3.6)$$

where:

P pressure in suspension bellows
m_t gross axle weight rating as measured at the ground (Design limit)
m_u usprung mass average−typically taken as 8% of m_t
r_t transmission ratio $= \frac{L_1}{L_1 + L_2}$ and
p pressure rating in suspension bellows usually in units of bar/kg load.

The pressure rating "*p*" will be given by the bellows designers but a value in the region of 0.002 bar/kg would be typical.

The denominator 2 occurs in Eq. (3.6) because there are 2 bags per axle.

Example E3.1 Calculate the pressure in a truck suspension bellows system under both full load and part load conditions.

Full load situation

$m_t = 9000$ kg
$m_u = 0.08m_t$
$L_1 = 500$ mm
$L_2 = 370$ mm
$p = 0.002$ bar/kg

Solution:
Unsprung mass

$$m_u = 0.08 \times 9000 = 720 \text{ kg}$$

Transmission ratio

$$r_t = \frac{500}{500 + 370} = 0.575$$

Therefore

$$P = \frac{(9000 - 720) \times 0.575 \times 0.002}{2} = 4.76\,bar$$

Part load situation

$m_t = 3000$ kg
$m_u = 0.08m_t$
$L_1 = 500$ mm
$L_2 = 370$ mm
$p = 0.002$ bar/kg

Solution:
As before, unsprung mass = 720 kg and transmission ratio = 0.575
Therefore

$$P = \frac{(3000 - 720) \times 0.575 \times 0.002}{2} = 1.31\ bar$$

Ratio of pressures between full load and part load:

$$\frac{Full\ Load}{Part\ Load} = \frac{4.76}{1.31} = 3.63 : 1$$

Compare with ratio of masses:

$$\frac{Full\ Load}{Part\ Load} = \frac{9000}{3000} = 3 : 1$$

With such systems (requiring pressurised air), it would be possible to vary the supply pressure from side to side to counter vehicle roll and possible roll-over. In addition fore to aft pressure could be controlled to reduce vehicle pitch during braking/acceleration.

3.4.2 Anti-roll Bars (Roll Stabilisers)

Anti-roll bars (ARB's) are a very simple type of spring and consequently very cheap to manufacture. In essence the design consists of a bar loaded in torsion when the two wheels connected by the ARB deflect by differing amounts. As such a torsion bar is both wear and maintenance free. Despite their simplicity they cannot easily be adopted for some of the more popular forms of suspension.

ARB's are used to reduce body roll and have an influence on a vehicle's cornering characteristics (in terms of understeer and oversteer). They do not contribute to overall suspension stiffness if both wheels bump/rebound the same amount. Figure 3.29a shows how a typical roll bar is connected to a pair of wheels. The ends of the U-shaped bar are connected to the wheel supports and the central length of the bar is attached to the body of the vehicle. Attachment points need to be selected to ensure that bar is subjected to torsional loading without bending.

If one of the wheels is lifted relative to the other, half the total anti-roll stiffness acts downwards on the wheel and the reaction on the vehicle body tends to resist

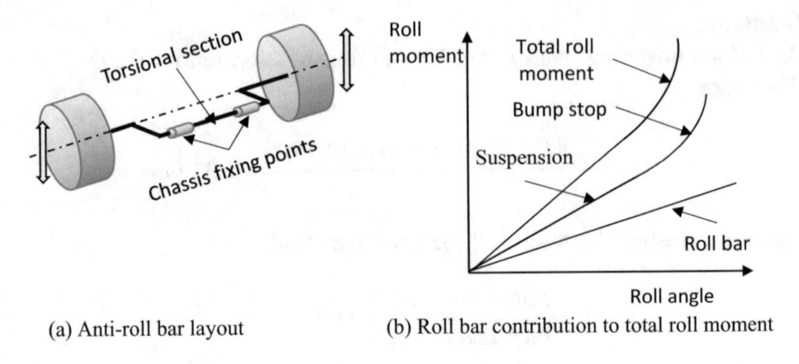

(a) Anti-roll bar layout (b) Roll bar contribution to total roll moment

Fig. 3.29 Anti-roll bar geometry and the effect on roll stiffness

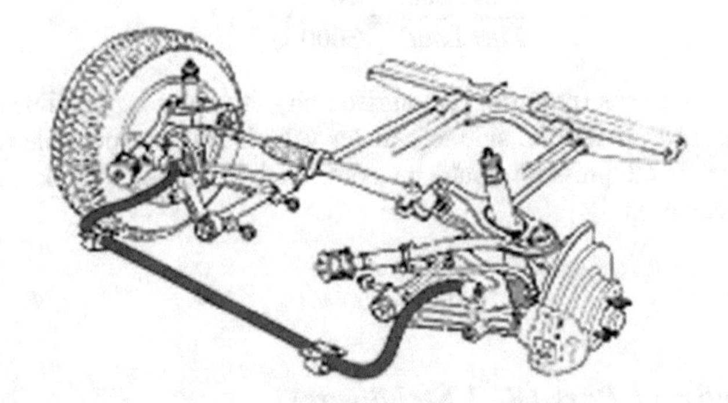

Fig. 3.30 Anti-roll bar added to suspension system. https://upload.wikimedia.org/wikipedia/commons/thumb/a/a9/Alfetta_front_suspension_antiroll.jpg/220px-Alfetta_front_suspension_antiroll.jpg

body roll. If both wheels are lifted by the same amount, the bar does not twist and there is no transfer of load to the vehicle body. If the displacements of the wheels are mutually opposed (one wheel up and the other down by the same amount), the full effect of the anti-roll stiffness is produced.

Roll stiffness is equal to the rate of change of roll moment with roll angle. Figure 3.29b shows the contributions to the total roll moment made by suspension stiffness and anti-roll bar stiffness.

The anti-roll bar as shown in Fig. 3.30 is an adaptation of the simple torsion bar. In this design the total bar construction would be of the same section to reduce costs. As such the torque arm section of the bar will be subject to bending which should be included in calculation of total stiffness.

The principle of operation of any ARB (Fig. 3.31) is to convert the applied wheel load F_W into a torque $F_W \times R$ producing twist in the twist section of the bar. A circular cross-section bar gives the lowest spring weight for a given stiffness and may be solid or tubular. In this case simple torsion of shaft theory can be used to determine the stiffness of the bar and the stresses within it. In general, stiffness is

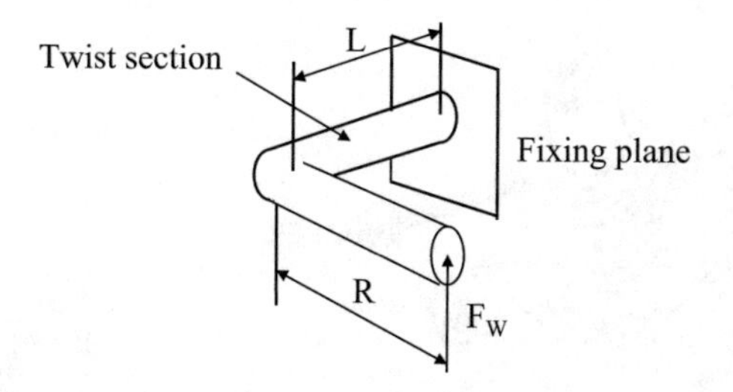

Fig. 3.31 Principle of operation of a torsion bar spring

related to diameter and length of the torsion bar and the shear modulus of the material. The torque arm (of length R as shown in Fig. 3.31) should be relatively stiff in comparison to the torsion section and may sometimes be neglected in stiffness calculations. A bending moment $F_W \times L$ is also induced in the twist section of the member and supports should be positioned to minimise this. As the torque arm rotates under load, the moment arm changes, requiring twist angle corrections (for large rotations) to be made in design calculations.

For the twist section shown in Fig. 3.31, torque $T = RF_W$
giving the strain energy in the section as:

$$U = \frac{T^2 L}{2GJ} = \frac{R^2 F_W^2 L32}{2G\pi d^4}$$

where $J = \frac{\pi d^4}{32}$ for a solid torsion section.

Since $\frac{dU}{dF} = x$, we have for the vertical deflection x in the direction of F_W:

$$x = \frac{R^2 F_W L32}{G\pi d^4}$$

Thus the effective vertical stiffness of the ARB is given by:

$$\frac{F_W}{x} = \frac{G\pi d^4}{32LR^2} \tag{3.7}$$

The torsional stiffness of the twist section is given by the standard equation for a solid circular shaft:

$$\frac{T}{\theta} = \frac{GJ}{L} = \frac{G\pi d^4}{32L} \tag{3.8}$$

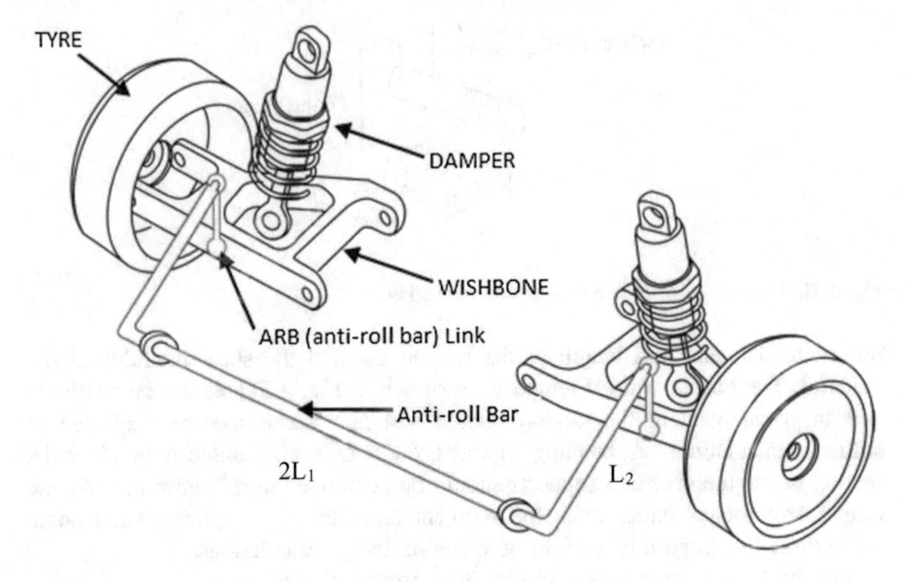

Fig. E3.2a Typical anti roll bar installation

Example E3.2

A suspension system is shown in Fig. E3.2a for which the following parameters are given:

Anti-roll bar:

Rod diameter	15 mm
$2L_1$	1000 mm
L_2	300 mm
Shear modulus (G)	79.3 GPa
Young's modulus	206 GPa

Spring:

Number of active coils	7
Wire diameter	12.7 mm
Mean coil diameter	120 mm
Shear modulus (G)	79.3 GPa

Given the above calculate:

(i) The stiffness of the coil spring.
(ii) The total vertical stiffness of the torsion bar at the wheel.
(iii) The total stiffness of the torsion bar and suspension spring only (not including the tyre).

Solution

(i) Consider coil spring

$$\text{Spring Rate } k_s = \frac{F}{x} = \frac{Gd^4}{8D^3n} = \frac{79.3 \times 10^3 \times 12.7^4}{8 \times 120^3 \times 7} = 21.3\,\text{N/mm}$$

(ii) Consider anti-roll bar torsional element $(2L_1)$

From Eq. (3.8):

$$\frac{T}{\theta} = \frac{GJ}{L_1} = \frac{G\pi d^4}{32L_1}$$

Note that only half the length of the torsional section is considered in calculating the stiffness for each wheel i.e. the mid-plane of the bar is considered a plane of symmetry as would be the case under equal and opposite movements of the 2 wheels.

But $T = FL_2$ and, for small angles $L_2\theta = x$, giving $\theta = \frac{x}{L_2}$

Substituting for T and θ in above gives:

$$k_T = \frac{F}{x} = \frac{G\pi d^4}{32L_1L_2^2} = \frac{79.3 \times 10^3 \times \pi \times 15^4}{32 \times 500 \times 300^2} = 8.75\,\text{N/mm}$$

Consider anti-roll bar bending element (L_2).
For a cantilever beam, end deflection under load is given as:

$$\delta = \frac{PL^3}{3EI}$$

where:

δ vertical deflection at end of cantilever
P load at end of cantilever

I second moment of area in bending
E Young's modulus (E)
L length of cantilever

In this case

$$x = \frac{FL_2^3}{3EI}$$

where $I = \frac{\pi d^4}{64}$ giving

$$k_B = \frac{F}{x} = \frac{3EI}{L_2^3} = \frac{3E\pi d^4}{64L_2^3}$$

Therefore

$$k_B = \frac{3 \times 206 \times 10^3 \times \pi \times 15^4}{64 \times 300^3} = 56.87\,\text{N/mm}$$

Torsion and bending stiffnesses of torsion bar act in series giving total vertical stiffness of the torsion bar at the wheel:

$$k_{TB} = \frac{k_T \times k_B}{k_T + k_B} = \frac{8.75 \times 56.87}{8.75 + 56.87} = 7.58\,\text{N/mm}$$

It should be noted that the stiffer the bending section of the anti-roll bar, the less influence it has on the total stiffness.

(iii) Consider total stiffness of the torsion bar and suspension spring only (not including the tyre).

Coil spring and torsion bar act in parallel giving total stiffness of the torsion bar and suspension spring:

$$k_{Total} = k_{TB} + k_S = 7.58 + 21.3 = 28.88\,\text{N/mm}$$

Example E3.3

A test engineer pushes on the centre of a car bonnet and records the front bounce frequency at 1.60 Hz. The engineer then pushes on the body over one front wheel, the body over the other wheel remaining stationary, and records a different frequency of 1.78 Hz.

The suspension configuration for this particular vehicle is generally as shown in Fig. E3.2a but with the relevant vehicle data as listed below:

Front sprung mass 500 kg

Suspension:

Front Bounce Frequency	1.60 Hz
Corner Bounce Frequency	1.78 Hz

Suspension spring:

Lever ratio	1:2
Wire diameter (d)	15 mm
Front mean coil diameter (D)	100 mm
Shear modulus (G)	79.3 GPa

Anti-roll (torsion) bar:

Lever ratio	1:1
L_1 (assume half total length)	600 mm
L_2	300 mm
Shear modulus (G)	79.3 GPa
Young's modulus (E)	206 GPa

The test engineer wishes to determine and confirm:

(i) The number of active coils in the spring.
(ii) Anti-roll bar stiffness.
(iii) Required diameter for the anti-roll bar rod.

(i) **To calculate number of active coils in spring**

Front bounce motion does not activate the ARB.
Front f = 1.6 Hz giving $\omega = 2\pi f = 10$ rad/s
But $\omega = \sqrt{(k/m)}$ or $k = m\omega^2$
The wheel rate is therefore given by:

$$k_w = (500/2) \text{ x } 10^2/1000 = 25\ N/mm$$

Lever ratio is 1:2 giving spring rate $k_s = 25/0.5^2 =$ **100 N/mm**
Knowing:

$$k_s = \frac{F}{x} = \frac{Gd^4}{8D^3 n}$$

gives:

$$n = \frac{Gd^4}{8D^3k} = \frac{79.3 \times 10^3 \times 15^4}{8 \times 100^3 \times 100} = 5 \ active\ coils$$

(ii) **To calculate torsion bar stiffness**

In this case, the coil spring & torsion bar act in parallel giving a combined stiffness k_c.

We have $f = 1.78\ Hz\ \ giving\ \omega = 2\pi f = 11.21\ rad/s$

But $\omega = \sqrt{\frac{k_c}{m}}\ \ or\ k_c = \omega^2 m = \frac{11.21^2 \times 250}{1000} = 32.42\ N/mm$

Noting lever ratio for anti-roll bar = 1:1

Combined stiffness **at the wheel** = $k_w + k_{TB}$ = **32.42 N/mm**

Giving k_{TB} = 32.42−25.00 = **7.42 N/mm**

(iii) **To calculate torsion bar rod diameter**:

Consider twist section of torsion bar:

$$\frac{T}{\theta} = \frac{GJ}{L} = \frac{G\pi d^4}{32L}$$

But $T = FL_2\ and\ \theta = \frac{x}{L_2}$ *Substituting in above gives*:

$$k_T = \frac{F}{x} = \frac{G\pi d^4}{32L_1L_2^2} = \frac{79.3 \times \pi \times 10^3 \times d^4}{32 \times 600 \times 300^2} = 144d^4 \times 10^{-6}\ N/mm$$

Consider bending element of torsion bar:

$$x = \frac{FR^3}{3EI}$$

where $I = \frac{\pi d^4}{64}$ giving

$$k_B = \frac{F}{x} = \frac{3E\pi d^4}{64L_2^3}$$

$$k_{TB} = \frac{F}{x} = \frac{3E\pi d^4}{64L_2^3} = \frac{3 \times 206 \times 10^3 \times \pi \times d^4}{64 \times 300^3} = 1223\ d^4 \times 10^{-6}\ N/mm$$

Torsion and bending sections of torsion bar are in series giving:

$$k_{TB} = \frac{k_T \times k_B}{k_T + k_B} = \frac{144d^4 \times 10^{-6} \times 1223\ d^4 \times 10^{-6}}{144d^4 \times 10^{-6} + 1223\ d^4 \times 10^{-6}}$$

$$= \frac{176,112d^8 \times 10^{-12}}{1367d^4 \times 10^{-6}} = 128.83\, d^4 \times 10^{-6} = 7.42\, N/mm$$

Hence:

$$d = 15.5\, mm\ (say\ 16\, mm\ to\ give\ standard\ size)$$

3.5 Dampers

3.5.1 Damper Types and Characteristics

Frequently called shock absorbers, dampers are the main energy dissipaters in a vehicle suspension. They are required to dampen vibration after a wheel strikes a pot-hole or similar. In addition they provide a good compromise between low sprung mass acceleration (related to ride) and adequate control of the unsprung mass to provide good road holding.

Suspension dampers are usually telescopic devices containing hydraulic fluid. They are connected between the sprung and unsprung masses and produce a damping force that is proportional to the relative velocity across their ends. The features of the two most common types of passive damper are shown in Fig. 3.32.

Figure 3.32a shows a dual tube damper in which the inner tube is the working cylinder while the outer cylinder is used as a fluid reservoir. The latter is necessary to store the surplus fluid that results from a difference in volumes on either side of the piston. This is a result of the variable rod volume in the inner tube.

In the monotube damper shown in Fig. 3.32b, the surplus fluid is accommodated by a gas-pressurised free piston. An alternative form of monotube damper (not shown in Fig. 3.32) uses a gas/liquid mixture as the working fluid to absorb the volume differences.

Comparing the two types of damper shown in Fig. 3.32, the dual tube design offers better protection against stones thrown up by the wheels and is also a shorter unit making it easier to package. On the other hand, the monotube strut dissipates heat more readily.

In dealing with road surface undulations in the bump direction (the damper in compression), relatively low levels of damping are required when compared with the rebound motion (the damper extended). This is because the damping force produced in bump tends to aid the acceleration of the sprung mass, while in rebound an increased level of damping is required to dissipate the energy stored in the suspension spring. These requirements lead to damper characteristics that are asymmetrical when plotted on force-velocity axes (Fig. 3.33). The damping rate (or coefficient) is the slope of the characteristic. Ratios of 3:1 for rebound to bump are quite common.

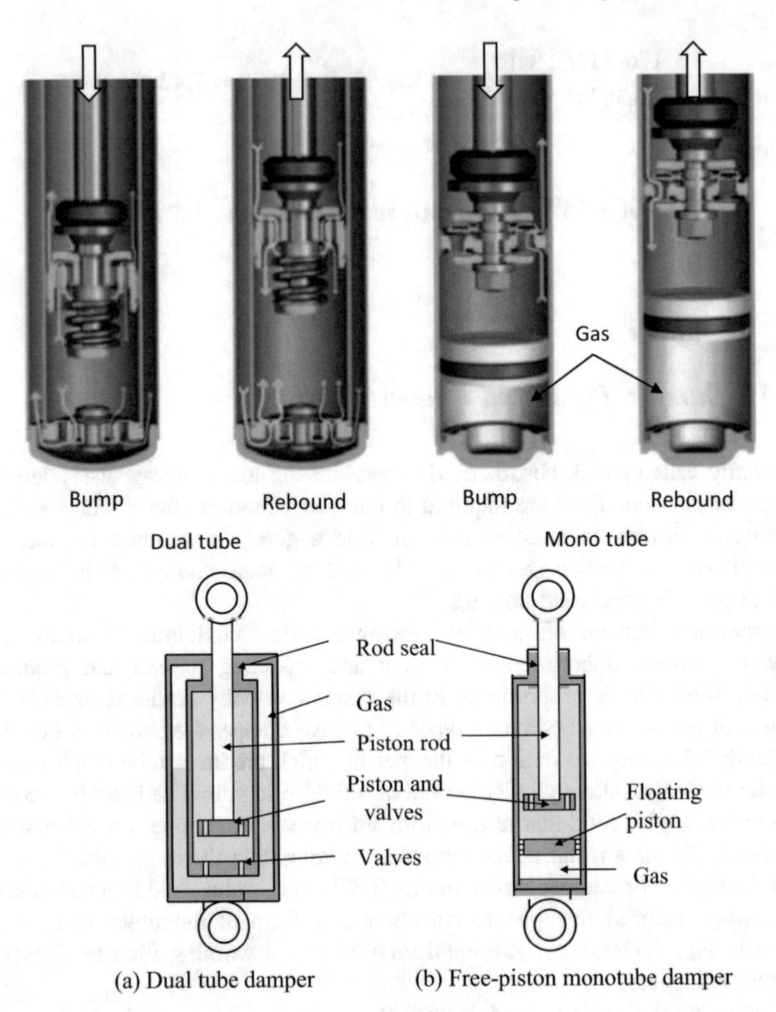

Fig. 3.32 Damper types.
Reproduced with kind permission of ZF © ZF Friedrichshafen AG

The characteristics called for in damper designs are achieved by a combination of orifice flow and flows through spring-loaded one-way valves. These provide plenty of scope for shaping and fine-tuning of damper characteristics. Figure 3.34 shows the principle of the combined orifice and valve control system and Fig. 3.35 the resultant characteristic. Bump and rebound would have different valves. At low relative velocities, damping is by orifice control until the fluid pressure is sufficient to open the pre-loaded flow-control valves. Hence the shape of the combined characteristic shown in Fig. 3.35. The problem with the design as shown in Fig. 3.34 is that by using a sphere in the cut-off valve the operation of the valve tends to be either open or closed (on/off).

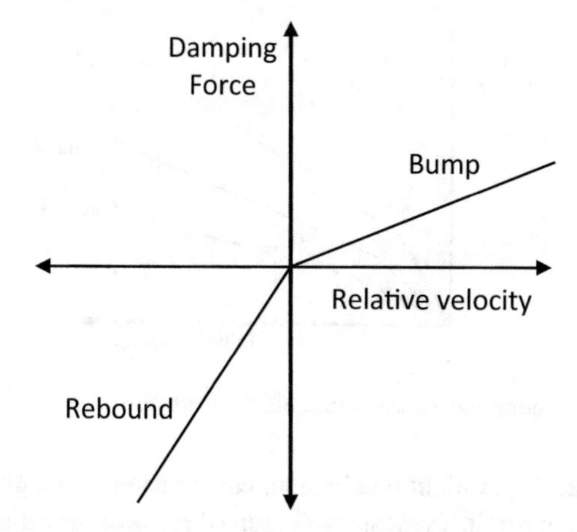

Fig. 3.33 Non-linear damper characteristics

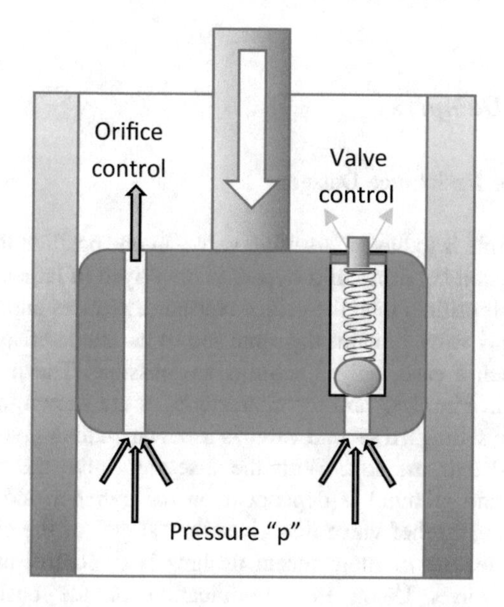

Fig. 3.34 Principle of the orifice and one way valve control arrangement

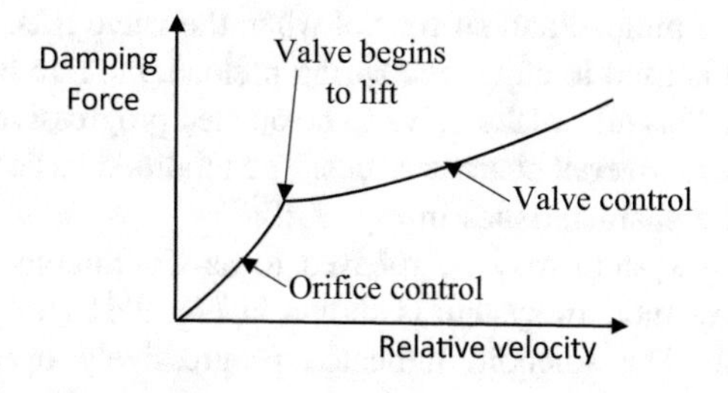

Fig. 3.35 Shaping of damper characteristics

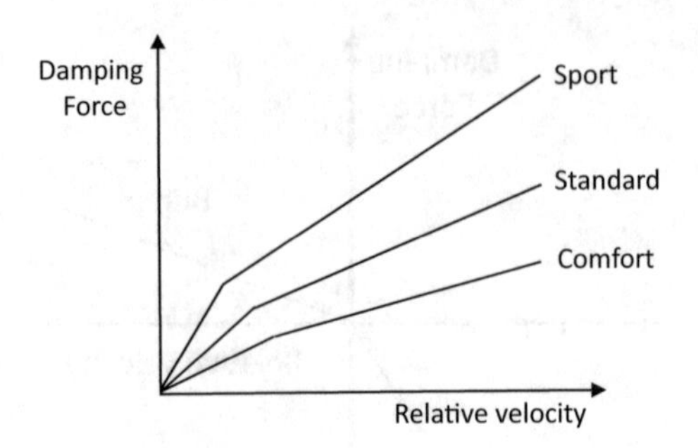

Fig. 3.36 Different operating modes for adjustable dampers

A driver-operated adjustment mechanism can be used to obtain several damping characteristics from one unit. Typical curves for a three position adjustable damper are shown in Fig. 3.36. A continuous electronically controlled adjustment forms the basis of one type of controllable suspension designed to improve both ride and handling.

3.5.2 Active Dampers

3.5.2.1 Variable Resistance Designs

The general principle is to have a 2-setting valve. In one position the orifice is small and so the suspension is "firm". If a bypass is employed (a larger orifice is opened alongside the small orifice) then the orifice resistance reduces and the suspension is "soft". The one-way valve remains the same and so its characteristic (slope) remains unchanged. In such a case, only 2 settings are possible. The modified system is indicated simply in Fig. 3.37 and the characteristics are shown in Fig. 3.38.

To obtain more settings, a second valve is used which then gives 4 settings, firm, medium, soft and extreme soft. With the 2-setting valve the ratio of damping between bounce and rebound is dependent on the cylinder and rod dimensions. There is an additional relief valve that gives the "slope" of the characteristic.

An additional feature of more recent designs is to control the valve opening using a needle valve. Using this modification of the basic principle, the force-velocity characteristics may be shaped by altering the needle design (taper and shape) giving more control of flow. A further advance would be to also provide for a variable spring preload and so control when the valve lifts. In these advanced designs a solenoid is used to adjust the spring preload, the rate being controlled by the current supply. This allows the valve to be opened progressively (rather than on/off) and so change the overall characteristics. The modified basic principle is shown in Fig. 3.39 and the characteristics in Fig. 3.40.

Such a damping system may be referred to as Continuous Damping Control (CDC) and a typical modern system is shown in Fig. 3.41 along with the internal valve arrangement. The solenoid indicated progressively opens the valve, the

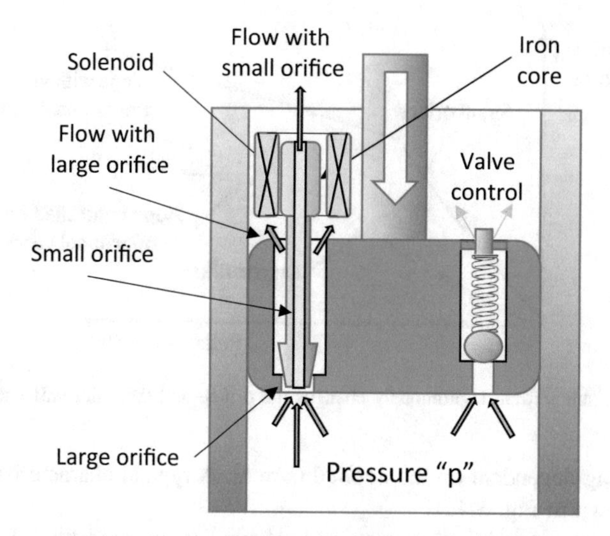

Fig. 3.37 The basic design modified to include a solenoid controlled orifice

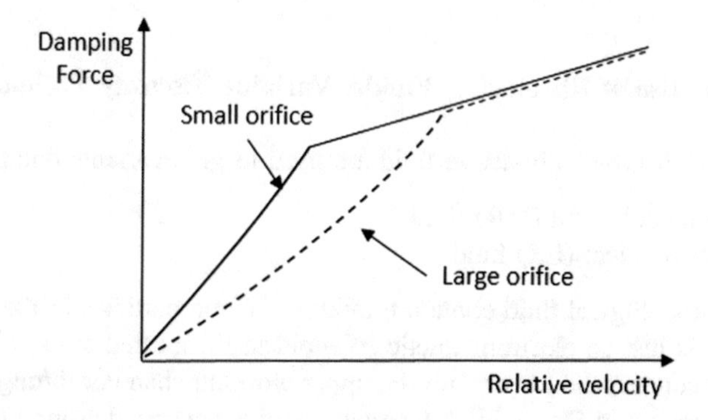

Fig. 3.38 Typical characteristics of a 2-setting valve

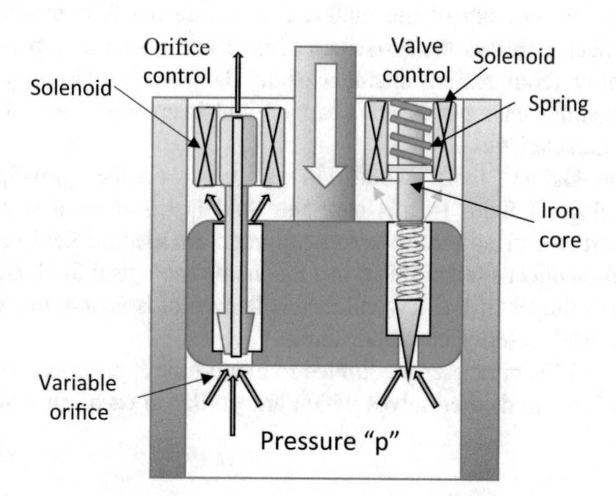

Fig. 3.39 Variable current continuously changes the relief valve setting and orifice opening

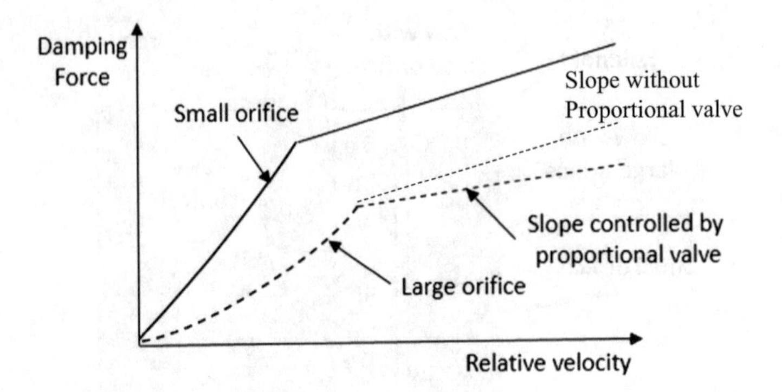

Fig. 3.40 Variable current continuously changes the orifice and the relief valve setting

opening being dependant on the applied current. A typical characteristic of such a valve is shown in Fig. 3.42.

If independent control of bounce and rebound is required then 2 proportional valves can be used, one each for bounce and rebound as shown in Fig. 3.43.

3.5.2.2 The Use of Rheological Fluids: Variable Viscosity Technology

Two types of rheologically-active fluid are used to give variable damping:

- Magneto-reheological (MR) fluid
- Electro-rheological (ER) fluid.

Magneto-rheological fluid contain a mixture of iron particles in the damper oil. The "valve" is just an electromagnetic solenoid itself, located at the piston, which has holes to connect the lower with the upper working chamber through which the oil flows as shown in Fig. 3.44. A magnetic field is generated around the valve, its strength changing the orientation of the iron particles within the valve area. This in turn changes the viscosity of the fluid and therefore the flow through the valve holes. The effect increases the resistance of flow and generates a pressure change inside the active (compressed) chamber of the damper. The viscosity of the fluid inside the working chamber is not changed and keeps the same physical compressibility characteristics.

Electro-rheological fluid works in a very similar principle to the magneto-rheological fluid. In this case the fluid is based on a silicon oil fluid containing particles of an electro-active polymer. An electric field is applied and this generates an effect similar to that in a magneto-rheological fluid. Because of the required high voltages of 1–2 kV and the challenges of isolation, this principle has not yet been used in series car applications.

Both MR and ER principles are limited to mono-tube applications because of the restrictions of the fluids themselves which are similar to each other in this regard.

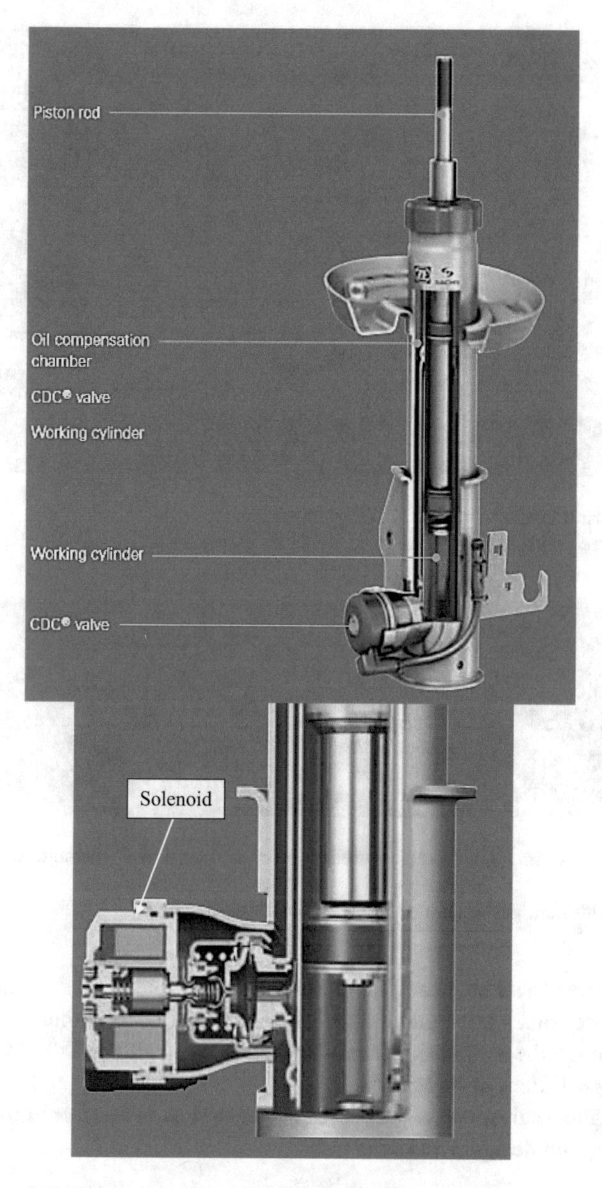

Fig. 3.41 External Continuous Damping Control (CDC). Reproduced with kind permission of ZF © ZF Friedrichshafen AG

3.6 Kinematic Analysis of Suspensions

Kinematics is the study of the geometry of motion, without reference to the forces involved. In the case of vehicle suspensions, which are simply examples of the many linkages and mechanisms encountered in mechanical engineering, study of their kinematic properties involve some important assumptions in relation to their

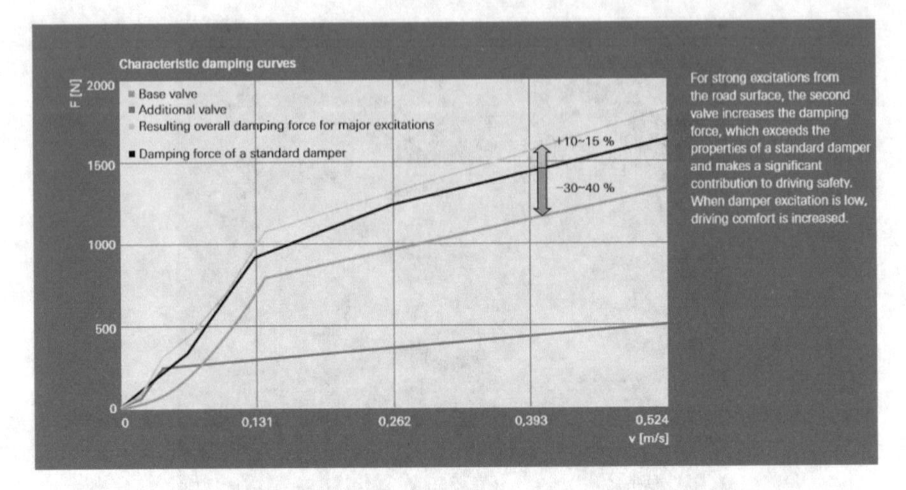

Fig. 3.42 General characteristic of a CDC damper.
Image reproduced with kind permission of ZF © ZF Friedrichshafen AG

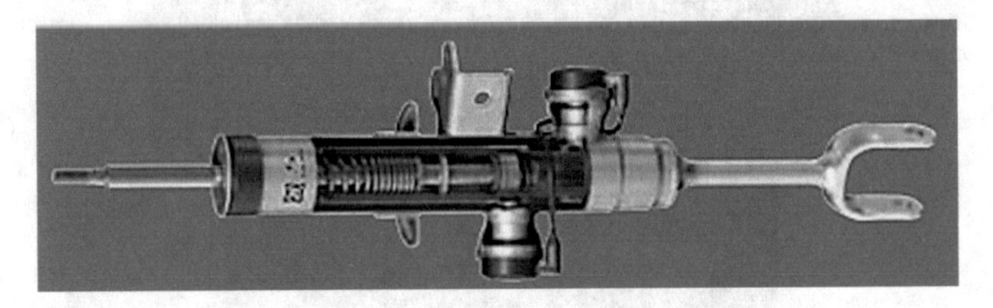

Fig. 3.43 Two external CDC valves used to control bump and rebound continuously and independently.
Reproduced with kind permission of ZF © ZF Friedrichshafen AG

practical operation. For example, elastic deformations are excluded from the analysis (since forces are excluded). This means that bush compliances cannot be included. Hence the suspension must be treated as if all the link connections are perfect pivots, sliders or ball joints.

A kinematic analysis of a proposed suspension is invariably the first analysis performed by the designer in order to:

- Check that the mechanism actually works over the full bump/rebound range and does not interfere with the body
- Check the way in which the wheel geometry is controlled over the suspension working range
- Calculate the effective ratios involved between spring and damper travel relative to wheel travel.

Suspensions are really three-dimensional mechanisms so that kinematics analyses should really be three-dimensional. However, the largest displacement

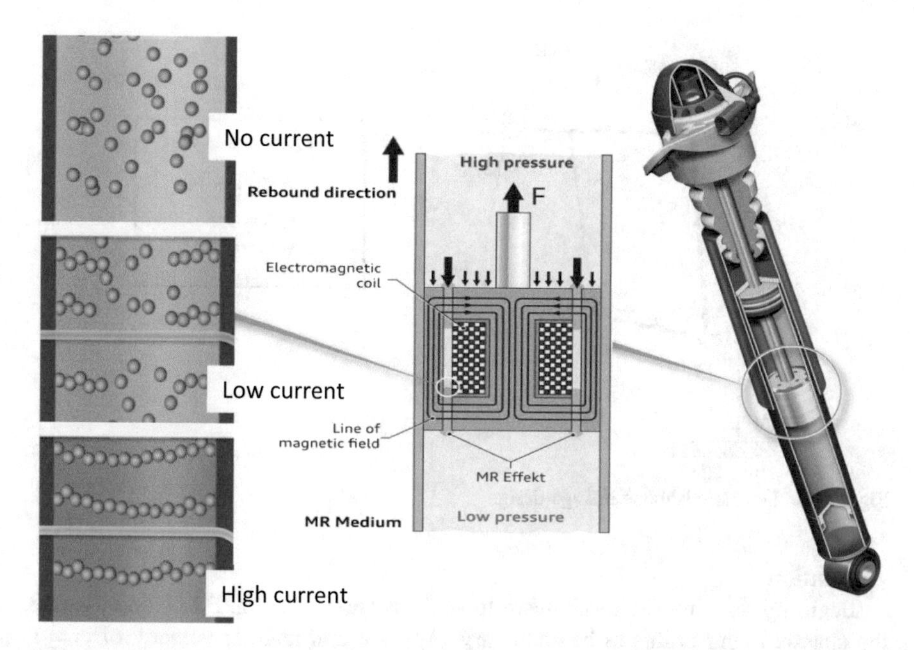

Fig. 3.44 The general principle of the control valve (piston head) using a magneto-rheological fluid.
Reproduced with kind permission of Audi © Audi AG

components occur in a transverse plane perpendicular to the longitudinal axis of the vehicle. It follows that much useful information can be obtained from a two-dimensional analysis.

There are three ways in which suspension kinematics can be analysed:

- Analytically using vector-based methods
- Using a computer package with a mechanisms analysis capability
- Graphically (only really relevant for a two-dimensional analysis).

It is the third approach that we are going to use in the following example.

Example E3.4 A scaled drawing of a double wishbone suspension is shown diagrammatically in Fig. E3.4a. The damper is mounted between E and F, and E can be assumed to be on the line AD. The centre of the tyre contact patch is at W and WV is the centre-line of the wheel.

Use a velocity diagram to determine:

(a) The scrub to bump derivative at the tyre-road interface (ratio of wheel horizontal to vertical velocities)
(b) The damper rate to bump rate (ratio of relative velocity across the damper to the vertical velocity of the wheel)
(c) The suspension ratio (the inverse of the damper rate to bump rate).

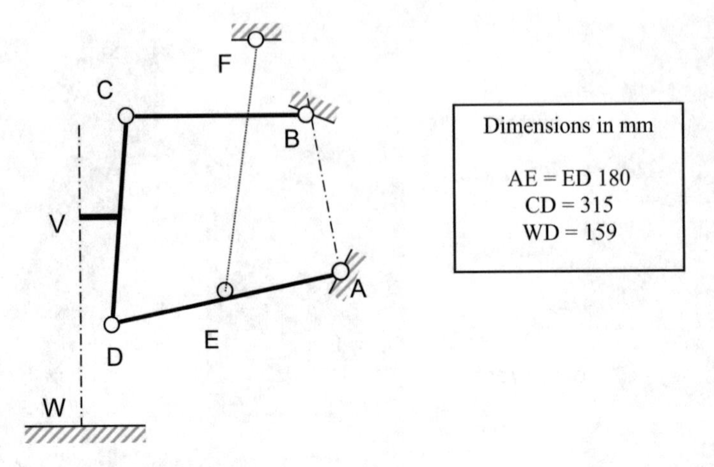

Fig. E3.4a Double wishbone linkage design

Solution

Begin by drawing the mechanism to scale as shown in Fig. E3.4b and assume the chassis fixing points to be stationary. Apply a unit angular velocity of $\omega = 1$ rad/s to link AD (the absolute magnitude is not important as the movements to be determined are all ratios). The magnitude of the velocity of D is $v_D = 360 \times 1 = 360$ mm/s and the vector $\underline{v}_D$ is perpendicular to AD as shown.

The velocity of C is given by: $\underline{v}_C = \underline{v}_D + \underline{v}_{CD}$ (a vector addition, where a symbol underlined denotes a vector). $\underline{v}_C$ is perpendicular to BC, its magnitude is unknown at this stage. Because CD can be considered as a rigid link, $\underline{v}_{CD}$ is perpendicular to CD but its magnitude is unknown at this stage. The directions of these vectors are also unknown (see dotted lines in Fig. E3.4b). Now draw a line from D to W; it will be assumed this imaginary link is rigid and forms part of a link CDW.

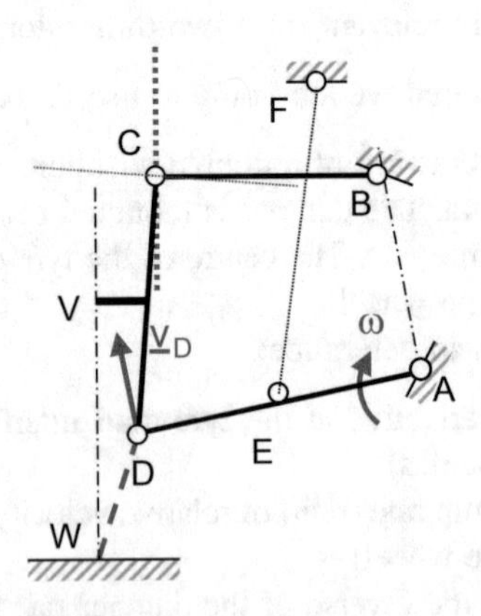

Fig. E3.4b Drawing of linkage to some arbitrary scale

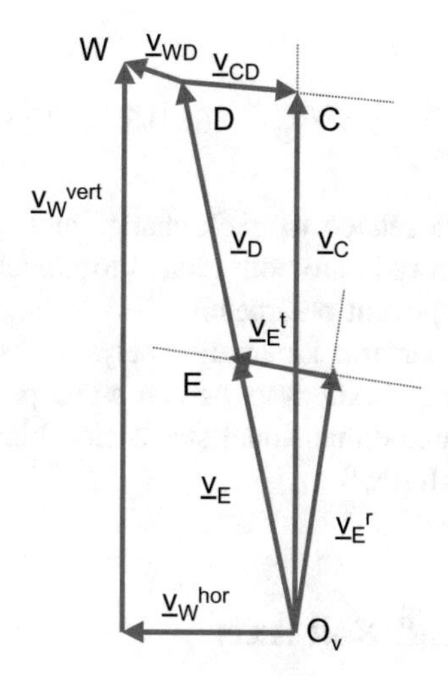

Fig. E3.4c Velocity diagram (to some arbitrary scale)

Construction of the velocity diagram (Fig. E3.4c) begins by first locating the pole O_v. Vectors drawn from here are "absolute" velocities.

1. There should be a *velocity image* CDW in the velocity diagram of similar shape to CDW in Fig. E3.4b. Determine W by proportion from Fig. E3.4b and hence add the vector $\underline{v}_{WD}$ as shown in Fig. E3.4c.
2. The (absolute) vertical and horizontal velocity components of W, $\underline{v}_W^{hor}$ and $\underline{v}_W^{vert}$ can be added to the diagram as shown in Fig. E3.4c.
3. The (absolute) velocity of E, $\underline{v}_E$, is half that of D and can now be added to the diagram. There are components parallel and perpendicular to EF which are related respectively to relative velocity across the damper and its angular velocity. Add the component vectors $\underline{v}_E^r$ and $\underline{v}_E^t$.
4. Relevant values can now be scaled from the diagram. These are: $\underline{v}_W^{vert} = 336$ mm/s, $\underline{v}_W^{hor} = 114$ mm/s and $v_E^r = 168$ mm/s.

Answers

(a) Scrub to bump derivative:

$$v_W^{hor}/v_W^{vert} = 114/366 = 0.393$$

(b) Damper rate: bump rate

$$v_E^r/v_W^{vert} = 168/366 = 0.459$$

(c) Suspension ratio

$$v_W^{vert}/v_E^r = 366/168 = 2.18$$

Bump to scrub ratio is related to track change and tyre wear. Damper rate to bump rate and suspension ratio are both related to modelling for ride analysis and the determination of component parameters.

Note: The facility to perform kinematic analysis of three-dimensional mechanisms is available in many CAD systems and some packages have been written specifically to analyse three-dimensional suspension kinematics, for example, the one embodied in MSC/ADAMS.

3.7 Roll Centres and Roll Axis

Roll centre and *roll axis* are important concepts for studying vehicle handling and are also used in the calculation of lateral load transfer for cornering operations.

There are two definitions of roll centre. One is based on forces and the other on kinematics. The first of these (the SAE definition) states that:

(1) *The roll centre is a point in the transverse plane through any pair of wheels at which a transverse force may be applied to the sprung mass without causing it to roll.*

The second is a kinematic definition which states that:

(2) *The roll centre is the point about which the body can roll without producing any lateral movement at either of the wheel contact areas.*

In general each *roll centre* lies on the vertical line produced by the intersection of the longitudinal centre plane of the vehicle and the vertical transverse plane through a pair of wheel centres. The roll centre heights in the front and rear wheel planes tend to be different as shown in Fig. 3.45. The line joining the centres is

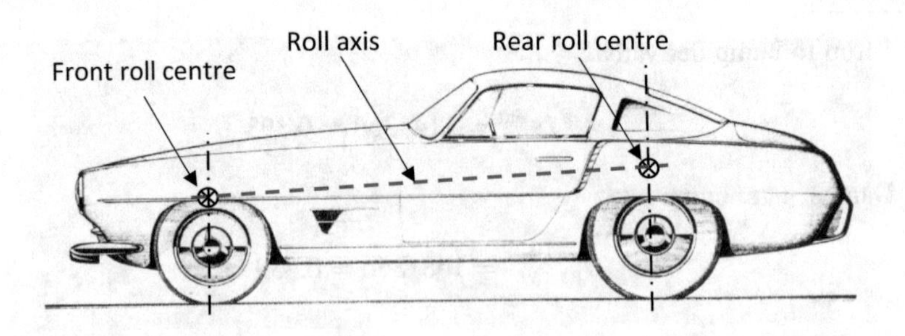

Fig. 3.45 Roll centres and roll axis.
http://www.carstyling.ru/resources/studios/1955_Ghia_Ferrari_375_MM_Drawing.jpg

called the *roll axis*, with the implication that a transverse force applied to the sprung mass at any point on this axis will not cause body roll.

3.7.1 Roll Centre Determination

For a given front or rear suspension, the roll centre can be determined from the kinematic definition by using the Aronhold-Kennedy theorem of three centres. This states: "*when three bodies move relative to one another they have three instantaneous centres all of which lie on the same straight line*".

To illustrate the determination of roll centre by this method consider the double wishbone suspension shown in Fig. 3.46. Consider the three bodies capable of relative motion as being the sprung mass, the left-hand wheel and the ground. The instantaneous centre of the wheel relative to the sprung mass $I_{wb,}$ lies at the intersection of the upper and lower wishbones, while that of the wheel relative to the ground lies at I_{wg}. The instantaneous centre of the sprung mass relative to the ground (the roll centre) $I_{bg,}$ must (according to the above theorem) lie in the centre plane of the vehicle and on the line joining I_{wb} and I_{wg}, as shown in the diagram.

I_{wg} always lies in the position shown in Fig. 3.46. Changing the inclination of the upper and lower wishbones can vary I_{wb}, thereby altering the location of I_{bg} and ultimately the load transfer between inner and outer wheels during cornering. In the case of the McPherson strut suspension (Fig. 3.47) the upper line defining I_{wb} is perpendicular to the strut axis. Figures 3.48, 3.49 and 3.50 illustrate the locations of roll centres for a variety of independent suspensions. The importance of roll centre location is discussed below.

For the swing axle suspension shown in Fig. 3.48 the wheel pivots about the inner swivel axis giving the roll centre as indicated.

For the trailing arm suspension (Fig. 3.49), the trailing arm pivots about a transverse axis (forward of the wheel centre). In the front view the wheel is

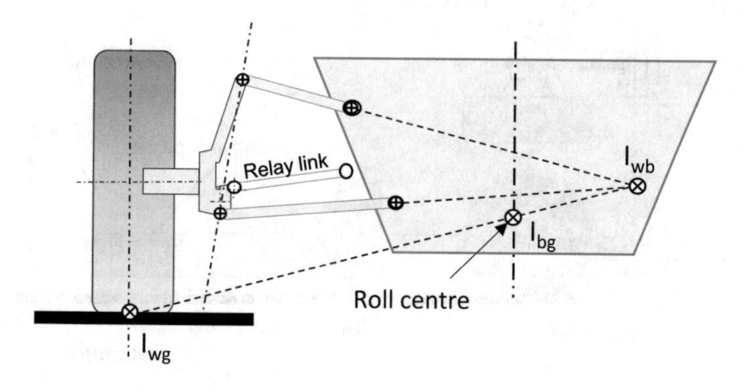

Fig. 3.46 Roll centre determination for a double wishbone suspension

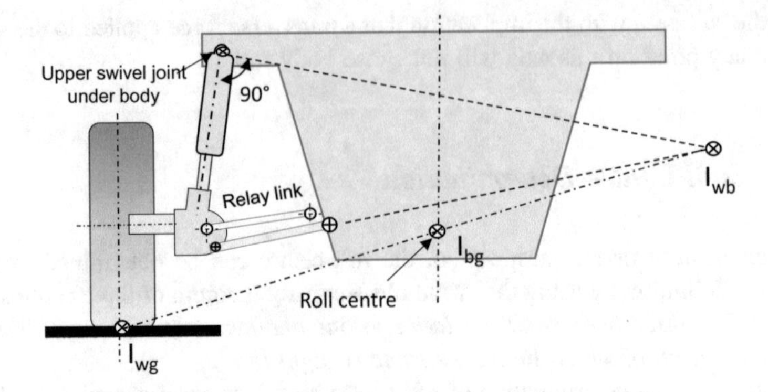

Fig. 3.47 Roll centre location for a McPherson strut

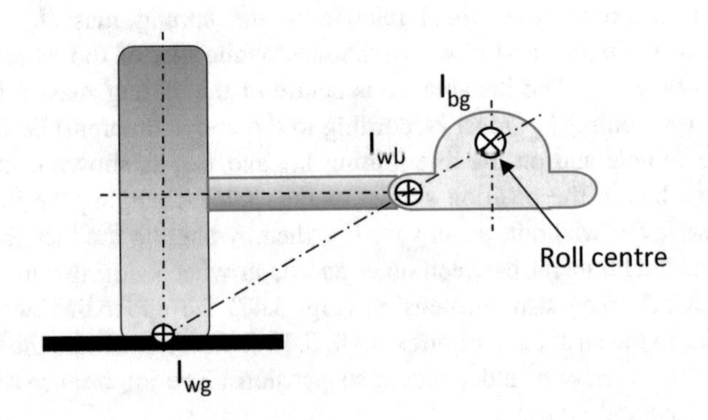

Fig. 3.48 Roll centre location for a swing axle suspension

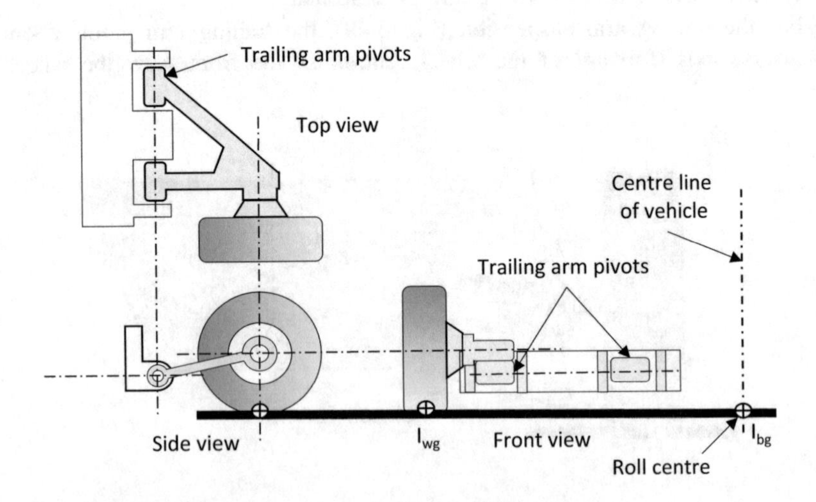

Fig. 3.49 Roll centre location for a trailing-arm suspension

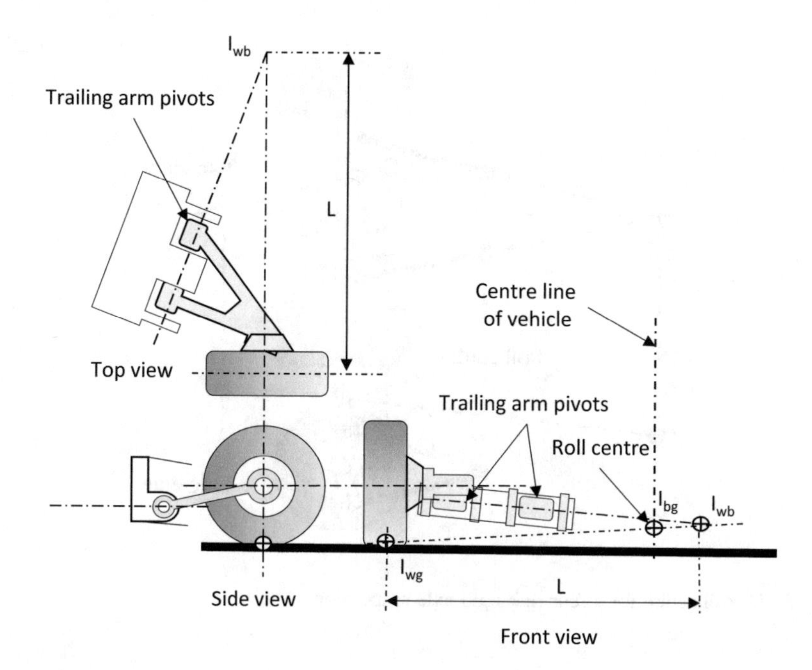

Fig. 3.50 Roll centre location for semi-trailing arm suspension

constrained to move in a vertical plane (with no transverse movement) and hence I_{wb} lies at infinity along the pivot axis (to the right). The roll centre therefore lies in the ground plane on the centre-line of the vehicle.

For the semi-trailing arm suspension (Fig. 3.50) the pivot axis is inclined and intersects the vertical lateral plane through the wheel centre at I_{wb} a distance L from the centre plane of the wheel. The roll centre I_{bg} lies on the line connecting I_{wb} with the instantaneous centre of the wheel relative to the ground I_{wg}.

Figure 3.51 shows the roll centre determination for a four-link rigid axle suspension. In this case the wheels and axle can be assumed to move as a rigid body. The upper and lower control arms produce instant centres at A and B respectively (projected to the vehicle centreline). Connecting these together produces a roll axis for the suspension. The intersection of this axis with the transverse wheel plane defines the roll centre.

The final example illustrating roll centre determination is the Hotchkiss rear suspension shown in Fig. 3.52. The analysis in this case is somewhat different to the previous examples. Lateral forces are transmitted to the sprung mass at A and B. The roll centre height is at the intersection of the line joining these points and the vertical transverse plane through the wheel centres. The roll centre is of course at this height in the centre plane of the vehicle.

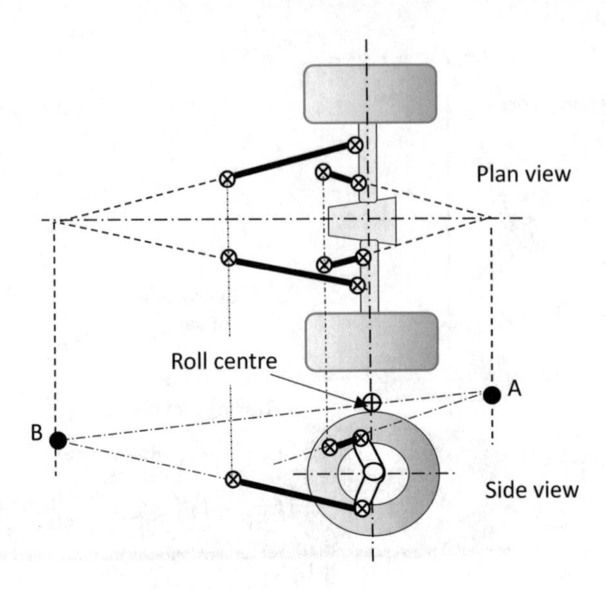

Fig. 3.51 Roll centre for a four link rigid axle suspension

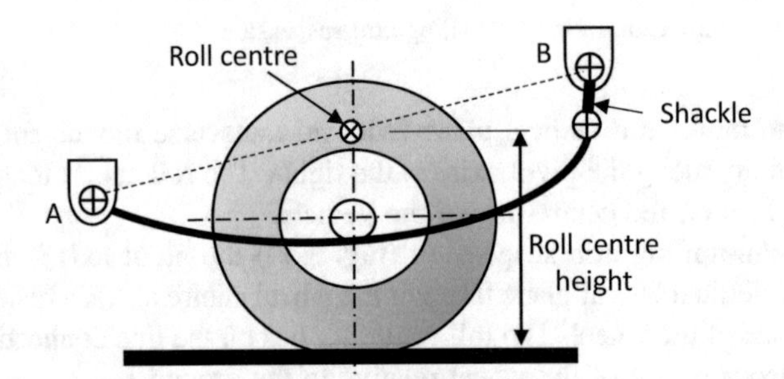

Fig. 3.52 Roll centre location for a Hotchkiss suspension

3.7.2 Roll Centre Migration

When a suspension deflects, the roll centre locations move. Even for modest suspension deflections, this migration of the roll centre can be quite significant as shown in Fig. 3.53, arising in this case from the vertical deflection of the vehicle body. Large roll centre migrations can have undesirable effects on vehicle handling. Hence attempts are made at the design stage to limit migration effects for the full range of suspension travel. Some suspension types are more amenable than others for controlling roll centre migration.

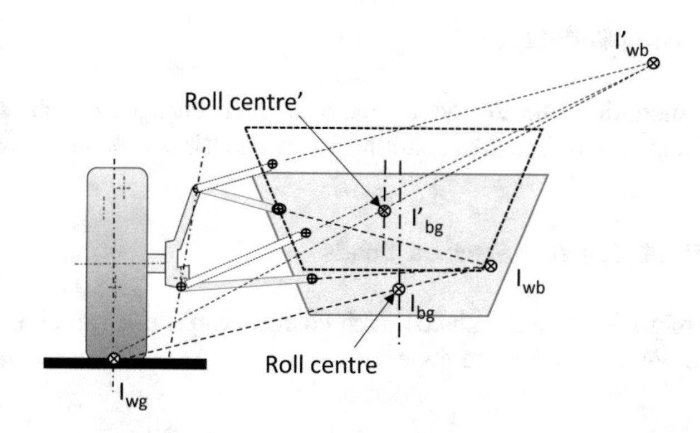

Fig. 3.53 Effect of suspension deflection on roll centre migration

A high roll centre produces low lateral load transfer (good anti-roll properties) but produces high levels of half-track changes. The high roll centre is typically closer to the centre of gravity and hence there is a reduced lever arm and lower roll moment. Lateral forces generated under cornering therefore produce less roll. However, the suspension geometry to give a high roll centre typically means that the wheel scrub is significant over the suspension travel, hence the larger half-track changes.

A low roll centre produces small half-track changes but poor anti-roll properties.

The migration of the roll centre needs to be limited to minimise the steering balance variation under different driving conditions as described below.

3.7.2.1 Steering Response

The balance between steady-state understeer/oversteer is largely dictated by the front to rear lateral load transfer ratio which is dependent on roll centre heights. This can be modified by changes in roll stiffness usually with anti-roll bars. Migration of roll centres can create unwanted variations in steering balance.

3.7.2.2 Straight Line Stability

Increasing front to rear lateral load transfer ratio produces an understeer tendency. Unsymmetrical road undulations induce half-track changes producing lateral force variations at the tyre contact patches with resultant steer effects. A positive half-track change produces a force towards the vehicle centreline and this effect, when combined with bump steer, will affect vehicle stability.

3.7.2.3 Braking Stability

Roll centre migration due to braking induced pitch changes can cause plunging front and rising rear roll centres resulting in an undesirable oversteer tendency.

3.7.2.4 Lift-off/Traction Steer on Bends

Roll centre migration due to induced pitch changes can cause variations in steering balance (see Steering Response above).

3.7.2.5 Ride

Half-track changes produce resisting forces having a vertical component. In some cases this may be large enough to affect ride quality. On asymmetric undulations, the lateral forces due to half-track changes may excite roll-rock. Vehicle roll is fundamentally linked to roll centre height, but can be modified by the use of anti-roll bars.

3.7.2.6 Jacking

Vehicle jacking is affected by roll centre height.

3.8 Lateral Load Transfer Due to Cornering

During cornering, centrifugal (inertia) forces act horizontally on the sprung and unsprung masses. These forces act above the ground plane through the respective mass centres causing moments to be generated on the respective masses. These in turn lead to changes in the vertical loads at the tyre which affect vehicle handling and stability. In general the vertical loads increase on the outer wheels while the loads on the inner wheels decrease.

The process of converting the transverse forces into vertical load changes is termed ***lateral load transfer***. Lateral load transfer due to cornering was considered for a rigid body vehicle model in Chap. 1. Here the effects of deflections in the vehicle suspension are considered to give more accurate predictions of lateral load transfer effects using D'Alembert's approach.

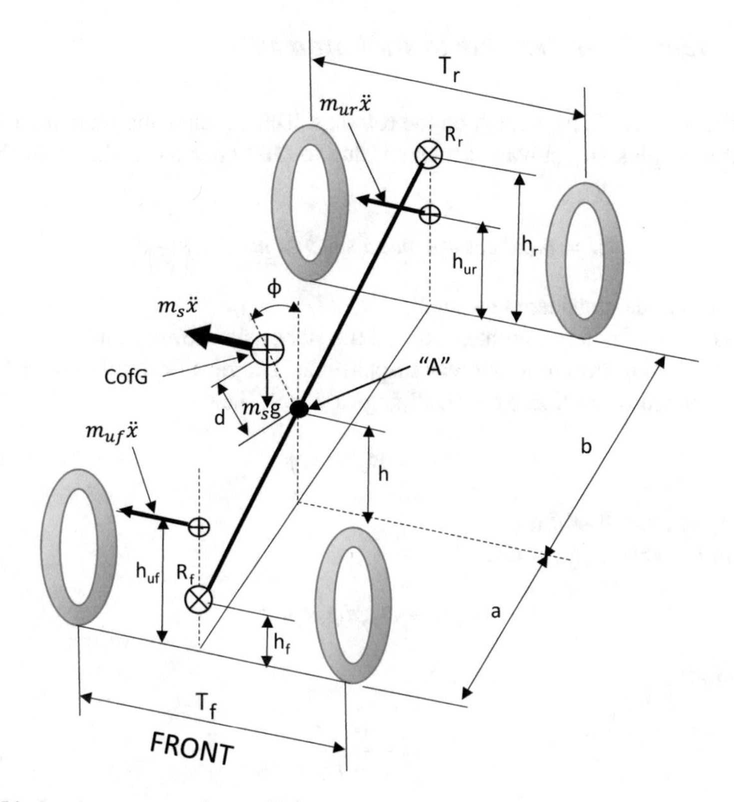

Fig. 3.54 Steady state cornering analysis

Assumptions:

- G is the sprung mass centre of gravity (CofG in Fig. 3.54).
- The transverse acceleration at G due to cornering is $\ddot{x}$.
- The sprung mass rolls through angle ϕ about the roll axis.
- The centrifugal (inertia) force on the sprung mass $m_s\ddot{x}$ acts horizontally through G.
- The gravity force on the sprung mass acts vertically downwards through G.
- The transverse inertia forces $m_{uf}\ddot{x}$ and $m_{ur}\ddot{x}$ act directly on the unsprung masses at the front and rear axles. Each transfers load only between its own pair of wheels.

The parameters in Fig. 3.54 are defined as follows:

m_s sprung mass
m_{uf} front unsprung mass
m_{ur} rear unsprung mass
R_f front roll centre
R_r rear roll centre
$\ddot{x}$ lateral acceleration during cornering.

Considering Fig. 3.54, the analysis is carried out in four stages as outlined below.

3.8.1 Load Transfer Due to Roll Moment

Refer the forces at G to point A on the roll axis. This requires the same forces to be applied at A plus an equivalent moment due to the two forces, M_s, about the roll axis:

$$M_s = m_s\ddot{x}d\cos\phi + m_s gd\sin\phi \approx m_s\ddot{x}d + m_s gd\phi \tag{3.9}$$

where ϕ may be considered as small.

M_s is reacted by a roll moment M_ϕ (at the suspension springs and anti-roll bars) and distributed to the front and rear suspensions. The relationship between M_s and M_ϕ is assumed to be linear for small angles of roll, i.e.

$$M_s = M_\phi = k_s\phi \tag{3.10}$$

where k_s is the roll stiffness.

From the above equations:

$$m_s xd + m_s gd\phi = k_s\phi$$

Giving:

$$\phi = \frac{m_s\ddot{x}d}{k_s - m_s gd} \tag{3.11}$$

M_ϕ can be split into components $M_{\phi f}$ and $M_{\phi r}$ at the front and rear axles and likewise the roll stiffness comprises front and rear components such that:

$$k_{sf} + k_{sr} = k_s$$

The load transfer due to the roll moment is then:

$$F_{fsM} = \frac{k_{sf}\phi}{T_f} = \frac{k_{sf}m_s\ddot{x}d}{T_f\left(k_{sf} + k_{sr} - m_s gd\right)} \tag{3.12}$$

Similarly the rear load transfer due to roll moment is:

$$F_{rsM} = \frac{k_{sr}\phi}{T_r} = \frac{k_{sr}m_s\ddot{x}d}{T_r(k_{sr} + k_{sr} - m_s gd)} \tag{3.13}$$

where T_f and T_r are the front and rear track widths of the vehicle.

3.8.2 Load Transfer Due to Sprung Mass Inertia Force

The sprung mass is distributed to the roll centres at the front and rear axles according to the position of the CofG. The respective masses are:

$$m_{sf} = \frac{m_s b}{(a+b)} = \frac{m_s b}{L}$$

And

$$m_{sr} = \frac{m_s a}{(a+b)} = \frac{m_s a}{L}$$

The centrifugal force at A is also distributed to the respective roll centres at the front and rear axles as follows:

$$F_{fs} = m_{sf}\,\ddot{x} \text{ and } F_{rs} = m_{sr}\ddot{x}$$

and the corresponding load transfers are:

$$F_{fsF} = \frac{m_{sf}\,\ddot{x}\,h_f}{T_f} \text{ and } F_{rsF} = \frac{m_{sr}\,\ddot{x}\,h_r}{T_r} \tag{3.14}$$

3.8.3 Load Transfer Due to Unsprung Mass Inertia Forces

The respective load transfers at the front and rear axles due to the unsprung mass inertia forces are:

$$F_{fuF} = \frac{m_{uf}\,\ddot{x}\,h_{uf}}{T_f} \text{ and } F_{ruF} = \frac{m_{ur}\,\ddot{x}\,h_{ur}}{T_r} \tag{3.15}$$

3.8.4 Total Load Transfer

Combine the load transfers due to roll moment with those due to inertia forces on the sprung and unsprung masses using relevant equations above, the load transfers for the front and rear wheels are:

$$F_f = F_{fsM} + F_{fsF} + F_{fuF} \tag{3.16}$$

and:

$$F_r = F_{rsM} + F_{rsF} + F_{ruF} \tag{3.17}$$

3.8.5 Roll Angle Gradient (Roll Rate)

This represents the relationship between the body roll angle and the lateral acceleration.

From Eq. (3.11), the roll angle gradient k_ϕ is given by:

$$k_\phi = \frac{d\phi}{d\ddot{x}} = \frac{m_s\ d}{k_s - m_s\ g\ d} \tag{3.18}$$

Example: E3.5

The following data is given for a luxury passenger car:

Total mass	1600 kg
Unsprung masses, front/rear	140/180 kg
Sprung mass CofG to front axle	0.46 × wheelbase
Track front & rear	1.46 m
Roll centre heights front/rear	70/320 mm
Spring mass CofG above ground	550 mm
Roll stiffness front/rear	350/150 Nm/deg
Unsprung mass CofG heights front/rear	280/300 mm

(a) Calculate the load transfers due to cornering with lateral acceleration 0.6 g.
(b) Calculate roll angle gradient.

Solution:

(a) Calculate the load transfers due to cornering with lateral acceleration 0.6 g

Determine roll angle from Eq. (3.11):

$$\phi = \frac{m_s \ddot{x} d}{k_s - m_s g d}$$

Sprung mass is total mass − unsprung mass = 1600 − (140 + 180) = 1280 kg
Calculate distance "d" from known roll centre heights as follows:

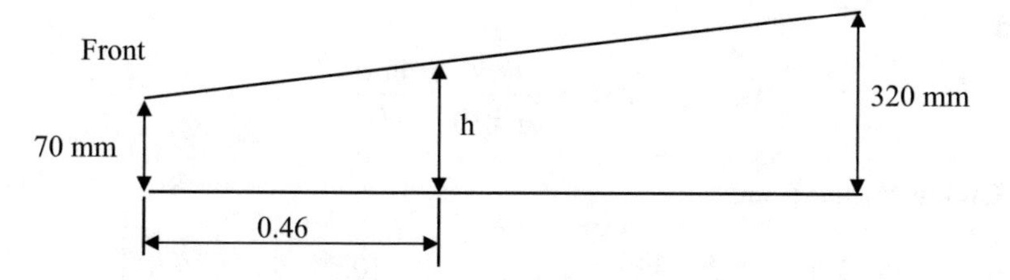

Fig. E3.5 Height of roll axis (h) at CofG

$$h = 70 + (320 - 70)0.46 = 185 \text{ mm}$$

giving

$$d = 550 - 185 = 365 \text{ mm}$$

Determine combined roll stiffness in Nm/rad:

$k_s = k_{sf} + k_{sr}$ = 350 + 150 = 500 Nm/deg or 500 (Nm/deg) × 57.3 (deg/rad) = 28,650 Nm/rad

Thus

$$\phi = \frac{1280 \times 0.6 \times 9.81 \times 0.365}{28,650 - 1280 \times 9.81 \times 0.365} = 0.114 \text{ rad} = 6.55 \text{ deg}$$

Determine load transfer due to roll moment:

$$F_{fsM} = \frac{k_{sf}\phi}{T_f} = \frac{350 \times 57.3 \times 0.114}{1.46} = 1566 \text{ N}$$

$$F_{rsM} = \frac{k_{sr}\phi}{T_r} = \frac{150 \times 57.3 \times 0.114}{1.46} = 671 \text{ N}$$

Determine load transfer due to sprung mass inertia forces from the following equations:

$$F_{fsF} = \frac{m_{sf}\,\ddot{x}\,h_f}{T_f} \quad and \; F_{fsF} = \frac{m_{sr}\,\ddot{x}\,h_r}{T_r}$$

where

$$m_{sf} = \frac{m_s b}{(a+b)} = \frac{m_s b}{L}$$

and

$$m_{sr} = \frac{m_s a}{(a+b)} = \frac{m_s a}{L}$$

Giving for the front:

$$F_{fsF} = \frac{m_{sf}\,\ddot{x}\,h_f}{T_f} = \frac{m_s b}{L} \times \frac{\ddot{x} h_f}{T_f} = (1280 \times 0.54)\left[\frac{0.6 \times 9.81 \times 0.07}{1.46}\right] = 195\,\mathrm{N}$$

and for the rear:

$$F_{rsF} = \frac{m_{sr}\,\ddot{x}\,h_r}{T_r} = \frac{m_s a}{L} \times \frac{\ddot{x} h_r}{T_r} = (1280 \times 0.46)\left[\frac{0.6 \times 9.81 \times 0.32}{1.46}\right] = 760\,\mathrm{N}$$

Determine load transfer due to unsprung mass inertia forces from the following equations:

$$F_{fuF} = \frac{m_{uf}\,\ddot{x}\,h_{uf}}{T_f} \; and \; F_{fuF} = \frac{m_{ur}\,\ddot{x}\,h_{ur}}{T_r}$$

Giving for the front:

$$F_{fuF} = \frac{m_{uf}\,\ddot{x}\,h_{uf}}{T_f} = \frac{140 \times 0.6 \times 9.81 \times 0.280}{1.46} = 158\,\mathrm{N}$$

and for the rear

$$F_{ruF} = \frac{m_{ur}\,\ddot{x}\,h_{ur}}{T_r} = \frac{180 \times 0.6 \times 9.81 \times 0.3}{1.46} = 218\,\mathrm{N}$$

Determine the total load transfers:

$$\text{Front} = 1556 + 195 + 158 = 1919\,N$$

$$\text{Rear} = 671 + 760 + 218 = 1649\,N$$

(b) Calculate roll angle gradient from Eq. (3.18):

$$k_\phi = \frac{m_s\ d}{k_s\ -\ m_s\ g\ d} = \frac{1280\ \times\ 0.365}{28650\ -\ 1280\ \times 9.81\ \times\ 0.365} = 0.0194\,\mathrm{rad/(ms^{-2})}$$

or

$$k_\phi = 0.0194\ \times\ 57.3\ \times\ 9.81\ = 10.9\ deg/g$$

3.9 Spring Rate and Wheel Rate

In the following sections, we discuss how dynamic loads can be quantified and analysed for some typical loading situations. The relationship between the suspension spring stiffness (spring rate) and the equivalent stiffness in the vertical plane of the wheel (wheel rate) is examined. This relationship is important in terms of determining suspension forces and the natural frequencies of vibration of the vehicle.

A generalised schematic of wheel load against wheel vertical displacement is given in Fig. 3.55. The figure identifies the critical points where the response changes, e.g. due to engagement of a bump stop. Whilst the main suspension spring and its associated spring rate may be linear (or intentionally designed to be non-linear), the wheel rate is almost always non-linear owing to the suspension geometry which defines how the spring rate translates to give the wheel rate.

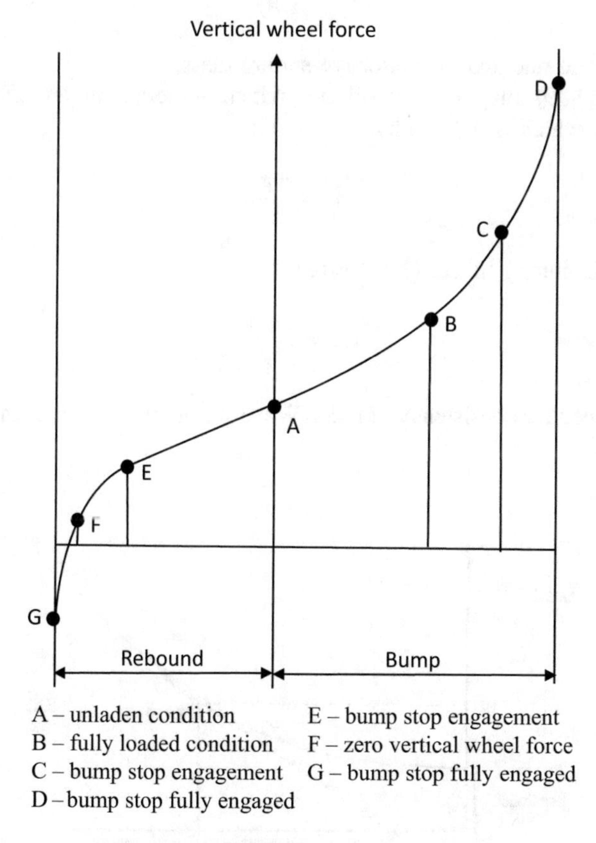

Fig. 3.55 Wheel load against vertical displacement

3.9.1 Wheel Rate Required for Constant Natural Frequency

Wheel rate is the vertical stiffness of the suspension in a wheel plane. One of the many problems facing the suspension designer is that as the body mass of a vehicle changes from the unladen to the laden condition, the natural frequency of the mass can change. This is undesirable and happens when the wheel rate is constant over the range of suspension travel. The problem can be overcome by arranging a wheel rate that increases with suspension travel. The following analysis shows that it is possible to determine how the wheel rate needs to vary in order to provide a constant natural frequency as the payload changes.

A vehicle body mass supported on its suspension can be modelled most simply as a mass on a spring for which the natural frequency is given by:

$$\omega_n = \left(\frac{k}{m}\right)^{\frac{1}{2}} \tag{3.19}$$

where k = wheel rate and m = effective sprung mass.

Equation (3.19) may first of all be written in terms of an effective static deflection, δ_s, which is defined by:

$$\delta_s = \frac{mg}{k} \tag{3.20}$$

Hence, substituting in Eq. (3.19) gives:

$$\omega_n = (\frac{g}{\delta_s})^{\frac{1}{2}} \tag{3.21}$$

Note the need for consistency of units. When g is in m/s^2 and δ_s is in m, the units of ω_n are rad/s.

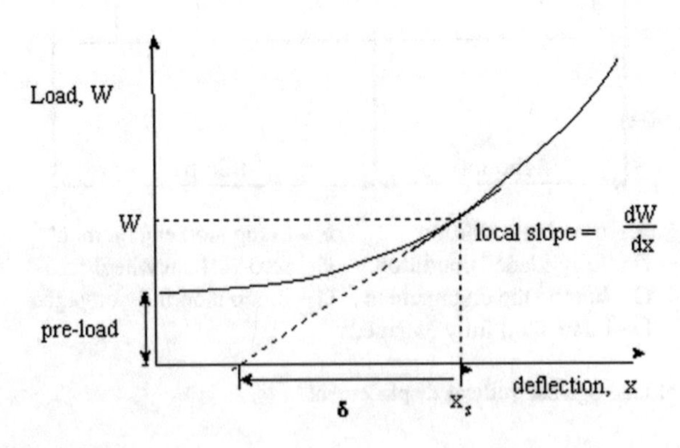

Fig. 3.56 Rising rate spring characteristics

A typical load-deflection curve for a rising-rate spring is shown in Fig. 3.56. The effective rate is the local slope of this curve at a particular point. To maintain a constant value of natural frequency requires a particular relationship between the local spring rate and the load to maintain in effect a constant static deflection i.e.

$$\frac{\text{Load}}{\text{Rate}} = \text{constant} \tag{3.22}$$

Giving:

$$\frac{W}{dW/dx} = \delta_s = \text{constant} \tag{3.23}$$

Equation (3.23) now needs to be re-arranged and integrated to obtain an expression for W as a function of x:

$$\frac{dW}{W} = \frac{dx}{\delta_s} \tag{3.24}$$

Hence by integration:

$$\log_e W = \frac{x}{\delta_s} + c \tag{3.25}$$

The integration constant, c, is given by substituting values at the static load condition as indicated in Fig. 3.56:

$$W = W_s \qquad \text{when } x = x_s \tag{3.26}$$

Therefore:

$$c = \log_e W_s - \frac{x_s}{\delta_s} \tag{3.27}$$

Hence Eq. (3.25) becomes:

$$\log_e\left(\frac{W}{W_s}\right) = \frac{(x - x_s)}{\delta_s} \tag{3.28}$$

Equation (3.28) defines the required variation of load W with suspension deflection x to give a constant natural frequency.

3.9.2 The Relationship Between Spring Rate and Wheel Rate

The rising-rate stiffness of the suspension in the wheel-plane will dictate the required characteristic for the suspension spring. The relationship between wheel rate and spring rate will depend on the kinematics of the suspension. In general, the displacement relationships between wheel travel and deflection across the suspension spring will be nonlinear. Some examples will illustrate how the relationship can be obtained.

3.9.2.1 Basic Example

This example is based on the swing (live) axle suspension shown in Fig. 3.57, a design that is not commonly used nowadays because it is prone to the suspension-jacking problem mentioned above. However, it serves as a uscful

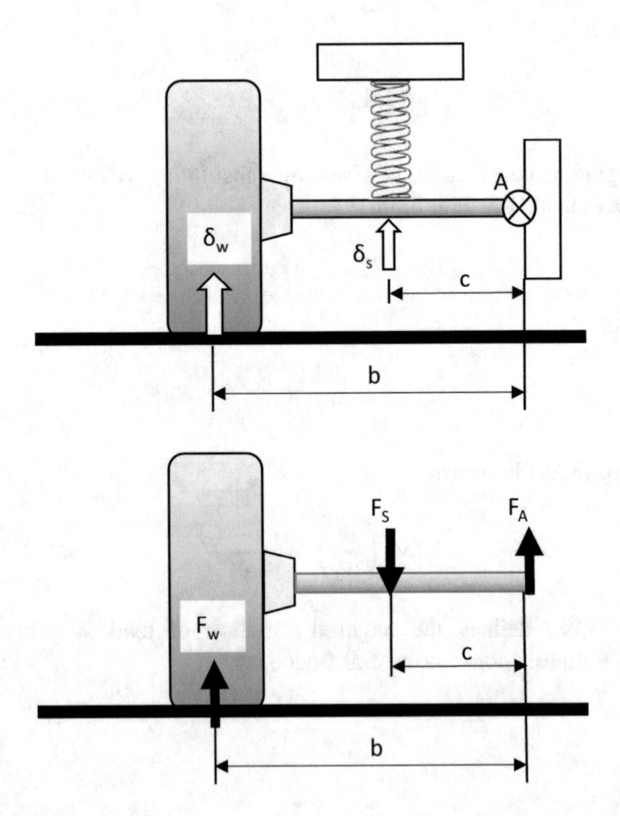

Fig. 3.57 Swing axle example

example to introduce spring/wheel rate calculations. For the case shown in Fig. 3.57, the wheel rate may be calculated as follows.

Assume the suspension spring has stiffness k_S.

A small displacement, δw, at the wheel results in a displacement at the spring of:

$$\delta s = \frac{c\delta w}{b} \tag{3.29}$$

Hence the spring force is:

$$\mathrm{F_s = k_s(\frac{c\ \delta w}{b})} \tag{3.30}$$

Taking moments about the pivot at A (Fig. 3.57) gives:

$$F_w b = F_s c \tag{3.31}$$

Expressing the wheel force in terms of an effective wheel rate, k_w, gives:

$$\mathrm{F_w = k_w\ \delta w} \tag{3.32}$$

Substituting in Eq. (3.31) then gives:

$$k_w = k_s(\frac{c}{b})^2$$

or

$$k_w R^2 = k_s \tag{3.33}$$

In this case, the rates are related by the **lever arm ratio (R = b/c) squared**. This is strictly valid only for small displacements around the mean position shown. For larger displacements the relationship is non-linear.

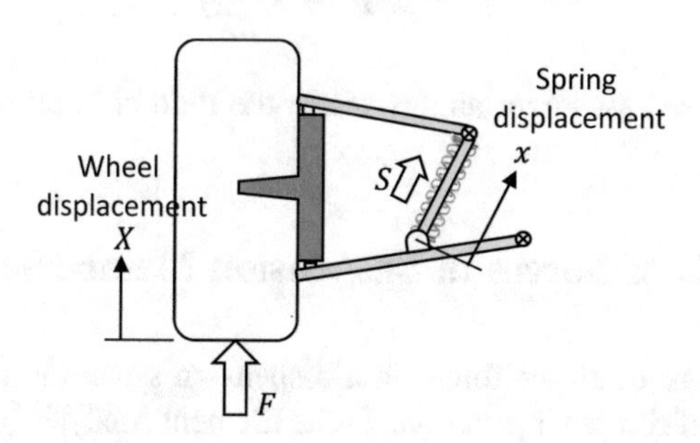

Fig. 3.58 Independent suspension with coil spring example

3.9.2.2 More General Example—Independent Suspension with Coil Spring

A more general relationship between spring rate and wheel rate can be derived using Fig. 3.58. The suspension ratio (R) is defined as:

$$R = \frac{Spring\,force}{Wheel\,force} = \frac{S}{F} \tag{3.34}$$

The spring stiffness is:

$$k_s = \frac{dS}{dx} = \frac{d(RF)}{dx} \tag{3.35}$$

$$k_s = R\frac{dF}{dX}\frac{dX}{dx} + F\frac{dR}{dX}\frac{dX}{dx} \tag{3.36}$$

where x is spring displacement and X is wheel displacement.

Also by virtual work:

$$Sdx = FdX \tag{3.37}$$

This leads to an alternative definition of the suspension ratio:

$$R = \frac{S}{F} = \frac{dX}{dx} \tag{3.38}$$

Noting that the wheel rate is:

$$k_W = \frac{dF}{dX} \tag{3.39}$$

Equation (3.36) may be rewritten as:

$$k_S = k_W R^2 + S(\frac{dR}{dX}) \tag{3.40}$$

This may be used for larger angles where the ratio changes as the wheel lifts.

3.10 Analysis of Forces in Suspension Members

A thorough analysis of all the forces in a suspension system is a complex problem requiring the use of computer packages. Finite Element Analysis (FEA) packages can be used for static analyses and stress distribution within members while Multi-Body Dynamics (MBD) packages are required for the analysis of three-dimensional dynamic loading. Analysing static forces is straightforward, but superimposed on these are dynamic loads due to vertical and longitudinal road irregularities, braking, accelerating, cornering etc. Many of these dynamic loads are of a transient nature and are very difficult to model. For this reason they tend to be modelled by dynamic load factors.

Despite all the complexities associated with practical situations, it is nevertheless possible to carry out some simple design calculations. These can be carried out graphically for a given load application and enable the loads in the suspension members and at body mounting points to be determined. Inevitably these calculations involve some simplifying assumptions. Some examples below illustrate the procedure.

3.10.1 Longitudinal Loads Due to Braking and Accelerating

Vertical, longitudinal and lateral tyre loads have been discussed briefly in Sect. 3.1.6. It is useful to examine in a little more detail the generation of longitudinal loads arising from vehicle control actions such as braking and accelerating. For these cases the loads are dependent on how braking and drive torques are reacted. They in turn depend upon whether the vehicle has inboard or outboard brakes and whether the drive is to a live axle or via an independent suspension system.

3.10.1.1 Braking—Outboard Brakes (Braking at the Wheel)

In this case the braking torque is reacted at the hub assembly. Consider the free-body diagram of the wheel in Fig. 3.59a. The corresponding force and moment on the hub carrier are shown in Fig. 3.59b. These provide the suspension loading. An equivalent suspension loading is shown in Fig. 3.59c.

3.10.1.2 Braking—Inboard Brakes (Braking Inboard on the Axle)

In this case (for example the De Dion suspension), the braking torque is reacted inboard at the vehicle body. The free-body diagram of the wheel is the same as in Fig. 3.60a. However, in this case, there is no moment exerted on the hub carrier. The suspension load is simply equal to the longitudinal tyre force acting through the centre of the hub, Fig. 3.60a. An equivalent (resultant) force system is shown in Fig. 3.60b.

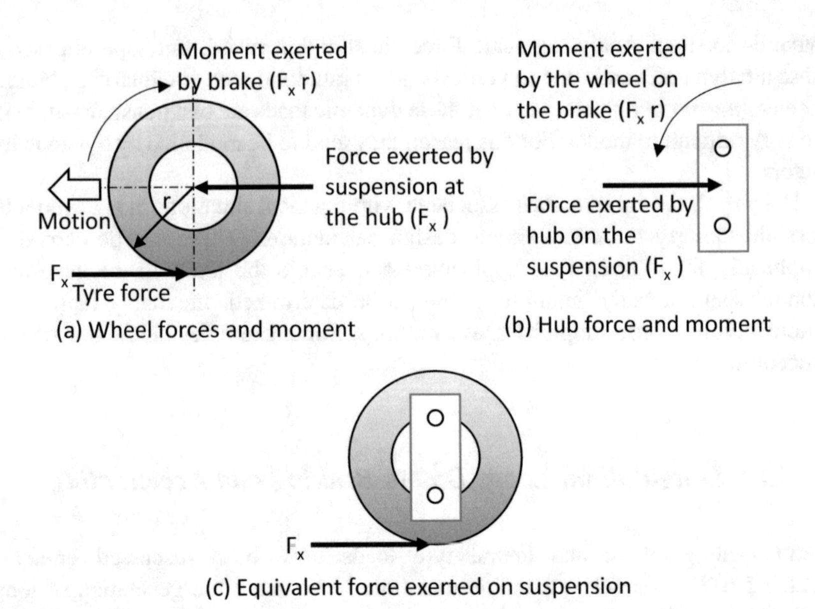

Fig. 3.59 Loads due to braking with outboard brakes

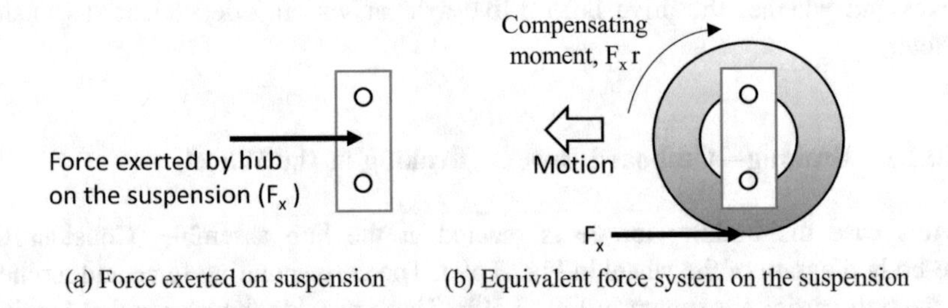

Fig. 3.60 Loads due to braking with inboard brakes

3.10.1.3 Drive to a Live Axle

In this case the drive torque is reacted outboard at the hubs. The analysis is similar to the case of outboard brakes. The equivalent force is shown in Fig. 3.61.

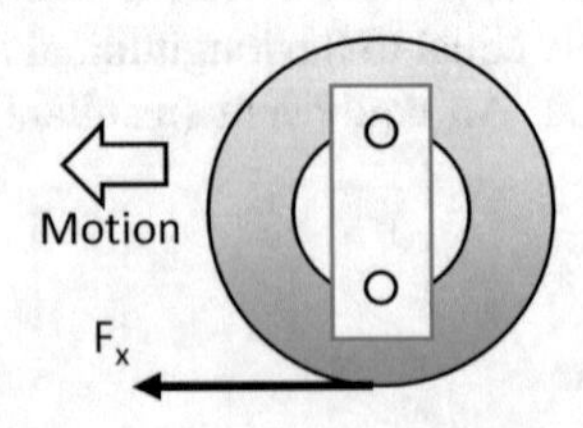

Fig. 3.61 Suspension load resulting from drive to a live axle

3.10.1.4 Drive to Independent Suspension

In this case the drive torque is reacted inboard. The analysis is similar to the case of inboard brakes. The equivalent force system acting on the suspension is shown in Fig. 3.62.

3.10.2 Vertical Loading

3.10.2.1 Double Wishbone Suspension Example

A diagram of a double wishbone suspension in which the spring acts on the lower arm is shown in Fig. 3.63. The following assumptions are made:

- Loads due to the masses themselves are ignored on the basis that they will be insignificant compared with loads arising from the static wheel load, F_w.
- All joints are treated as simple pin joints so that effects of friction, compliance in bushes etc. are ignored.
- The problem can be treated in two dimensions.

The static wheel load, F_W, would typically be calculated from the vehicle data. The problem then is to calculate the forces transmitted to the body mounting points. Clearly this is a statics problem since all the members are in equilibrium and the simplest approach is a graphical one. The free body diagrams and associated force vector diagrams are shown in Fig. 3.63.

Note the following features in obtaining these results:

- The force in the upper arm DA must be along the link since it is pin-jointed at both ends.
- Consider the wheel plus kingpin assembly first and construct the FBD. The directions of F_W and F_D are known so the direction of F_C can be found (it must pass through G for equilibrium).
- Next, the FBD for the lower arm can be drawn.

A full numerical solution to this problem requires that (i) the position diagram of the linkage is drawn to scale to define all the geometry and (ii) the force vector

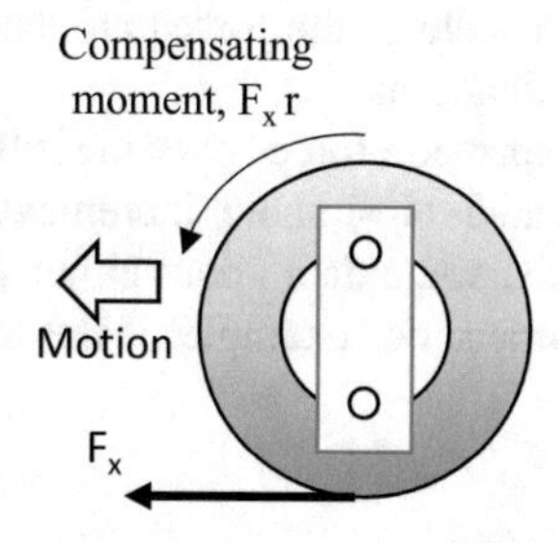

Fig. 3.62 Suspension load resulting from drive to an independent suspension

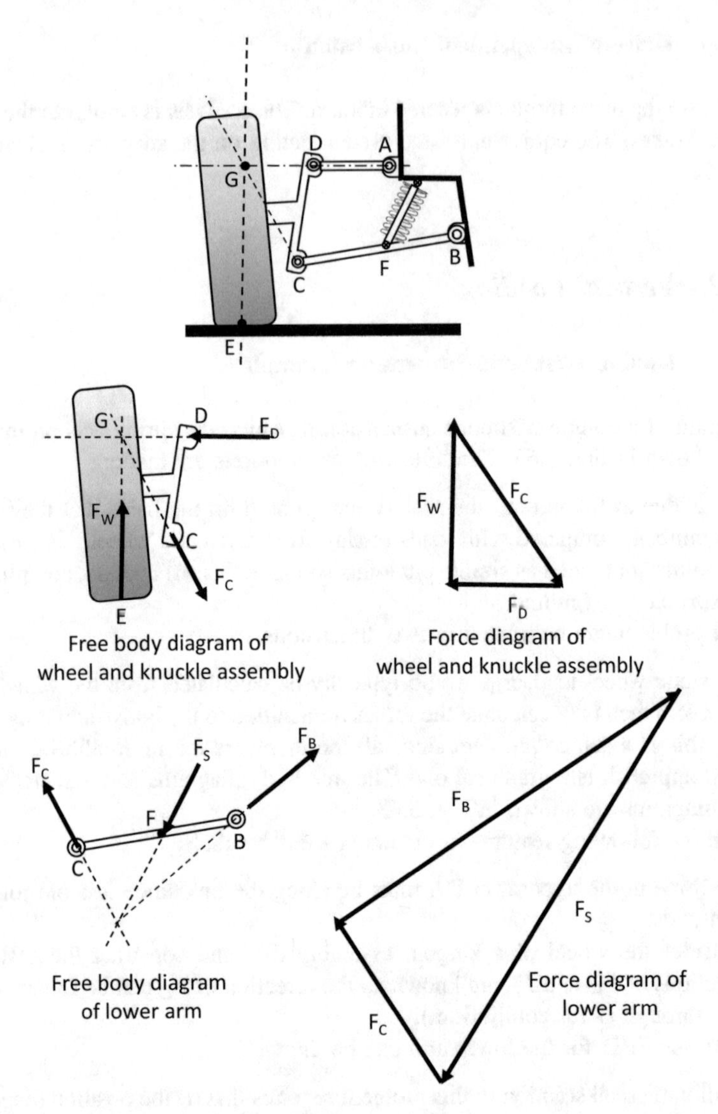

Fig. 3.63 Force analysis of a double wishbone suspension

diagrams are either drawn to scale or the forces are calculated using the geometry defined by the force vector diagrams.

In order to analyse the suspension forces over the full range of suspension travel, the procedure can be repeated at suitable increments between full bump and rebound positions. To do this, some data about either the spring rate or the wheel rate must be known or assumed. For example, if the load-deflection curve for the

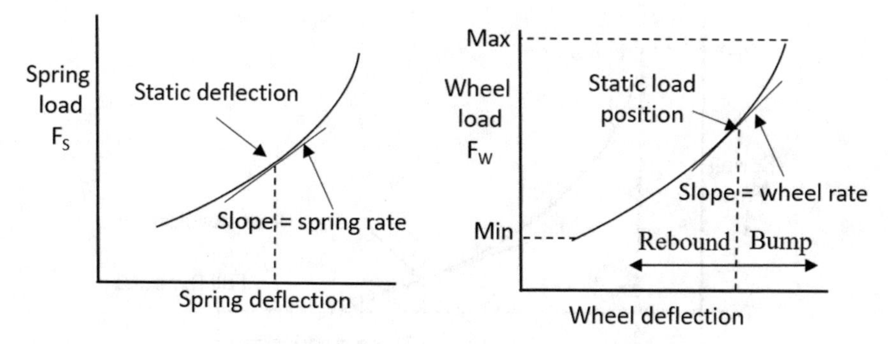

Fig. 3.64 Typical load versus deflection characteristics for spring and wheel

spring is known, then the load-deflection curve for the wheel can be plotted and the effective wheel rate calculated (Fig. 3.64). Note that because of geometrical effects, the relationship between wheel deflection and spring deflection is not linear and so even a linear spring will result in a non-linear wheel rate as defined by Eq. (3.40) for example. An alternative calculation is to assume a particular wheel rate and hence calculate the spring load-deflection curve (Fig. 3.65). This defines the spring characteristics necessary to achieve a desired wheel rate, which again may be linear or non-linear.

3.10.2.2 McPherson Strut Example

A diagram of a McPherson strut suspension is shown in Fig. 3.66. The same assumptions as in the previous section are made and in addition the strut is treated as a piston which is free to slide within a cylinder. The free body and force vector diagrams are constructed as follows:

- The force in the arm, *CB,* is along its longitudinal axis since it is pin-jointed at both ends and there are no loads applied along its length.
- For the wheel-axle-strut assembly, *DCE*, the resultant force at D must pass through point *G* for equilibrium.
- The force at *D* is reacted by two components: the spring force in the direction of *AC* and a side force on the cylinder. Note that this is only true if the spring is concentric with the strut *AC*. The side force is undesirable since it produces friction and wear on the piston in the strut. Mounting the spring more nearly in line with AG can reduce this side force and this is commonly done in McPherson strut designs.

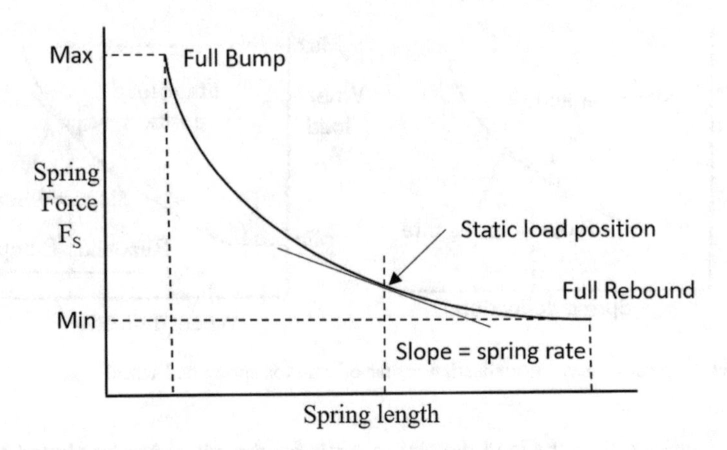

Fig. 3.65 Spring rate calculation

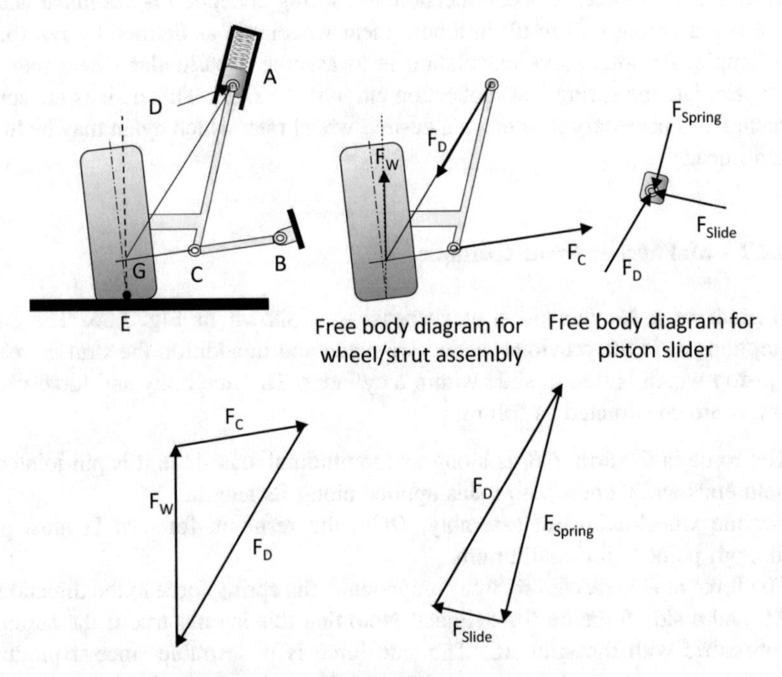

Fig. 3.66 Analysis of a McPherson strut suspension

3.10.3 Lateral, Longitudinal and Mixed Loads

When a lateral tyre force is applied in addition to a vertical load, the analysis for the example in Fig. 3.63 is modified to that shown in Fig. 3.67. The resultant wheel force is first calculated from the vector combination of F_W and F_Y. This then defines

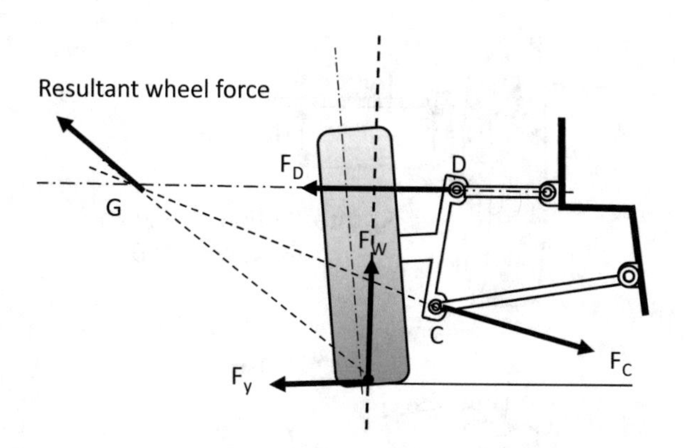

Fig. 3.67 Combined lateral and vertical loads on double wishbone suspension

a new point for G and the force acting on the lower link is changed in both magnitude and direction compared with the previous analysis.

For loads in the fore-aft direction, the situation is shown in Fig. 3.68. A fore-aft load may be applied at either points *E* or *O* (i.e. at right angles to the plane of the diagram.) A force at the tyre-ground contact point, *E*, would arise from braking with conventional outboard brakes. The resulting braking torque is transmitted through the stub axle/kingpin assembly which may be treated as a single rigid member. If the fore-aft force is due to aerodynamic drag, shock loading, braking with inboard brakes or traction, then its effective point of application is at the wheel centre, *O*.

In general any force at a given point may be treated as the same force at another point plus a couple around that point. This is a convenient way of analysing the effect of these fore-aft forces. If the fore-aft force (F_X) is moved to the kingpin axis then a moment of either $F_X d_1$ or $F_X d_2$ acts around the kingpin axis. This moment is, in practice, reacted by the steering arm and track rod so that the force F_{TR} in the track rod is either:

$$F_{TR} = \frac{F_X d_1}{d_3} \tag{3.41}$$

or

$$F_{TR} = \frac{F_X d_2}{d_3} \tag{3.42}$$

where d_3 is the effective moment arm of the steering arm around the kingpin axis.

In the side view (Fig. 3.69), the torque due to this longitudinal force is reacted by the forces in the upper and lower arms. It may then be possible to proceed to analyse all the forces on the upper and lower arms. However, the analysis is now starting to become complicated. Remember also we are not taking proper account of all the three dimensional effects. Despite this, our simple analysis does provide a good indication of the primary forces in the suspension. The subtle effects arising from a full three-dimensional treatment are of secondary importance.

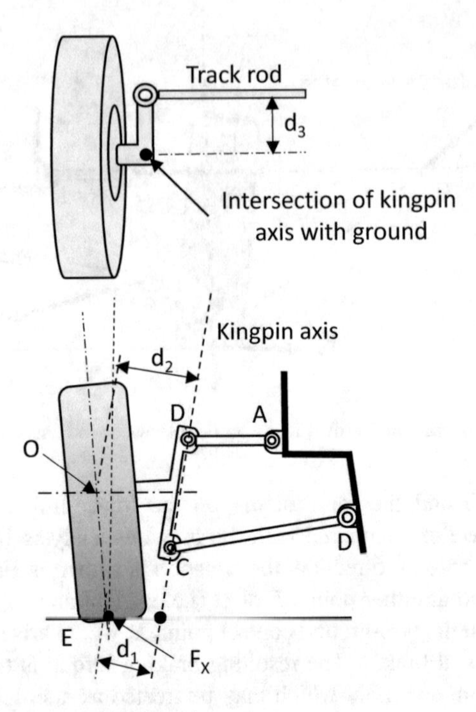

Fig. 3.68 Defining the fore-aft loading points

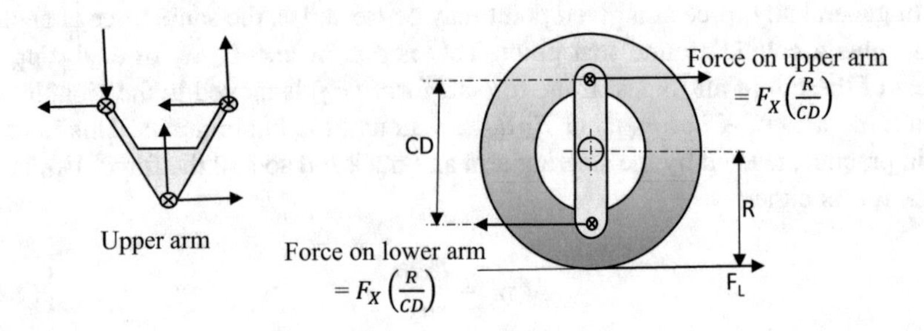

Fig. 3.69 Longitudinal forces reacted by suspension arms

3.10.4 Limit or Bump Stops

Suspension travel is limited by bump stops. Normally these are made of rubber and produce a rapidly increasing spring rate with deflection. This obviously prevents damage from metal to metal contact and provides a rapidly increasing force to react the extreme motion of the wheel.

There are two philosophies involved in bump stop design. The first, and perhaps more conventional approach is to make them relatively short and stiff; this means that contact is relatively infrequent and restricted to extreme events. The forces involved are high. The second approach is to make them relatively long and compliant; this means that they come into contact much more often, but much lower forces are involved and they may be viewed as a secondary spring. This effectively helps to stiffen the suspension for large deflections. Whichever approach is adopted, contact with the bump stops changes the force system in the suspension members.

An example is shown in Fig. 3.70. One method of tackling this problem is by superposition, i.e. by separating the forces arising from the main suspension spring at full deflection from those due to the bump stop spring. Both sets of suspension loads are then added together using the principle of superposition.

The procedure is as follows:

(a) Draw the suspension in the required position (in contact with the bump stop).
(b) Ignoring the force due to the bump stop compression, determine the forces and wheel load F_W'.
(c) Calculate the forces as described previously—this will give nominal loads F_c' and F_s'.
(d) Calculate the excess wheel load carried by the bump stop, $F_{ex} = F_W - F_W'$
(e) Ignoring the spring force, draw the FBD of the upper arm as shown in Fig. 3.70 and determine the suspension forces arising from the bump stop wheel load F_W'.
(f) Determine the total suspension forces by adding the results from steps (c) and (e).

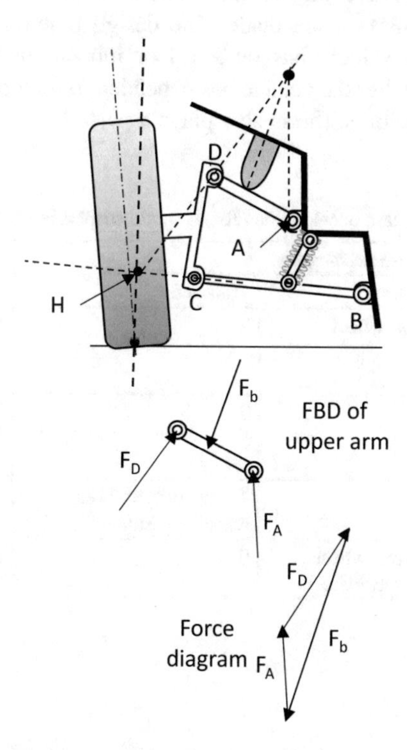

Fig. 3.70 Effect of bump stop on suspension force analysis

3.10.5 Modelling Transient Loads

In order to safeguard the robustness of a suspension it is particularly important that the effects of worst-case loading scenarios are analysed. These tend to arise from the transient (shock) loads due to wheels striking potholes and kerbs as well as from panic braking.

Because every set of events is likely to be different, designers tend to cover worst-case scenarios by using load factors based on the static wheel load. Different manufacturers tend to adopt their own set of design factors. Alternatively, the manufacturers work to a set of worst-case accelerations along the different vehicle axes; a typical example is shown in Table 3.1. The force loadings can then be obtained using an appropriate mass for the wheel and unsprung mass.

3.11 Suspension Geometry to Combat Squat and Dive

Squat (nose up) and dive (nose down) are changes in vehicle body attitude arising respectively from acceleration and braking. It is possible to design a suspension that will combat squat or dive; however, the geometry required for anti-squat will be different to that for anti-dive. Again, this is one of those areas of suspension design that requires compromises to be made. The design requirements to combat dive depend on whether a vehicle has outboard or inboard brakes, while anti-squat design depends on whether the vehicle has dependent or independent suspension. In the latter case, suspension stiffness also plays a part.

Table 3.1 Typical worst case accelerations for dynamic suspension loads

Load case	Worst case acceleration		
	Longitudinal	Transverse	Vertical
Front/rear pothole bump	3 g at the wheel affected	0	4 g at the wheel affected, 1 g at other wheels
Bump during cornering	0	0	3.5 g at wheel affected, 1 g at other wheels
Lateral kerb strike	0	4 g at front and rear wheels on side affected	1 g at all wheels
Panic braking	2 g at front wheels, 0.4 g at rear wheels	0	2 g at front wheels, 0.8 g at rear wheels

3.11.1 Anti-dive Geometry

The free body diagram for a car during braking is shown in Fig. 3.71. Note the pseudo (D'Alembert) force, $m\ddot{x}$, in which $\ddot{x}$ is the deceleration. Braking forces, B_f and B_r, are applied at the front and rear wheel pairs and the relationship between them is defined by the fixed braking ratio:

$$k = \frac{B_f}{B_f + B_r} \tag{3.43}$$

Under these conditions, the vertical loads on the axles are different from their static values, and can be calculated as follows.

Taking moments about the rear tyre contact patch:

$$N_f l - m\ddot{x}h - mbg = 0 \tag{3.44}$$

Giving:

$$N_f = \frac{mgb}{l} + \frac{m\ddot{x}h}{l} \tag{3.45}$$

The first term is the static load and the second term is the load transfer due to braking. Hence, the load at the rear is:

$$N_r = \frac{mga}{l} - \frac{m\ddot{x}h}{l} \tag{3.46}$$

This load transfer term will cause an increased deflection at the front axle and a reduced deflection at the rear axle. Hence the change of attitude of the vehicle body.

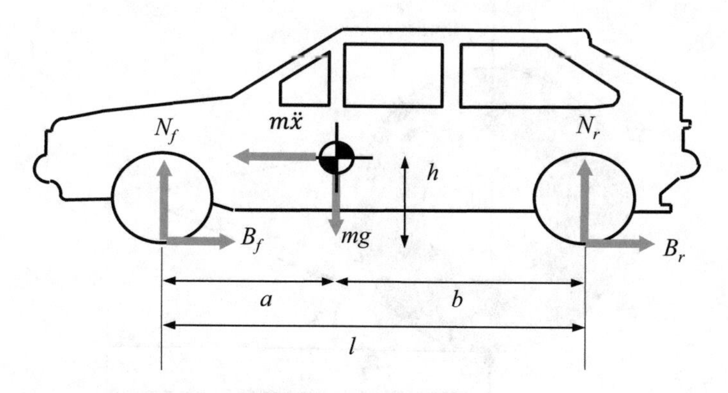

Fig. 3.71 Free body diagram of vehicle during braking

3.11.1.1 Outboard Brakes (Braking at the Wheel)

Consider now the diagram of the front suspension shown in Fig. 3.72. The axis of the link pivots are inclined such that the effective centre of rotation of the wheel in side view is at O_f.

The spring force may be expressed as static load plus a perturbation due to braking:

$$S_f = S_f' + \delta S_f \tag{3.47}$$

Under static conditions ($B_f = 0$) the static spring load is:

$$S_f' = \frac{mgb}{l} \tag{3.48}$$

For the braking condition, taking moments about O_f gives:

$$N_f d - S_f d - B_f e = 0 \tag{3.49}$$

Substituting for N_f from Eq. (3.45) and S_f from Eq. (3.47):

$$d\left(\frac{mgb}{l} + \frac{m\ddot{x}h}{l}\right) - d\left(S_f' + \delta S_f\right) - B_f e = 0 \tag{3.50}$$

Removing the static load and setting δS_f to zero to represent the case in which the suspension load does not change (no dive) gives:

$$\frac{m\ddot{x}hd}{l} - B_f e = 0 \tag{3.51}$$

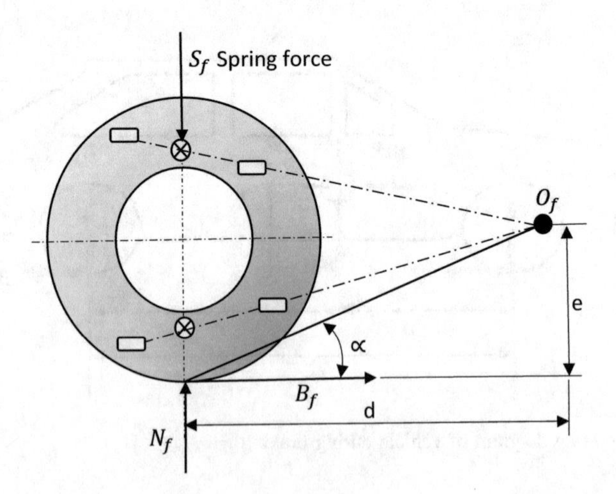

Fig. 3.72 Free body diagram for front suspension

Substituting for the braking force:

$$B_f = mk\ddot{x} \tag{3.52}$$

results in:

$$\frac{e}{d} = \frac{h}{lk} = \tan \alpha \tag{3.53}$$

Hence, if the instantaneous centre of the suspension O_f lies anywhere along the line defined by Eq. (3.53) then the condition for no front suspension deflection (zero dive) is satisfied. If the instantaneous centre is designed to lie below this line at an angle of α' say, then the percentage anti-dive is defined as

$$(\frac{\tan \alpha'}{\tan \alpha}) \times 100\% \tag{3.54}$$

A similar analysis can be applied at the rear (Fig. 3.73) to give

$$tan\beta = \frac{e}{d} = \frac{h}{l(1-k)} \tag{3.55}$$

Again, if the instantaneous centre lies along the line defined by Eq. (3.55), then will be no change in spring force and hence no lift will occur. Therefore, the condition for 100% anti-dive is that the front and rear instantaneous centres lie along the two lines shown in Fig. 3.74.

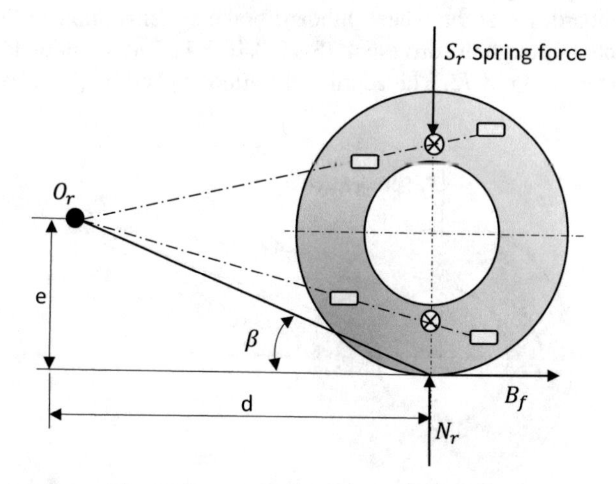

Fig. 3.73 Free body diagram for rear suspension

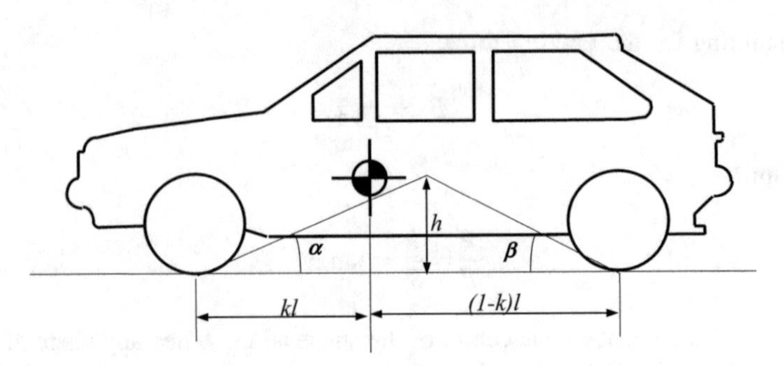

Fig. 3.74 Condition for 100% anti-dive

It is common practice to aim for only around 50% anti-dive. The reasons are:

- Zero pitch during braking appears to be undesirable from a subjective viewpoint.
- Full anti-dive conflicts with anti-squat so that some compromise must be reached.
- Full anti-dive can cause large castor angle changes because all the brake torque is reacted through the suspension links. Such castor changes may tend to make the steering unacceptably heavy during braking.

3.11.1.2 Inboard Brakes (Braking Inboard on the Axle e.g. de Dion Suspension)

All the foregoing analysis applies to the most common case in which the brakes are mounted outboard, i.e. at the wheel. Inboard brake systems differ in that the brake torque is reacted through the driveshaft (Sect. 3.10.3.1). The situation for say a front wheel is shown in Fig. 3.75. The additional torque applied by the driveshaft to the

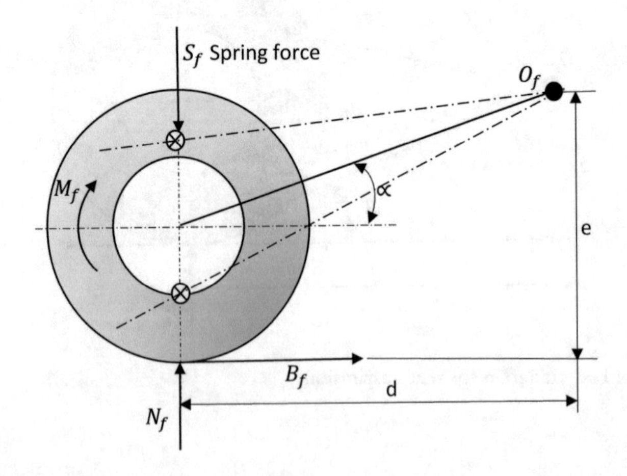

Fig. 3.75 FBD for front suspension with inboard brakes

wheel unit must be included in the free body diagram. Taking moments about the wheel centre (radius "r"):

$$M_f = B_f r = m\ddot{x}kr \tag{3.56}$$

Now, taking moments about O_f gives:

$$N_f d - S_f d - B_f e + m\ddot{x}kr = 0 \tag{3.57}$$

Comparing this with Eq. (3.49), it can be seen that an additional term is now included. Using the same analysis as before leads to:

$$\frac{(e-r)}{d} = \frac{h}{kl} = \tan\alpha \tag{5.58}$$

Note from the above expression that the angle α is now defined from the wheel centre (e–r) rather than the ground level. The analogous situation occurs at the rear to give:

$$\frac{(\mathrm{e}-\mathrm{r})}{\mathrm{d}} = \frac{\mathrm{h}}{\mathrm{l}(1-\mathrm{k})} = \tan\beta \tag{3.59}$$

3.11.2 Anti-squat Geometry

The analysis for anti-squat or anti-pitch design during acceleration follows a very similar process to that adopted in the previous section. In Fig. 3.76, the pseudo force is now reversed in direction. Tractive forces may be applied at either axle, or

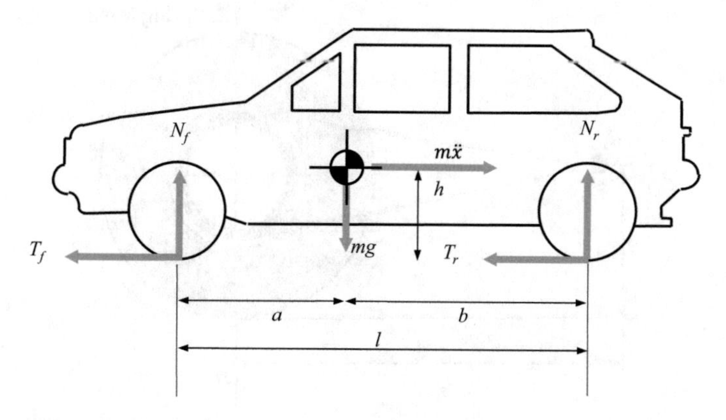

Fig. 3.76 FBD for an accelerating car (all-wheel drive)

both in the special case of four-wheel drive. There are two general cases to consider: solid axle drive or independent suspension/driveshaft designs.

3.11.2.1 Solid Rear Drive Axle

This configuration is shown in Fig. 3.77 with the driven rear axle assumed to be mounted on twin trailing arms, with an instantaneous centre of rotation at O_r. This point is sometimes referred to as the *virtual reaction point* (i.e. the point at which the forces in the suspension arms can be resolved and assumed to act on the vehicle body). Note that this treatment of the suspension is exactly the same as if a trailing arm of length, d, connected rigidly to the rear axle acting about a pivot at O_r. Note that the drive torque is reacted through the axle casing and the suspension members (thus making the analysis analogous to the case of outboard brakes). The analysis assumes that the axle mass is small relative to the vehicle mass.

The vertical wheel load is:

$$N_r = \frac{mga}{l} + \frac{m\ddot{x}h}{l} \tag{3.60}$$

where $\frac{m\ddot{x}h}{l}$ is the load transfer onto the rear axle due to the acceleration.

The tractive force is:

$$T_r = m\ddot{x} \tag{3.61}$$

Taking moments about O_r gives:

$$T_r e + S_r d - N_r d = 0 \tag{3.62}$$

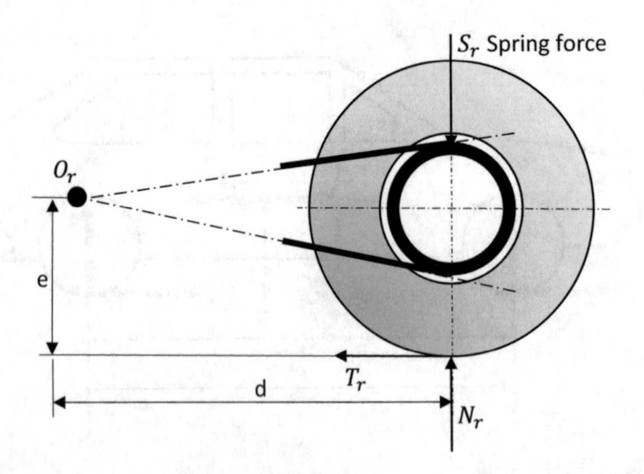

Fig. 3.77 Rear axle mounted on twin trailing arms

From Eq. (3.47), we have:

$$S_r = S_r' + \delta S_r$$

Also:

$$N_r = \frac{mga}{l} + \frac{m\ddot{x}h}{l} = S_r' + \frac{m\ddot{x}h}{l}$$

Substituting for all the forces and eliminating the static spring force gives:

$$m\ddot{x}e + \left(S_r' + \delta S_r\right)d - \left(S_r' + \frac{m\ddot{x}h}{l}\right)d = 0$$

$$\delta S_r = m\ddot{x}\left(\frac{h}{l} - \frac{e}{d}\right) = k_r\delta_r \tag{3.63}$$

where δ_r = rear suspension deflection, k_r = rear spring rate.

The front suspension must also deflect under these conditions due to the load transfer term, $m\ddot{x}h/l$, reacted there. Note that it is not possible to design the front suspension to directly accommodate this. It was possible to do this for the braking case because part of the braking force was reacted there. Hence, the spring force change at the front is given by:

$$\delta S_f = k_f\delta_f = -\frac{m\ddot{x}h}{l} \tag{3.64}$$

The pitch angle, θ, of the vehicle is simply:

$$\theta = \frac{(\delta_r - \delta_f)}{l} \tag{3.65}$$

$$\theta = \frac{\frac{m\ddot{x}}{k_r}\left(\frac{h}{l} - \frac{e}{d}\right) + \frac{m\ddot{x}h}{lk_f}}{l} \tag{3.66}$$

$$\theta = \frac{m\ddot{x}}{l}\left(\frac{h}{lk_f} + \frac{h}{lk_r} - \frac{e}{k_r d}\right) \tag{3.67}$$

Hence the condition for zero pitch may be expressed as:

$$\frac{e}{d} = \frac{\left(1 + \frac{k_r}{k_f}\right)h}{l} \tag{3.69}$$

3.11.2.2 Independent Suspension—Rear Wheel Drive

The free body diagram for the rear wheel is now altered to include the moment provided by the driveshaft (Fig. 3.78). Note that this is similar to the case of the inboard mounted brakes. Thus Eq. (3.62) becomes:

$$T_r e + S_r d - N_r d - M_r = 0 \tag{3.70}$$

where

$$M_r = T_r r$$

and

$$r = radius\ of\ tyre$$

Carrying this small difference through the analysis results in the condition for zero pitch:

$$\frac{(e-r)}{d} = \frac{\left(1 + \frac{k_r}{k_f}\right)h}{l} \tag{3.71}$$

A similar analysis could be carried out for a front wheel drive independent suspension system which is the most common configuration for small to medium passenger cars.

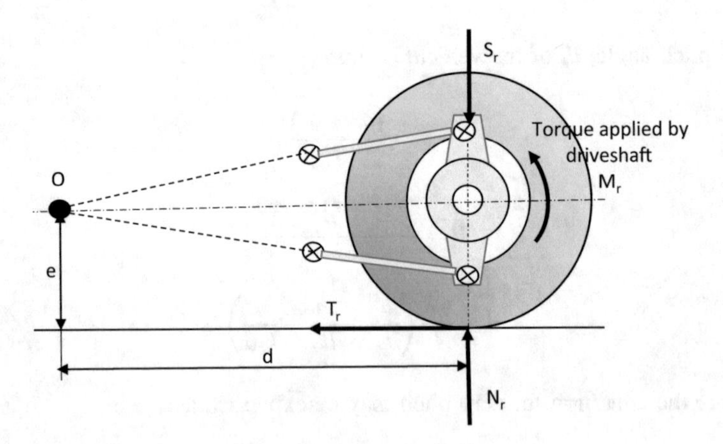

Fig. 3.78 Independent rear suspension

Example E3.6

A multi-purpose vehicle has a live rear axle with outboard brakes. The pivot centre of the rear suspension is located at O_r which is positioned relative to the underside of the wheels by the dimensions d and e, as shown in Fig. E3.6a.

Relevant vehicle data are as follows:

Vehicle mass m	2.1 tonne
Wheelbase L	2.8 m
Vehicle CofG height above road h	0.85 m
Front spring-rate (both sides) k_f	52.5 kN/m
Rear spring-rate (both sides) k_r	35.2 kN/m
Braking ratio (front: rear) k	70: 30

(a) Determine the ratio e/d for zero pitch during acceleration.

(b) What then is the lift at the rear suspension when the vehicle is braked with a constant deceleration of 0.3 g.

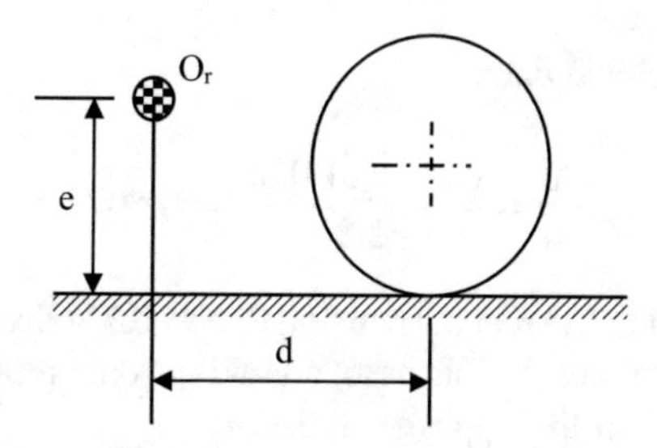

Fig. E3.6a Position of rear suspension pivot point

Solution

(a) The free body diagram of wheel and axle during acceleration is shown in Fig. E3.6b:

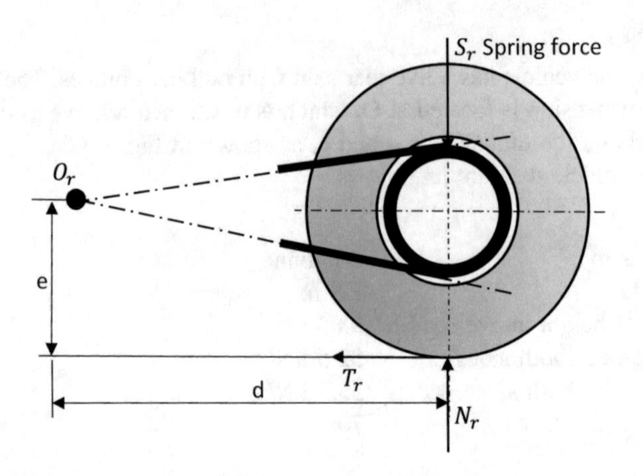

Fig. E3.6b General arrangement of rear wheel loading

It has been shown that for zero pitch during acceleration:

$$\frac{e}{d} = \frac{(1 + \frac{k_r}{k_f})h}{l}$$

Using the given vehicle data:

$$\frac{\mathrm{e}}{\mathrm{d}} = \frac{\left(1 + \frac{35.2}{52.5}\right) 0.85}{2.8} = 0.507$$

To calculate the lift at the rear suspension due to 0.3 g deceleration, note that the vehicle has outboard brakes. In this case, a braking force replaces the traction force in Fig. E3.6b and acts in the opposite direction.

The moment equation about $\mathrm{O_r}$ is: $-\mathrm{B_r}\,\mathrm{e} + \mathrm{S_r}\,\mathrm{d} - \mathrm{N_r}\,\mathrm{d} = 0$

We have:

$$N_r = \frac{mga}{l} - \frac{m\ddot{x}h}{l} \qquad \mathrm{S_r} = \mathrm{S'_r} + \delta\mathrm{S_r}, \qquad B_r = (1-k)m\ddot{x}$$

Substituting and cancelling out terms gives:

$$\delta_{sr} = m\ddot{x}(1-k)\frac{e}{d} - \frac{m\ddot{x}h}{l} = m\ddot{x}\left[(1-k)\frac{e}{d} - \frac{h}{l}\right]$$

Change of spring force:

$$\delta\mathrm{S_r} = 2100 \text{ x } 0.3 \text{ x } 9.81 \left(0.3 \text{ x } 0.507 - \frac{0.85}{2.8}\right) = -936 \text{ N}$$

Note that the negative sign indicates a force reduction.
Lift at rear suspension is given by:

$$\delta_r = \frac{\delta S_r}{k_r} = -\frac{936}{35.2} = -26.6 \text{ mm}$$

The negative sign indicates lift at the rear axle.

3.12 Vehicle Ride Analysis

Ride comfort is one of the most important characteristics defining the quality of a vehicle. It is principally related to vehicle body vibration, the dominant source of which is due to road surface irregularities. In order to design vehicles that have good ride properties, it is essential to be able to model ride performance in the early stages of vehicle development. This requires an understanding of road surface characteristics, human response to vibration and vehicle modelling principles. These are introduced in the following sections.

3.12.1 Road Surface Roughness and Vehicle Excitation

When a vehicle traverses over a road profile, the irregularities in the profile are converted into time-varying vertical displacement excitation inputs at each tyre contact patch. In general, road surfaces have random profiles which means that the excitations that they generate are also of a random nature.

The spatial random profile of a road surface can be frequency analysed and shown to comprise a set of harmonic components having different amplitudes and wavelengths as shown in Fig. 3.79. In general the components having the longest wavelengths have the largest amplitudes. While only three components are shown in Fig. 3.79, there are in reality an infinite number. For a given frequency component it is possible to define *spatial frequency* "n" as the number of cycles per metre. It can then be shown that the frequency characteristics of the profile are described as a function of "n", by the *spatial power spectral density* S(n). The SI units of S are m^3/cycle.

From large amounts of measured road data it has been established that S and n are related approximately by:

$$S(n) = \kappa\, n^{-2.5} \tag{3.72}$$

where κ is the road roughness coefficient. Typical values of κ for a motorway, a principal road and a minor road are 0.25×10^{-6}, 4×10^{-6} and 15×10^{-6} m^2/(cycle/m) respectively for $0.01 < n < 10$ cycle/m.

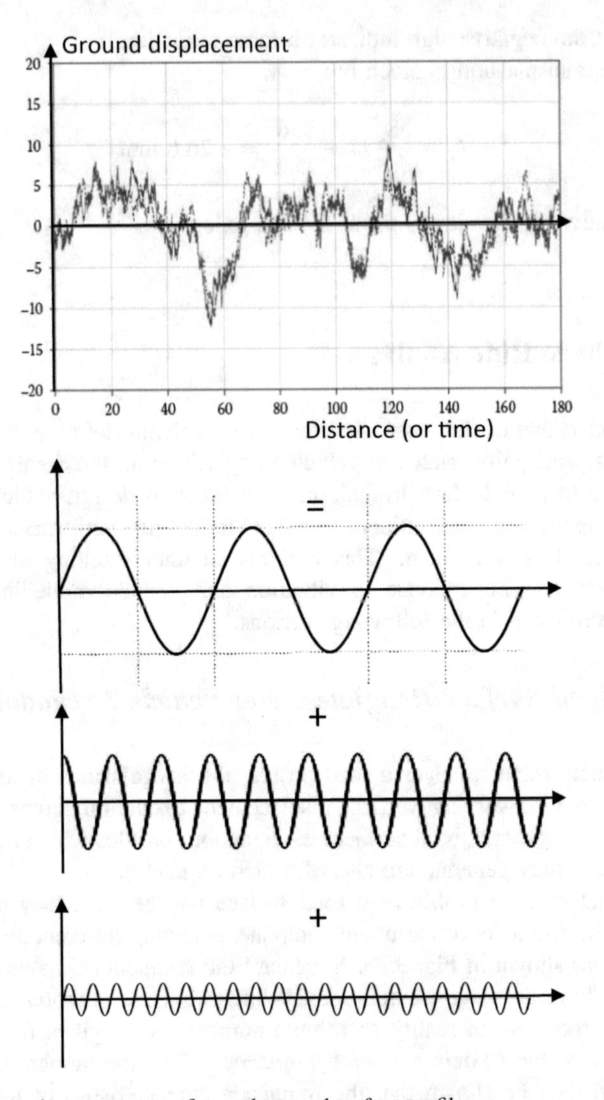

Fig. 3.79 Harmonic components of a random road surface profile

When a vehicle travels at a velocity V m/s over a random road profile described by S(n), the resulting random excitation to the vehicle is described by a temporal *spectral density* S(f), where f = n V is the excitation frequency in Hz.

It may be shown that:

$$S(f) = \frac{S(n)}{V} \tag{3.73}$$

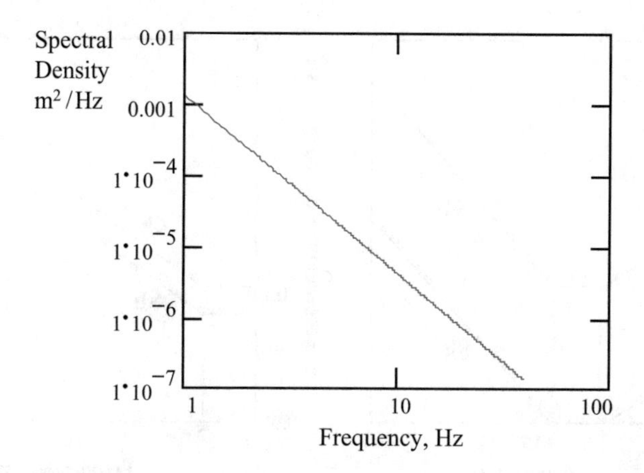

Fig. 3.80 Spectral density of a road input as a function of vehicle speed (poor minor road)

As

$$n = \frac{f}{V}$$

it follows from Eq. (3.72) that:

$$S(f) = \kappa\, V^{1.5}\, f^{-2.5} \tag{3.74}$$

The SI units for S(f) are m^2/Hz. The variation of S(f) for a vehicle traversing a poor minor road at 20 m/s is shown in Fig. 3.80. It is important to note the distribution of frequency components in this spectrum in relation to the natural frequencies of a suspended vehicle.

3.12.2 Human Perception of Ride

Ride is dominated by the level of road induced vibration in a vehicle body. In assessing ride performance of a vehicle it is necessary to take into account the sensitivity of the human body to vibration.

The human body is sensitive to vibration and has a number of its own natural frequencies. It follows that, when the human body is subjected to excitation at any of these frequencies, resonance and discomfort will result. From the previous section, it is obvious that road induced vibration will be broadband and hence some of the components of the excitation will tend to produce human body resonance. From an understanding of human body resonances, it is possible to weight the vehicle body response in order to obtain a single figure for assessment of ride performance.

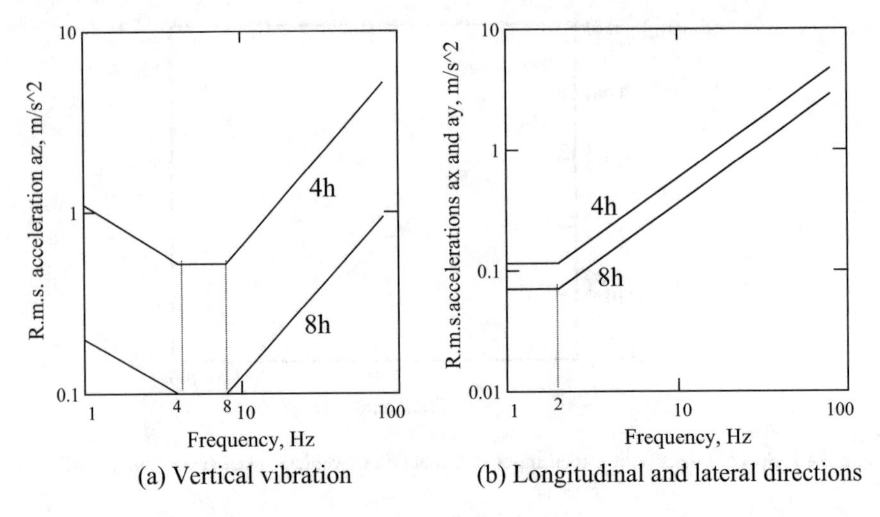

Fig. 3.81 Equal discomfort curves for different time durations

From a vibration point of view, the human body can be considered to be a complex assemblage of mass, elastic and damping elements that result in a number of natural frequencies in the range from 1 to 900 Hz. From a ride point of view, we are concerned with whole-body vibration of a seated person in the range from 0.5 to 15 Hz. There are two whole-body natural frequencies in this range. They are associated with the head-neck (1–2 Hz) and the thorax-abdomen (4–8 Hz) resonances. Head-neck resonance can be excited by vehicle pitching and rolling motions while resonance of the thorax-abdomen can be produced by vehicle bounce motion.

In general the tolerance to whole-body vibration decreases with time, so a high level of vibration for a short time can produce the same level of discomfort as a low level for a long time. Equal discomfort curves have been produced in standards, e.g. International Standard ISO 2631 and British Standard BS. 6841:1987. Some examples are given in Fig. 3.81 for vertical and longitudinal/lateral excitation. It should be noticed that the human body is acceleration-sensitive and the minimum values of the graphs correspond to the resonances discussed above.

The shapes of the graphs are used to establish weighting curves that can be applied to vehicle body accelerations determined from simulations of vehicle response to road induced vibration. This enables a single figure estimate of human response to be obtained. A typical weighting curve for vertical vibration is shown in Fig. 3.82. This weighting would be applied to the measured/calculated vibration level at the passenger seat position.

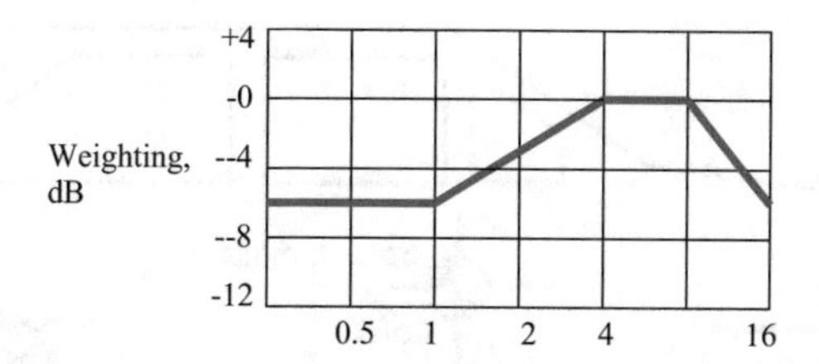

Fig. 3.82 Weighting curve for vertical whole-body acceleration

3.13 Vehicle Ride Models

Ride performance is assessed at the design stage by simulation of vehicle response to road excitation. This requires the development of a vehicle model and analysis of its response. Models of varying complexity are used. For a passenger car, the most comprehensive of these has seven degrees of freedom (DOFs) as shown in Fig. 3.83. These comprise three DOFs for the vehicle body (pitch, bounce and roll) and a further vertical DOF at each of the four unsprung masses. This model allows the pitch, bounce and roll performance of the vehicle to be studied.

The suspension stiffness and damping rates in the model are derived from the individual spring and damping units using the kinematics approach discussed in Sect. 3.3. Various tyre models have been proposed. The simplest of these uses a point-contact model to represent the elasticity and damping in the tyre with a simple spring and viscous damper. Since tyre damping is several orders of magnitude lower than suspension damping, it can be neglected in basic vehicle models.

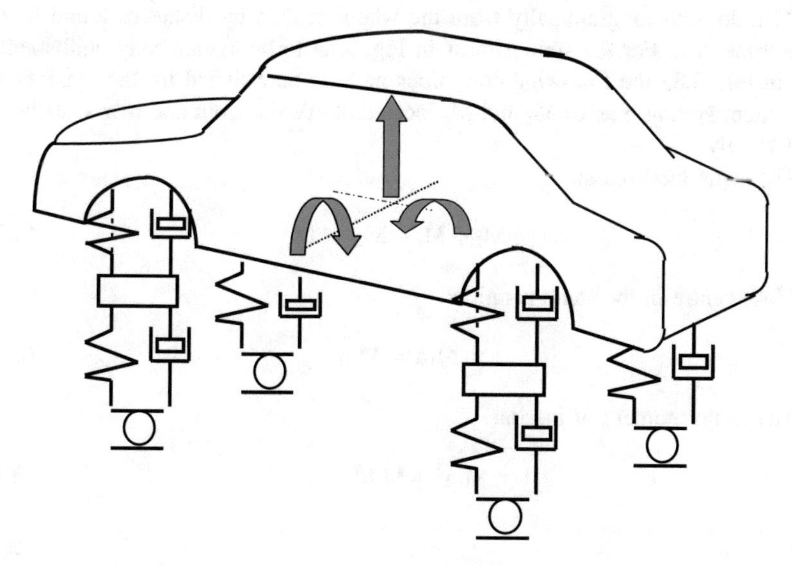

Fig. 3.83 Full vehicle model

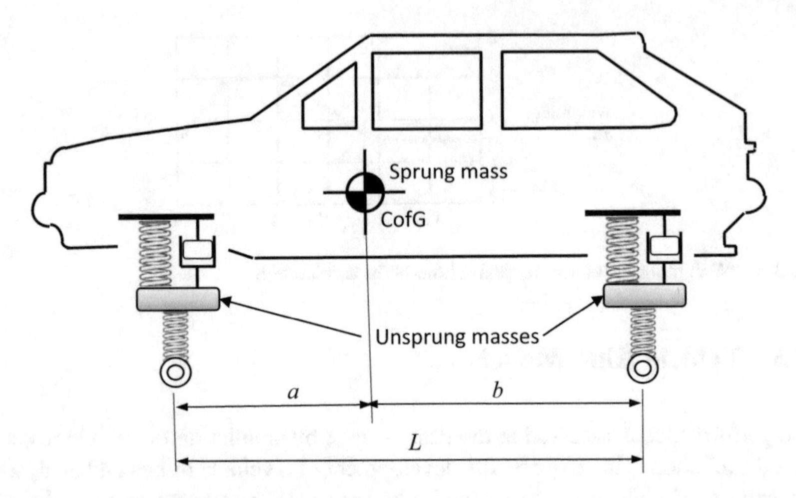

Fig. 3.84 Half-vehicle model

Much useful information can be derived from simpler vehicle models than shown in Fig. 3.83. For a normal road surface profile the components having the longer wavelength components are in phase (coherent) across the left and right tracks of a vehicle and there is therefore no tendency to excite body roll. This then justifies the use of a half vehicle model as shown in Fig. 3.84. This has four DOFs: a body mass translation and rotation, plus a translation for each of the unsprung axle masses.

The half vehicle model can be simplified still further to a quarter vehicle model if the body mass satisfies certain conditions. Let the total sprung mass in Fig. 3.83 be denoted by M and its moment of inertia about its centre of gravity CofG by "I". The CofG is located longitudinally from the wheel centres by distances a and b. The wheelbase is L. For the sprung mass in Fig. 3.84 to be dynamically equivalent to that in Fig. 3.85 the following conditions have to be satisfied for the dynamically equivalent sprung masses M_f and M_r located above the front and rear suspensions respectively.

The same total mass:

$$M_f + M_r + M_G = M \tag{3.75}$$

Mass centre in the same position:

$$M_f a = M_r b \tag{3.76}$$

The same moment of inertia:

$$M_f a^2 + M_r b^2 = I \tag{3.77}$$

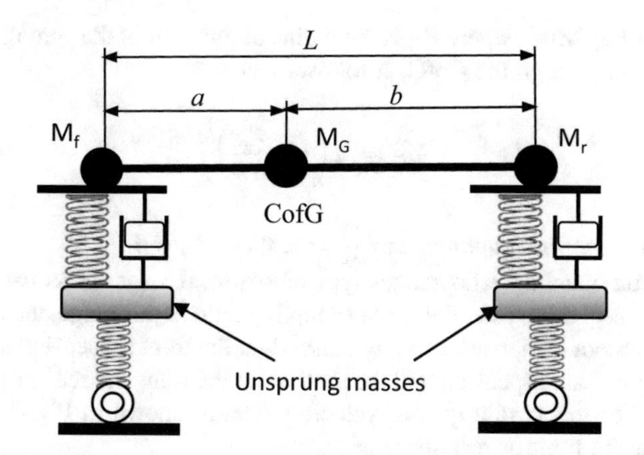

Fig. 3.85 Half vehicle model with dynamically equivalent sprung mass

From Eqs. (3.76) and (3.77):

$$M_f a^2 + M_f \frac{a}{b} b^2 = I$$

Re-arranging:

$$M_f a\,(a + b) = I$$

But:

$$a + b = L$$

Hence:

$$M_f = \frac{I}{a\,L} \tag{3.78}$$

and from Eq. (3.76):

$$M_r = \frac{I}{b\,L} \tag{3.79}$$

Substituting into Eq. (3.75):

$$M_G = M - \frac{I}{a\,L} - \frac{I}{b\,L} = M - I\,\frac{(a + b)}{a\,b\,L} = M - \frac{I}{a\,b}$$

Denoting I by MR_G^2, where R_G is the radius of gyration of the sprung mass about the lateral axis through the CofG, it follows that:

$$M_G = M\left(1 - \frac{R_G^2}{a\,b}\right) \tag{3.80}$$

Thus if the *inertia coupling ratio* $\frac{R_G^2}{a\,b} = 1$, then $M_G = 0$.

The inertia coupling ratio ranges typically from 0.8 for sports cars to 1.2 for some front wheel drive cars. For inertia coupling ratio in this range, the dynamically equivalent system comprises a sprung mass M_f at the front suspension and a sprung mass M_r at the rear suspension and the motions at the front and rear suspensions are uncoupled. For this case a quarter vehicle model as shown in Fig. 3.86 can represent either the front or rear suspension.

3.13.1 Vibration Analysis of the Quarter Vehicle Model

The model shown in Fig. 3.86 consists of linear elements, i.e. the springs and dampers have linear characteristics. These are valid approximations for small oscillations about the static equilibrium position of the system. Further details of the analysis methodology for a 2 DOF system are given in the Appendix. Let x_0, z_1 and z_2 be time varying displacements from the equilibrium position.

To understand the modes and frequencies of vibration of the system shown in Fig. 3.86, it is instructive to consider the free vibration response of the corresponding undamped system. For this case it is assumed that suspension damping $C_s = 0$ and the free vibration of the model ($x_0 = 0$) is then given by:

$$\begin{bmatrix} M_u & 0 \\ 0 & M_s \end{bmatrix} \begin{Bmatrix} \ddot{z}_1 \\ \ddot{z}_2 \end{Bmatrix} + \begin{bmatrix} (K_t + K_s) & -K_s \\ -K_s & K_s \end{bmatrix} \begin{Bmatrix} z_1 \\ z_2 \end{Bmatrix} = \begin{Bmatrix} 0 \\ 0 \end{Bmatrix} \tag{3.81}$$

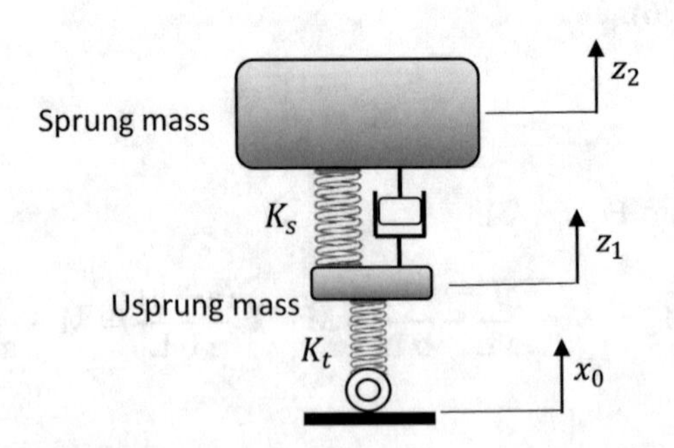

Fig. 3.86 Quarter vehicle model

Assume solutions $z_1 = A_1 \sin \omega t$ and $z_2 = A_2 \sin \omega t$. Substitute these expressions and their second differentials into Eq. (3.81) and cancel the time varying sinusoidal terms:

$$\begin{bmatrix} (K_t + K_s - M_u\,\omega^2) & -K_s \\ -K_s & (K_s - M_s\,\omega^2) \end{bmatrix} \begin{Bmatrix} A_1 \\ A_2 \end{Bmatrix} = \begin{Bmatrix} 0 \\ 0 \end{Bmatrix} \tag{3.82}$$

The non-trivial solution of Eq. (3.82) is given by:

$$\begin{vmatrix} (K_t + K_s - M_u\,\omega^2) & -K_s \\ -K_s & (K_s - M_s\,\omega^2) \end{vmatrix} = 0 \tag{3.83}$$

This is called the frequency or characteristic determinant, i.e. it describes the free vibration characteristics of the undamped system.

Consider a typical set of data for the rear corner of a mid-range saloon car:

Sprung mass M_s	317.5 kg
Unsprung mass M_u	45.4 kg
Suspension stiffness (wheel rate) K_s	22 kN/m
Tyre stiffness K_t	192 kN/m

Substituting into Eq. (3.83), expanding and then simplifying the result gives:

$$\omega^4 - 4786\,\omega^2 + 293 \times 10^3 = 0$$

The solution to this quadratic equation in ω^2 is:

$$\omega_1^2,\ \omega_2^2 = \frac{4786 \mp \sqrt{(4786)^2 - 4 \times 293 \times 10^3}}{2}$$

from which the natural frequencies are $\omega_1 = 7.87$ rad/s ($f_1 = 1.25$ Hz) and $\omega_2 = 68.73$ rad/s ($f_2 = 10.9$ Hz). These results are very typical of what one would expect for a car suspension. Corresponding to each natural frequency, there is a mode of vibration. These describe the relative amplitudes of free vibration of the masses at each of the natural frequencies. They are determined by substituting first ω_1^2 and then ω_2^2 into Eq. (3.82).

First mode of vibration

Substituting ω_1^2 into the second row of Eq. (3.82) and adding a second suffix 1 to A_1 and A_2 to denote that they are the amplitudes at the first natural frequency:

$$-K_s\,A_{11} + (K_s - M_s\,\omega_1^2)\,A_{21} = 0 \tag{3.84}$$

For the given numerical data: $\left(\frac{A_1}{A_2}\right)_1 = 0.104$

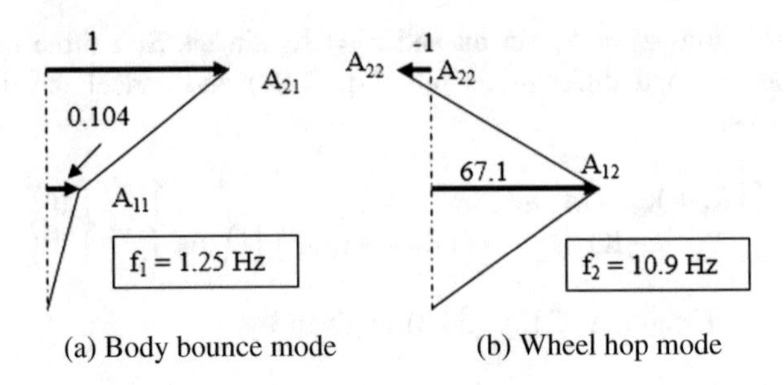

Fig. 3.87 Vibration modes for a quarter-vehicle suspension

This indicates that since the result is positive, both masses move in the same direction at any instant and the amplitude at the unsprung mass is 0.104 times that of the sprung mass. This mode of vibration is called the *body bounce* mode and can be represented graphically as shown in Fig. 3.87a. Note amplitudes are plotted horizontally for clarity.

Second mode of vibration

Substituting ω_2^2 into the second row of Eq. (3.82) and adding a second suffix 2 to A_1 and A_2 to denote that they are the amplitudes at the second natural frequency:

$$-K_s\, A_{12} + \left(K_s - M_s\, \omega_2^2\right) A_{22} = 0 \tag{3.85}$$

For the given numerical data: $\left(\frac{A_1}{A_2}\right)_2 = -67.1$

The negative ratio indicates the masses are moving in opposite directions at any instant and the amplitude at the unsprung mass is 67.1 times that of the sprung mass. This mode of vibration is called the *wheel hop* mode and can be represented graphically as shown in Fig. 3.87b.

From the above analysis, it is apparent that for the body bounce mode the amplitude of the unsprung mass is very small compared to that of the sprung mass, so in this case the system could be approximated as shown in Fig. 3.88a. Also, for the wheel hop mode, the amplitude of the sprung mass is very small in comparison to that of the unsprung mass. For this mode, the system could be approximated as shown in Fig. 3.88b. The corresponding approximate natural frequencies and damping ratios for a lightly damped system are as shown in Fig. 3.88. Note that the springs in wheel hop mode act in parallel so the effective stiffness is $K_s + K_t$.

Example E3.7

A road car has the following parameters:

General:

Curb Mass (m)	1140 kg
Weight Distribution	42% front and 58% at the rear
Wheelbase (WB)	2718 mm
Front corner un-sprung mass	40 kg

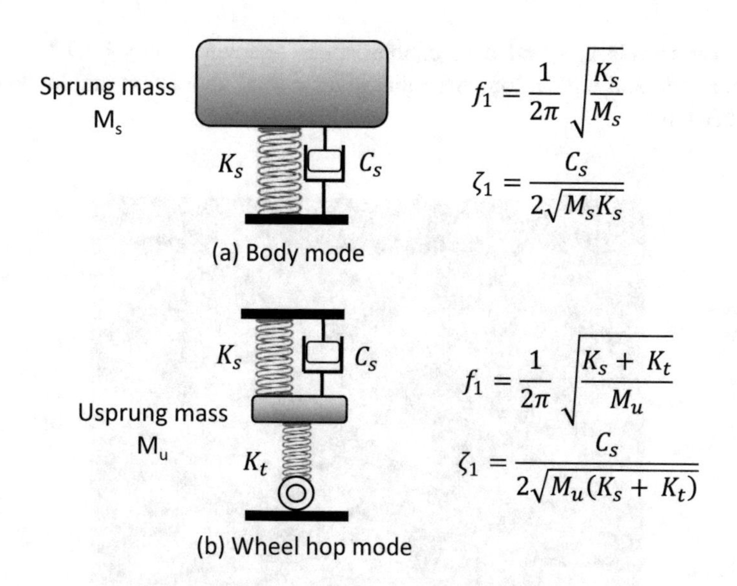

Fig. 3.88 Approximate models for body mode and wheel-hop mode

Tyres:

Front Tyre Spring Rate	300 N/mm
Rear Tyre Spring Rate	350 N/mm

Suspension:

Body bounce Frequency Front	1.43 Hz
Body bounce Frequency Rear	1.80 Hz

Suspension spring:

Wire diameter (d)	12.7 mm
Front mean coil diameter (D_f)	120 mm
Rear mean coil diameter (D_r)	100 mm
Shear modulus (G)	79.3 GPa

Assume tyre stiffness behaves linearly and the coil spring rate is given by:

$$Spring\ Rate = \frac{Gd^4}{8nD^3}$$

where n is the number of coils.

The car travels at speed over cobble stones—shown in Fig. E3.7.1—that are 200 mm wide and have a top curve that gives a total undulation amplitude of 2A mm (±A mm).

Fig. E3.7.1 Typical cobblestone road

It is required to calculate:

(a) The number of active coils on the front and rear coil springs.
(b) The safe speed the car can travel and the maximum amplitude of the undulation without the front tyres losing road contact due to wheel hop. Neglect any variable load due to spring/tyre deflection resulting from the undulations and any damper effects.

Solution:

(a) **Use the body bounce frequencies to determine the number of active coils in the front and rear suspensions**

$$\text{Front f} = 1.43\text{ Hz giving } \omega = 2\pi f = 8.986\text{ rad/s}$$
$$\text{Rear f} = 1.80\text{ Hz giving } \omega = 2\pi f = 11.31\text{ rad/s}$$

As $\omega = \sqrt{(K/m)}$ then $K = m\omega^2$
Giving:

$$\text{K front} = (0.42 \times 1140/2) \times 8.986^2 \times 10^{-3} = 19.3\text{ N/mm}$$
$$\text{K rear} = (0.58 \times 1140/2) \times 11.31^2 \times 10^{-3} = 42.3\text{ N/mm}$$

For body mode the tyre stiffness is not considered (Fig. 3.88a) and the overall spring rate is that due to the coil spring given by:

$$k_{spring} = \frac{Gd^4}{8\,n\,D^3}$$

Therefore the number of active coils in the front suspension is given by:

$$n = \frac{Gd^4}{8\,k_{spring} D^3} = \frac{79.3 \times 10^3 \times 12.7^4}{8 \times 19.3 \times 120^3} = 7.73\ coils$$

and the number in the rear suspension by:

$$n = \frac{Gd^4}{8k_{spring} D^3} = \frac{79.3 \times 10^3 \times 12.7^4}{8 \times 42.3 \times 100^3} = 6.10\ coils$$

(b) **Use the front wheel hop frequencies to determine the safe speed before front wheels experience wheel hop.**

For the wheel hop mode, the tyre spring rate and coil spring rates now act in parallel:

$$f = \frac{1}{2\pi}\sqrt{\frac{k_s + k_t}{M_u}}$$

$$f = \frac{1}{2\pi}\sqrt{\frac{(20.63 + 300)10^3}{40}}$$

$$f = 14.25\,\text{Hz}$$

Input frequency due to cobbles speed/(size of cobble) = v/0.2 = f = 14.25 Hz for resonance.

Hence maximum speed v = 2.85 m/s = 10.26 kph

To determine the amplitude of the undulations when the front wheels lose contact with the road.

The front tyres will lose contact with the road when the vertical inertia force at the wheel hop frequency equals the front tyre load.

Vertical load on each front tyre = 0.5 × 0.42 × 1140 × 9.81 = 2349 N

Inertia force due to cobbles = m A ω^2 = 40 × A × $(2\pi\ 14.25)^2$ = 2349 N for loss of contact.

Giving maximum amplitude of cobbles before loss of contact:

$$A = \pm 7.32\,\text{mm}$$

Note: The tyre stiffness will comprise the core stiffness which will be predominately dictated by the tyre pressure, and the tyre rubber crown or tread. The rubber stiffness will be significantly lower than the core and as the core and rubber are in series the core stiffness will not have a significant influence on overall stiffness. The rubber will compress appreciably more than the core and it is this contact area that provides the lateral tyre grip. The amplitude calculated above indicates the lift of the core and rubber components; the rubber will possibly still be in contact with the road but with a much lower loading and therefore much reduced ability to provide lateral grip.

3.14 Concluding Remarks

The chapter provides a foundation for understanding suspension systems and the processes needed to undertake first order analyses of suspension system dynamics and kinematics. It provides information on current designs and leads the reader to consider more advanced systems as superior vehicle refinement becomes increasingly in demand. Other areas that the designer needs to consider are determining the safe working space or travel of the suspension system. This will include determining the working stroke of the spring and damper before bump stops are engaged. It is necessary for the designer to also ensure the coil springs never bottom under full load as increased stresses will be induced in the coils with a shortening of life and performance. Finally there is a need to determine the forces induced onto the suspension system either from road conditions or unusual impacts in order to ensure the system will be sufficiently robust to endure such loads yet not result in too heavy a vehicle. These issues are further considered in Chap. 4 that follows. Most important is the realisation that no design area stands alone and that chassis suspension design needs to influence the design decisions taken on other vehicle systems (and *vice versa*).

Chapter 4
Vehicle Structures and Materials

Abstract This chapter commences with a review of chassis structures for the different classes of road vehicles including mass-produced passenger cars, high performance vehicles, small sports cars and commercial vehicles. It proceeds to consider the different materials used in vehicle structures with a focus on reducing vehicle weight and therefore emissions through the use of high strength steel, aluminium and composite materials. The following section outlines different methods of analysis of vehicle structures including both traditional theoretical methods and modern computational techniques. Crash safety of vehicles under impact loading is then considered and a particular case study of the crashworthiness of a small space frame sports car presented in detail. The final section of the chapter deals with the durability assessment of vehicle structures and again includes a detailed case study of the fatigue assessment and optimisation of a suspension component.

4.1 Review of Vehicle Structures

The purpose of any road vehicle structure is to support all the major components and sub-assemblies making up the complete vehicle (i.e. engine, transmission, suspension, etc.) and also carry the passengers and/or payload in a safe and comfortable manner.

In the early years of motor vehicles, both passenger cars and commercial vehicles were manufactured in the traditional way with a separate chassis frame onto which a non-structural body shell was attached. This form of construction has survived in commercial vehicles and also in specialist car brands such as Morgan. Since the chassis frame carries all the applied loads (i.e. dead-weight loads due to self-weight of vehicle and payload as well as "live" loads due to aerodynamic and dynamic tyre loads), it must be sufficiently strong and rigid. Most chassis frames are of ladder form, that is two longitudinal members connected by a number of transverse or cross-members which may not all be perpendicular to the longitudinal members but may take a diagonal or cruciform shape. The body shell serves mainly as a protection from the elements; it is generally isolated from the chassis via

© Springer International Publishing AG 2018

D. C. Barton and J. D. Fieldhouse, *Automotive Chassis Engineering*,
https://doi.org/10.1007/978-3-319-72437-9_4

flexible mountings (usually rubber) and therefore contributes very little to the overall vehicle stiffness.

The principal drawback of the separate chassis frame is that it is essentially a 2D structure and the members therefore have to be of high section modulus and are relatively heavy. Moreover it invariably leads to mounting problems due to the large difference in stiffness between the frame and the body shell. However, this type of construction is still a popular form of construction in commercial vehicles since a variety of different forms of body shell can be often mounted on a common ladder-type frame and the weight of the vehicle structure is of less concern than its overall load-carrying capability, see Fig. 4.1 for a typical example.

Following World War II, the drive to produce a more efficient structure for passenger cars led to the development of semi-integral forms of construction. These retain a strong chassis structure but, by mounting the body shell in a more rigid manner, the latter is required to carry a proportion of the applied loads. Of course, the body shell then has to be designed to withstand these loads and even more attention has to be paid to the mounting points between the frame and body because of the significant force transmission between the two structures. Moreover, semi-integral vehicles remain relatively heavy and have to be carefully assembled to tight tolerances since small misalignments can greatly increase stress concentrations near mounting points.

Fig. 4.1 Typical small truck chassis.
Photo Christopher Ziemnowicz, licensed under [CC BY-SA 2.5], via Wikimedia Commons from "Wikimedia Commons"

The final shape in the development of the structure of mass-produced passenger cars was the emergence of unitary or integral forms of construction. As the name suggests, such vehicles have no discernible separate chassis and the whole body is designed as an integral unit capable of reacting the applied loads and providing the necessary stiffness to the vehicle. This form of construction produces a structure which is truly 3D in the way it deforms and carries load. Thus it can be designed to be significantly lighter than traditional chassis-framed vehicles because of the relatively large depth of the fabricated structures used to resist the bending and torsional loads. However, since these structures are conventionally manufactured from relatively thin sheet steel or aluminium, they very often have to be stiffened with reinforcements or made as box sections. This requires complex tooling and assembly techniques, the cost of which can only be justified for mass produced vehicles. Figure 4.2 shows a modern aluminium integral body shell on a robotic assembly line. The complexity of the body shell as well as the robotic manufacturing processes employed in its construction are obvious.

Alongside the mass-produced integral steel and (more recently) aluminium structured vehicles, a number of alternative forms of construction exist for more specialised, low volume passenger vehicles. These range from the conventional chassis plus separate body used on traditional sports cars such as the Morgan, to the high-tech aluminium plus composite construction employed in the latest high performance sports vehicles such as manufactured by Lotus. Composites consisting

Fig. 4.2 Aluminium integral body shell during manufacture.
Reproduced with kind permission of Jaguar Land Rover © Jaguar Land Rover Ltd

of a thermo-setting epoxy or polyester resin reinforced with glass fibre mat have of course for a long time been used in the construction of body parts for niche sports cars (or as replacements for corroded steel parts in mass-produced vehicles). Recently, more advanced composites using a tougher thermoplastic matrix and stronger, stiffer fibres such as carbon or Kevlar have been introduced as load-bearing members in sports cars and other specialised vehicles. Of course, the body structure of F1 Grand Prix cars and similar relies almost exclusively on the use of carbon fibre composite materials to give the exceptional stiffness, strength and lightness of these vehicles.

An important class of vehicle in the small sports car category utilises the so-called space-frame chassis form of construction. A true space-frame is a collection of tubular members usually welded together and triangulated in such a way that the members carry load in tension and compression only and do not suffer bending or torsional to any significant degree. A good example of this form of construction is the Caterham Super Seven (formerly the Lotus Seven) illustrated in Fig. 4.3. The main members are predominantly square Sect. 16 swg or 18 swg steel tubes (with some round tube at certain locations) which are MIG welded together to produce a strong and stiff frame to which all other components of the vehicle (engine, suspension, body panels, etc.) are mounted directly. Such space-frame chassis are widely employed in the kit car market as well as in the construction of low cost race cars such as in Formula Student/SAE due to the low cost of tooling for such space frames.

Considering briefly commercial vehicle structures, most of these employ a separate chassis frame generally manufactured from rolled steel sections of channel or I-beam form. Two main longitudinal members usually extend the full length of

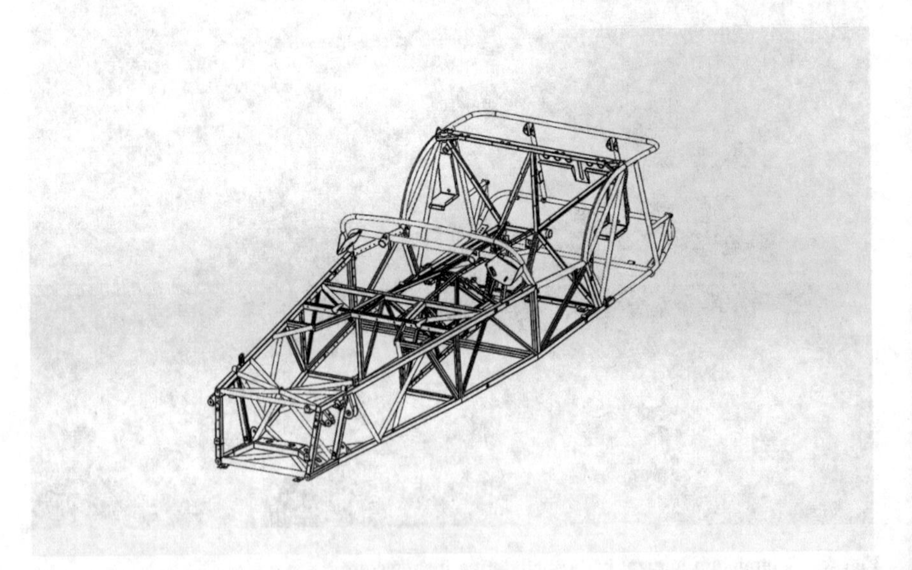

Fig. 4.3 Typical space-frame chassis for small sports car.
Reproduced with kind permission of Caterham Cars © Caterham Cars Ltd

the vehicle. These are joined together by cross members to form a ladder and/or cruciform shape as illustrated in Fig. 4.1. The power unit, driving controls, axles and the cab unit are mounted directly on the chassis frame and often form a self-contained unit. A separate loading platform or a superstructure is also mounted on the frame according to the usage for which the unit is intended. A strong structural panel (or bulkhead) normally extends upwards to the full height of the cab to give protection to the driver and to prevent the load from moving forward under heavy loading. Articulated vehicles which consist of a tractor unit and a separate trailer employ a similar form of construction except that only the tractor unit needs to accommodate the cab and powertrain. Finally, bus structures usually employ a sturdy 3D frame which is mounted rigidly onto the chassis and to which the body panels are attached. Roll-over and side impact protection are obviously of great importance in bus and coach structural design whereas, for heavy good vehicles, rear and side guards to prevent other vehicles from "submarining" beneath the chassis structure during crash scenarios are mandatory.

4.2 Materials for Light Weight Car Body Structures

The mandatory requirements for vehicle manufacturers to reduce their fleet average emissions of CO_2 have led not only to developments in powertrains such as down-sized IC engines and hybrid/electric drives but also to a strong motivation to reduce vehicle weights. Reduced vehicle mass reduces fuel consumption and gives enhanced acceleration/deceleration/handling performance. It can also reduce damage to road surfaces and generally enhances road safety because the kinetic energy of vehicles is reduced. Even commercial vehicles can benefit from lightweight chassis construction because hauliers can then increase the payload for a specified Gross Vehicle Weight (GVW).

In terms of integral forms of construction, interstitial-free steels (alloyed with small amounts of titanium to remove carbon from interstitial sites) have been introduced to reduce the thickness, and hence weight, of body panels. High strength low alloy (HSLA) steels (alloyed with manganese to increase strength and with niobium/vanadium to provide grain refinement) have also been widely used. More recently advanced high strength steels (AHSS) have been introduced because of their enhanced mechanical properties including good impact energy absorbing behaviour (except for the very high strength steels which are relatively brittle). These AHSS's typically include phases of hard steel such as martensite, bainite or austenite in addition to the more usual ferrite/pearlite microstructure. The resulting enhanced yield stress enables panel thicknesses to be further reduced but also causes manufacturing difficulties due to the reduced ductility and the high amounts of elastic energy stored that can exacerbate problems such as "springback" following forming operations (springback is the tendency of a pressed panel to return to its original shape after the forming operation).

One way to mitigate against these forming difficulties is the use of tailor-welded blanks (TWBs). These consist of multiple sheets of different shapes and thicknesses that are welded together and then pressed to the required shape. This means that the thickness can be enhanced where necessary for structural stiffness but elsewhere can be made thinner to improve formability and further reduce weight. Optimisation of the benefits of TWB's requires detailed understanding and analysis of body shell in-service loading. Knowledge of the relevant formability properties of the materials as well as their costs is also required.

In recent years, much attention has been paid at the high end of the passenger car market to the replacement of steel body shells with aluminium alloy which has a much lower density (about 2700 kg/m^3 compared with 7800 kg/m^3 for steel). As with steel, there are many different grades of aluminium alloy and different heat treatments are possible to give a wide range of material properties with yield strengths varying from about 70 MPa up to 700 MPa after age hardening. As with any metallic material, higher yield strength alloys have lower ductility and are therefore more difficult to form. For this reason, advanced body shells often employ a mixture of pressed sheet, extruded and cast aluminium components, such as in the Audi design shown in Fig. 4.4 which utilises steel, carbon fibre composite and even magnesium components as well as aluminium to form a hybrid structure which can be regarded as a space-frame. Both aluminium and magnesium alloys can be

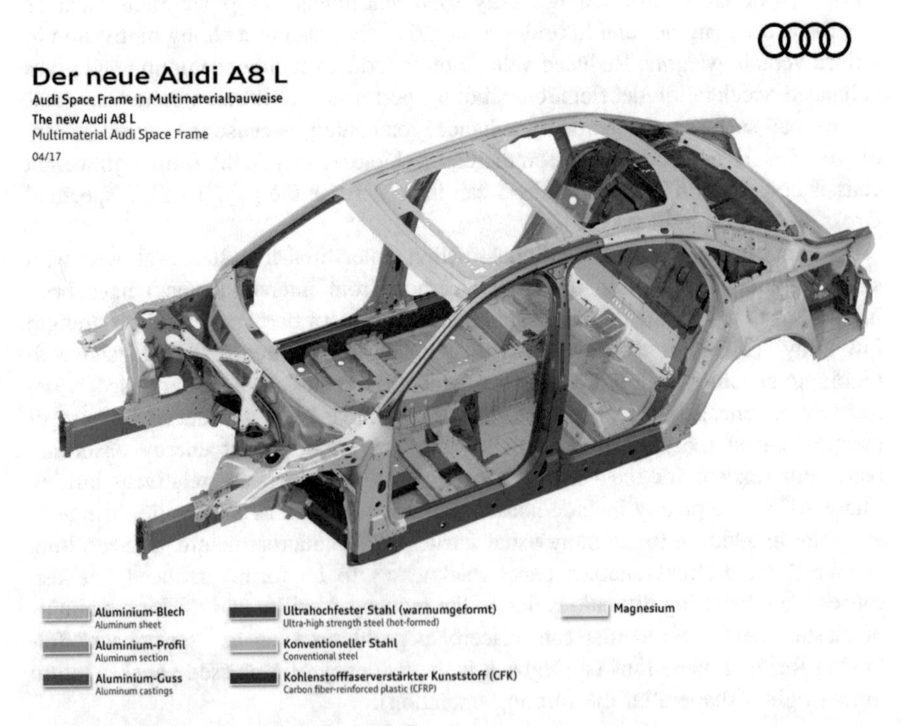

Fig. 4.4 Audi A8 multi-material body shell.
Reproduced with kind permission of Audi © Audi AG

pressure die cast (with or without vacuum assistance) which produces intricate parts with good properties especially in relation to dynamic loading.

Aluminium alloys can be laser, spot or metal inert gas (MIG) welded and are also very suitable for adhesive bonding. Higher strength alloys (e.g. the 6000 series commonly used in aerospace applications) are being increasingly used, often with quick hardening compositions so that age hardening can occur during the coating or painting processes. Developments in the properties of cast alloys have led to the consideration of aluminium for demanding applications such as the frontal impact longitudinal rails shown in blue in Fig. 4.4 and even in brake rotors where heavy ferrous materials are currently used.

Apart from its higher cost in comparison to steel, a further issue with aluminium is that it is more difficult to recycle. Again special alloys and processes have been developed to facilitate recycling e.g. the RC5754 alloy used extensively by Jaguar Land Rover in body shells such as shown in Fig. 4.2 contains up to 50% recycled material. In fact Jaguar aims to be using 75% recycled material in its aluminium body shells in the near future.

Alongside the developments in high performance metallic materials, polymeric matrix composites are being increasingly used in chassis structures. Over the past 10 years or so, the low cost, low tech glass fibre and the more expensive, high tech carbon fibre technologies have converged so that high performance composites have become an affordable option for body parts on normal road cars. There is now a wide choice of reinforcement types (short or continuous, randomly or fully aligned carbon, glass or aramid fibres) in both thermosetting (epoxy, polyester, vinyl ester) and thermoplastic (PP, PEEK) polymer matrices. Composite production processes have become more automated with much shorter cycle times. Prominent amongst these processes is Resin Transfer Moulding (RTM) whereby fibre reinforcement preforms are infiltrated by liquid polymer under carefully controlled conditions to achieve the near final net shape within a short cycle time.

Since complete composite body shells are likely to remain the reserve of high performance sports and race cars, an issue for more mass produced vehicles is the joining of a composite component to its metallic neighbours within a hybrid body shell. Since composites generally cannot be welded, the obvious joining technique is adhesive bonding but such joints need to be carefully designed and tested to demonstrate their integrity. An alternative is to incorporate metal fixing devices within the composite during the manufacturing process.

The modern vehicle structure designer is faced with a wide choice of different materials and manufacturing techniques. Not only must the complete body shell be lightweight and sufficiently stiff and strong, it should also be cheap and relatively easy to manufacture. It most also aim to protect the vehicle occupants and other road users from serious injury during road traffic accidents. From an environmental point of view, it is important to reduce the vehicle weight whilst at the same time improving the carbon footprint and recyclability of the materials used in its construction. Advanced computer simulation techniques exist which enable designers to predict both the manufacturability and on-road performance of the materials and components selected. Overall, with this seemingly never-ending choice of materials and manufacturing techniques, it is an exciting and challenging time for the vehicle structure designer.

4.3 Analysis of Car Body Structures

4.3.1 Structural Requirements

The structural requirements of any vehicle structure can be summarised as follows:

1. The structure must be sufficiently stiff to react the static loads (i.e. mainly due to dead-weight) and dynamic loads (i.e. mainly due to driving over rough terrain and handling manoeuvres) without excessive deformation.
2. The structure must be sufficiently strong to withstand many cycles of the applied loading without suffering from fatigue or other forms of material failure.
3. The structure should deform in such a manner under impact load conditions so as to minimise the risk of injury to the occupants and other road users.

The first job of any vehicle structure designer is to ensure that requirement 1 is met since this is critically dependant on the overall structural layout and determines how the vehicle will handle during normal and extreme manoeuvres. Basically, in order for the suspension system to perform the tasks it is designed for, its mounting points should remain as stationary with reference to the vehicle axis system as possible. There are three main categories of dynamic loading which tend to deform the vehicle structure about these mounting points:

i. Torsional loads such as when one wheel hits a bump in the road and the additional suspension load at that corner of the car is transmitted through the vehicle structure.
ii. Bending loads such as when both wheels on an axle hit a bump or a kerb simultaneously.
iii. Longitudinal/lateral loads due to inertia effects under traction, braking and/or cornering conditions, including the in-plane effects of minor collisions and impacts.

Although in principle all three of these loading types can act simultaneously, the first of these is the most significant and the latter relatively unimportant in terms of the chassis structure itself. Thus the torsional stiffness of the structure is perhaps the most important parameter to be considered in the initial stages of a chassis design. It is normally measured or calculated by fixing the suspension mounting points at three corners of the car and applying a vertical load at the fourth corner as indicated

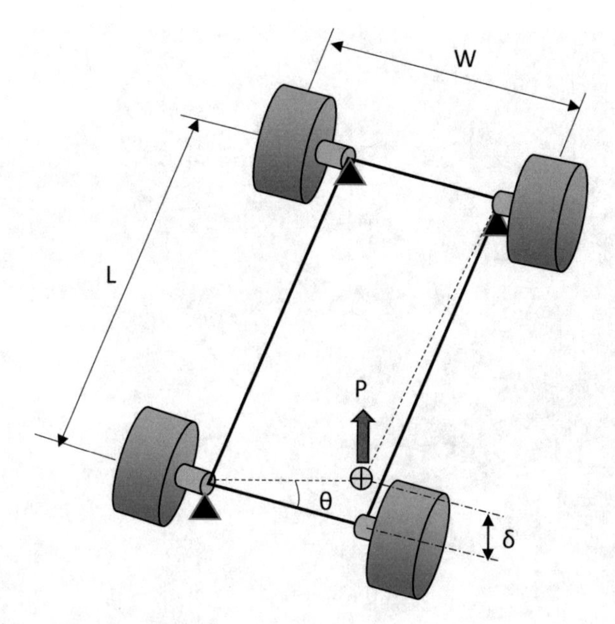

Fig. 4.5 Calculation of chassis torsional stiffness

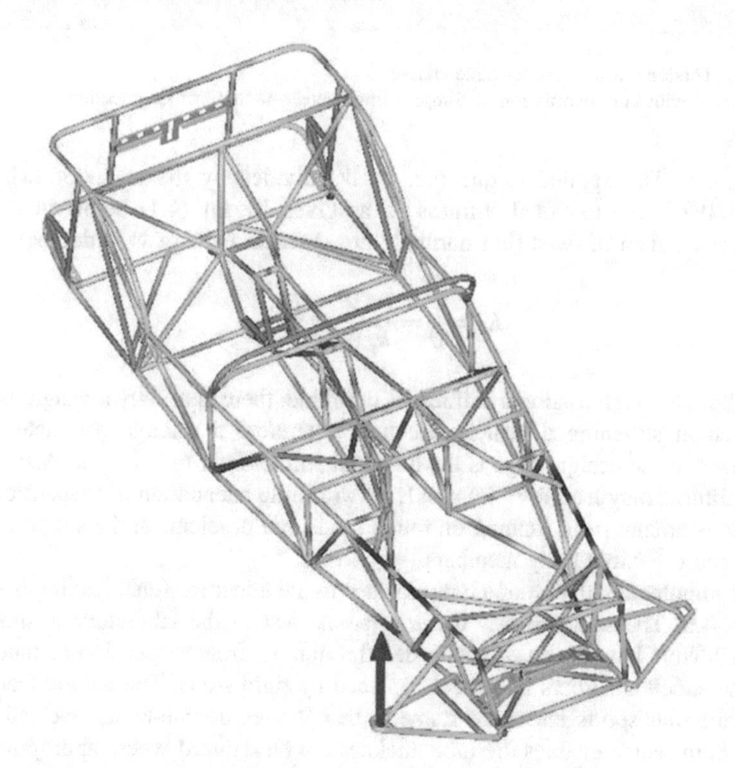

Fig. 4.6 Computer simulation of space-frame chassis under torsional loading. Reproduced with kind permission of Simpact Engineering © Simpact Engineering Ltd

Fig. 4.7 Torsional test of space-frame chassis.
Reproduced with kind permission of Simpact Engineering © Simpact Engineering Ltd.

in Fig. 4.5. The applied torque (i.e. $P \cdot W$) divided by the angle of twist (i.e. $\theta = \delta/W$) is the torsional stiffness K_t as given by Eq. (4.1) below in units of torque per radian of twist (but normally presented in units of N m/degree).

$$K_t = \frac{T}{\theta} = \frac{P \cdot W}{\delta/W} = \frac{P \cdot W^2}{\delta} \tag{4.1}$$

Although a high torsional stiffness is desirable, there is usually a weight penalty involved in stiffening a vehicle structure. Therefore a suitable parameter to be optimised at the design stage is the torsional stiffness per unit weight. Also a very high stiffness may lead to NVH problems with little attenuation of suspension load variations arising from running on rough roads and problems at the support points of the more flexible body members.

A computer simulation of a typical space frame under torsional loading is shown in Fig. 4.6. The same chassis under torsional test in the laboratory is shown in Fig. 4.7. Note how the chassis is loaded through the front suspension members but that the shock absorbers have been replaced by rigid struts. The unique feature of this particular sports car space-frame is that it uses double-butted welded joints throughout which enables the tube thickness to be reduced where appropriate and the overall chassis weight to be minimised, albeit at greater material and manufacturing cost.

Generally speaking, if a car structure is sufficiently stiff in torsion, it will be adequate in bending. Also, stresses in individual structural members during normal driving conditions will be low. However, requirement 2 in the list above (i.e. to maximise the fatigue life) remains a priority for the designer of the structure since stress concentrations can lead to material cracking or even failure, particularly at spot-welded or bolted connections. This fatigue mode of failure is considered in Sect. 4.5 below. The third and final requirement (i.e. to ensure the vehicle is crashworthy) often causes conflicts because a strong, stiff structure may not possess the necessary energy absorbing capabilities. Vehicle crashworthiness is the subject of Sect. 4.4 of this chapter.

4.3.2 Methods of Analysis

4.3.2.1 Simple Bending Analysis

Whether the purpose of the analysis is to determine the torsional stiffness or the stresses in the structure due to the applied loads, it is important to establish the load path through the structure, i.e. how the loads are input to the structure and transmitted from one member to the next. For example, for the analysis of static vertical loading, the distribution of loads and centres of mass of major items, e.g. engine, passengers, payload, etc., need to be determined or estimated. Usually the weight of the body shell, that is the sprung weight, is assumed to be uniformly distributed over the length of the vehicle. Then, assuming the vehicle to be simply supported at the wheels, the type of load diagram shown in Fig. 4.8 can be constructed leading to derivation of the shear force (SF) and bending moment (BM) diagrams as shown. It can be seen from these diagrams that the maximum SF occurs at A and the maximum BM at B.

Note that the static loads are often multiplied by factors of 2 or 3 to allow for dynamic effects as the vehicle traverses rough ground or hits obstacles in the road. However this is very much a "rule of thumb" and more precise methods of estimating the effect of dynamic loading are now available (see Sect. 4.5 below).

Once the maximum bending moment has been estimated, the maximum bending stress in the case of a chassis framed vehicle can then readily be calculated from the standard equation for a simple beam in bending:

$$\sigma = \frac{M \cdot y}{I} \tag{4.2}$$

where

σ is the maximum bending stress, which should be less than the design stress,
M is the applied bending moment,

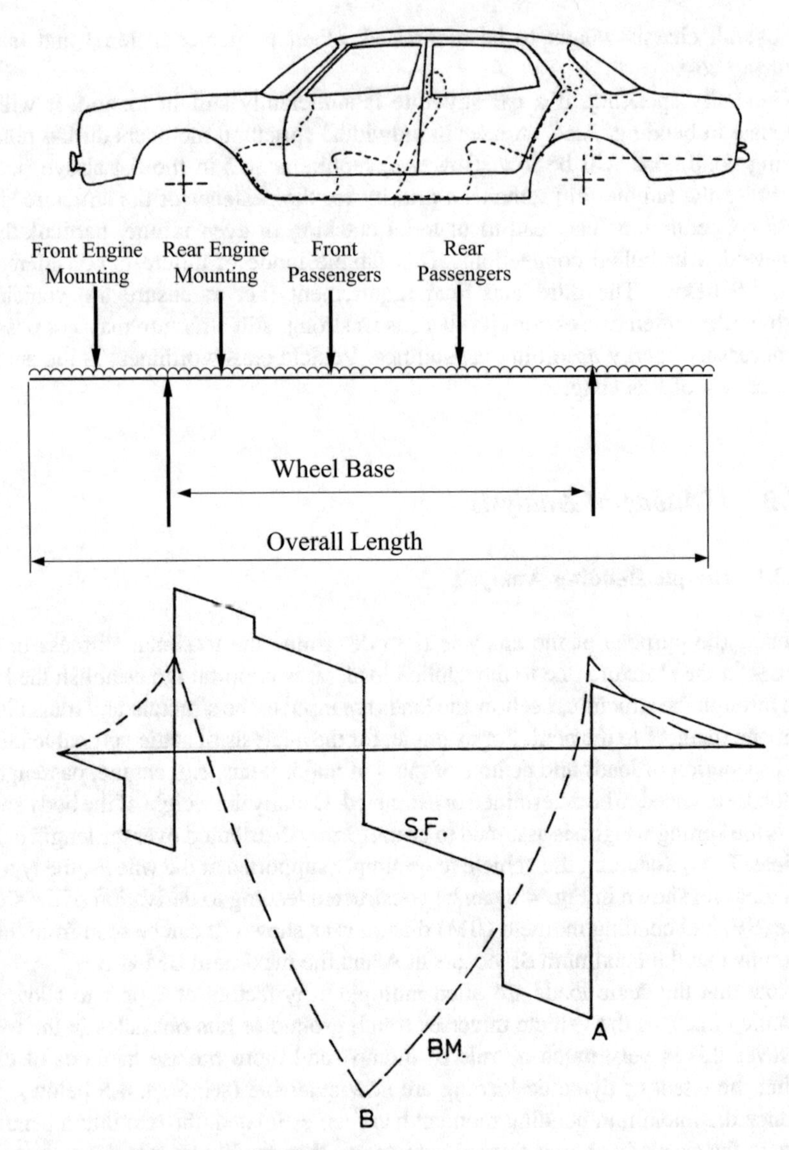

Fig. 4.8 Indicative static load diagram for saloon car: "A" indicates maximum shear force (S.F.), "B" indicates maximum bending moment (B.M.)

I is the second moment of area of the frame longitudinal member at that point y is the maximum distance from the neutral axis of the longitudinal member to its upper or lower surface.

For a vehicle of integral construction, the calculation is more complex because many parts of the structure are involved in reacting the applied loads. As a first approximation, the bending moment can be assumed to be carried solely by the sills, the longitudinal members usually of folded sheet steel construction which connect the front and rear of the car underneath the doors. Even with this assumption, the calculations are not straightforward because the sill section is often unsymmetrical and therefore subject to warping as well as bending. It should also be remembered that the torsional stiffness of beam sections with longitudinal cut-outs to allow for wiring runs e.t.c., is much reduced compared to that of the corresponding fully closed section. The theory of unsymmetrical sections and open sections in bending and/or torsion is outside the scope of the present book.

4.3.2.2 Simple Structural Surfaces (SSS) Method

An important concept in the design of integral structures is that of the shear panel. This is an idealisation of the large, approximately flat, thin steel panels used in the construction of many vehicles such as vans. It is assumed that a shear panel transmits load primarily by developing in-plane shear stresses due to shear forces applied parallel to the edges of the panel. To illustrate how this concept works, consider the simple box structure subject to a torsional load, T as shown in Fig. 4.9a. Free body diagrams can be drawn for each panel and the only loads

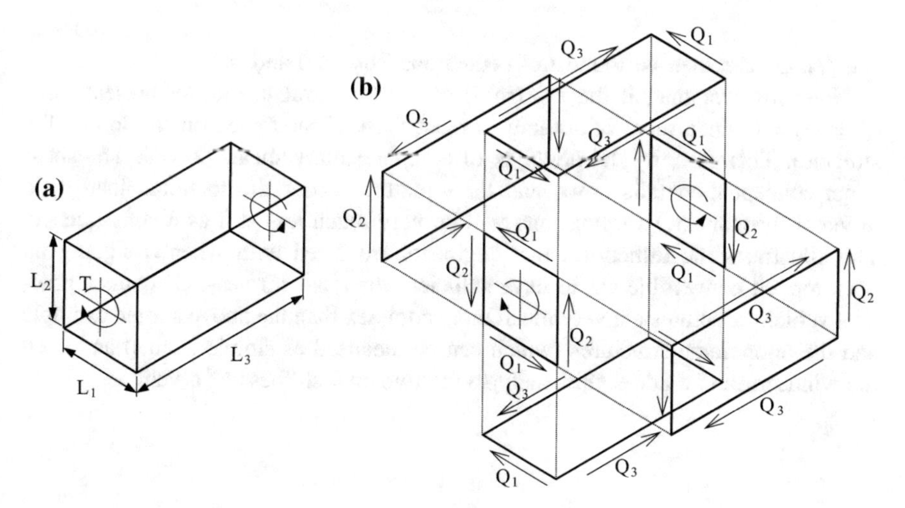

Fig. 4.9 Simple structural surfaces analysis of simple box structure **a** full structure **b** structure subdivided into panels with edge shear forces indicated

involved are assumed to be the shear forces, Q_i acting at the panel edges as shown in Fig. 4.9b. Then, for static equilibrium of the end, sides and top/bottom panels respectively, we can write:

$$Q_1L_2 + Q_2L_1 = T \tag{4.3}$$

$$Q_3L_2 - Q_2L_3 = 0 \tag{4.4}$$

$$Q_3L_1 - Q_1L_3 = 0 \tag{4.5}$$

From Eq. (4.3):

$$Q_3 = \frac{L_3}{L_1}Q_1 \tag{4.6}$$

Then, from Eq. (4.2):

$$Q_2 = \frac{L_2}{L_3}Q_3 = \frac{L_2}{L_3}\cdot\frac{L_3}{L_1}Q_1 = \frac{L_2}{L_1}Q_1 \tag{4.7}$$

Finally, from Eq. (4.1):

$$Q_1 = \frac{T}{L_2} - Q_2\cdot\frac{L_1}{L_2} = \frac{T}{L_2} - \frac{L_2}{L_1}\cdot Q_1\cdot\frac{L_1}{L_2}$$

Therefore:

$$Q_1 = \frac{T}{2L_2} \tag{4.8}$$

and Q_2, Q_3 can then be found from equations Eqs. (4.6 and 4.7).

Note however that, if the top (roof) of the box structure is not present, then $Q_1 = Q_3 = 0$ since there is nothing to react these shear forces on the top of the structure. Therefore, by the principle of complementary shear, $Q_2 = 0$. The shear panel concept then falls down and the structure carries the torsional load by a different mechanism (warping) and will be very much less stiff as a consequence. This illustrates the difficulties that engineers are faced with when designing an open-topped convertible car to have sufficient torsional stiffness. Of course, practical vehicle structures are very much more complex than the above simple example and the door frame structures (which can be idealised as simple beams) and even the windscreen can add significantly to the torsional stiffness of a vehicle.

4.3.2.3 Finite Element Analysis (FEA)

It will be appreciated from the above general discussion that the analysis of vehicle structures using traditional theoretical methods is complex and often very approximate in view of the number of simplifying assumptions that need to be made. An alternative approach which has become almost universally adopted in the industry is to use finite element analysis (FEA) to more accurately predict deformations and stresses in vehicle structures. FEA essentially involves breaking the structure down into a large number of non-overlapping small regions ("elements"). The problem is formulated in terms of the "degrees of freedom" (usually displacements) at certain discrete points known as "nodes" which always occur at element vertices but may sometimes also be located elsewhere.

There are many different types of elements but the ones most commonly employed in vehicle body structural analysis are beam and shell elements. Since both these types only explicitly model the mid-surface of the structure, the section properties (e.g. second moment of area, shell thickness) are specified separately. A beam element model of a space-frame sports car chassis is shown in Fig. 4.10 whereas a basic shell element model of a pick-up vehicle structure is shown in Fig. 4.11.

Note that in Fig. 4.10, the engine and final drive have been modelled with solid elements, the properties of which are calculated to give the correct mass distribution. The chassis members are modelled with beam elements, shown as straight lines in Fig. 4.10, which have 3 translational and 3 rotational degrees of freedom at each node. If the connections at the nodes can be considered to be pin-jointed, then the members carry loads in tension/compression only. However, if the connections are sufficiently reinforced to be considered rigid, the members can be subjected to bending/torsion as well as tension/compression. In practice it usual to make the rigid-joint assumption since this will generally overestimate the stiffness and the stresses in the members and so lead to conservative (i.e. safe) designs.

The shell element model of the pick-up truck in Fig. 4.11 is actually derived for a dynamic modal analysis of the body shell and is only indicative of the much more detailed model that would typically be used for structural assessment. The front windscreen is included because it does have a stiffening effect on both the static and

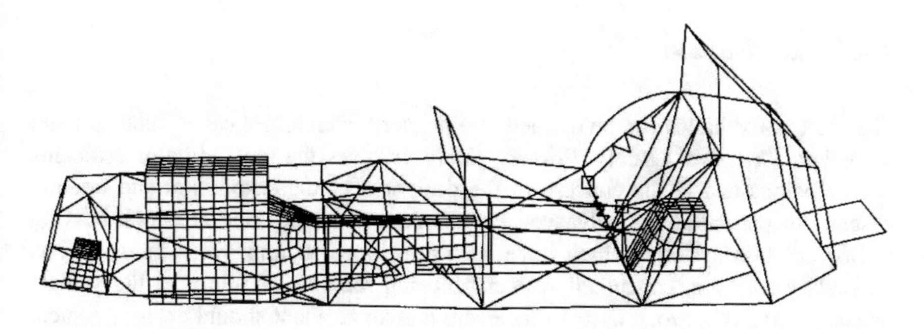

Fig. 4.10 Beam FE model of spaceframe

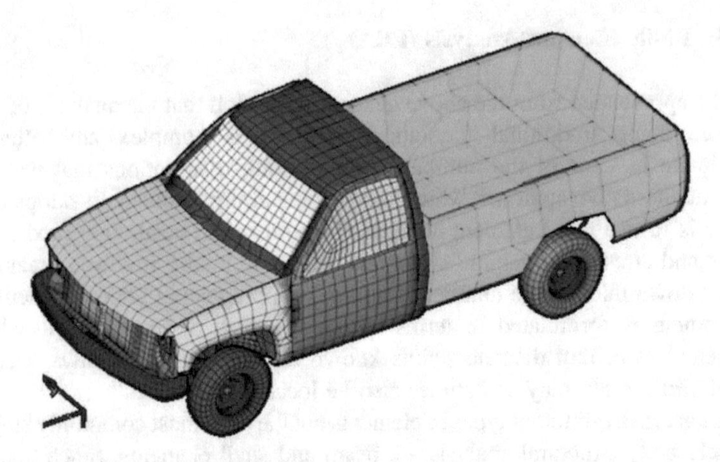

Fig. 4.11 Shell FE model of pick-up vehicle structure.
Source Advances in Vibration Engineering and Structural Dynamics, Intech 2012.
http://www.intechopen.com/books/advances-in-vibrationengineering-and-structural-dynamics

dynamic behaviour but the side windows and any sun-roof would normally be excluded from a structural model as they may be open during critical manoeuvres.

Occasionally, full 3D solid elements are used in place of shells e.g. in the detailed analysis of critical regions such as spot welded connections where stress concentrations may lead to early fatigue failures. Normally, isotropic linear elastic material properties are assumed for steel or aluminium components (only Young's modulus and Poisson's ratio required) but occasionally non-linear plasticity characteristics will be specified in heavily stressed regions where the material may be subject to plastic (permanent) deformation. Composite materials are more difficult to analyse because their properties are often anisotropic (different in different directions of loading). Modern FEA packages have specialist composite material models available but it is outside the scope of the present book to describe these in detail.

4.4 Safety Under Impact

4.4.1 Legislation

The safety of vehicles can be divided into primary (sometimes called "active") and secondary ("passive") safety. Primary safety involves the prevention of accidents and collisions and is the concern of the designers of the suspension and braking systems in particular e.g. advanced chassis control (ACC) and anti-lock braking (ABS) are two systems which have been developed to improve primary safety. Secondary safety is concerned with minimising the risk of injury to the vehicle occupants and other road users in the event that an accident should occur. As such, it is very much the responsibility of the vehicle structure designer to ensure that the vehicle has sufficient energy absorbing capability and general “crashworthiness” to meet the ever increasing legislation and customer-driven demands.

The most significant legislative codes affecting the international design of vehicles are the Federal Motor Vehicle Safety Standards (FMVSS) which apply to all vehicles sold in the USA and the various European Council (EC) directives on vehicle safety which are mandatory for volume-produced vehicles sold in the EU. Non-regulatory assessments also exist in many jurisdictions under the heading of NCAPs (New Car Assessment Programmes). Although passing the various NCAP tests is non-mandatory, obtaining good (ideally 5 star) ratings are important performance targets for vehicle manufacturers.

There are many parts within these standards relating to occupant protection and component design for frontal impact safety. The original and most well-known requirement is contained in FMVSS 208 which specifies a maximum allowable deceleration of 60 g measured on anthropomorphic test dummies in frontal impacts at 30 mph with a rigid wall at angles between 0° and 30° to the forward direction of motion. This original requirement was criticised as being unrepresentative of true crash situations and has been supplemented by tests involving a deformable barrier (manufactured from aluminium honeycomb to represent the front-end stiffness of a typical saloon car) and at various degrees of overlap (typically 40 or 60%) with the front-end of the test vehicle. A computer simulation of such an offset barrier frontal impact test is shown in Fig. 4.12.

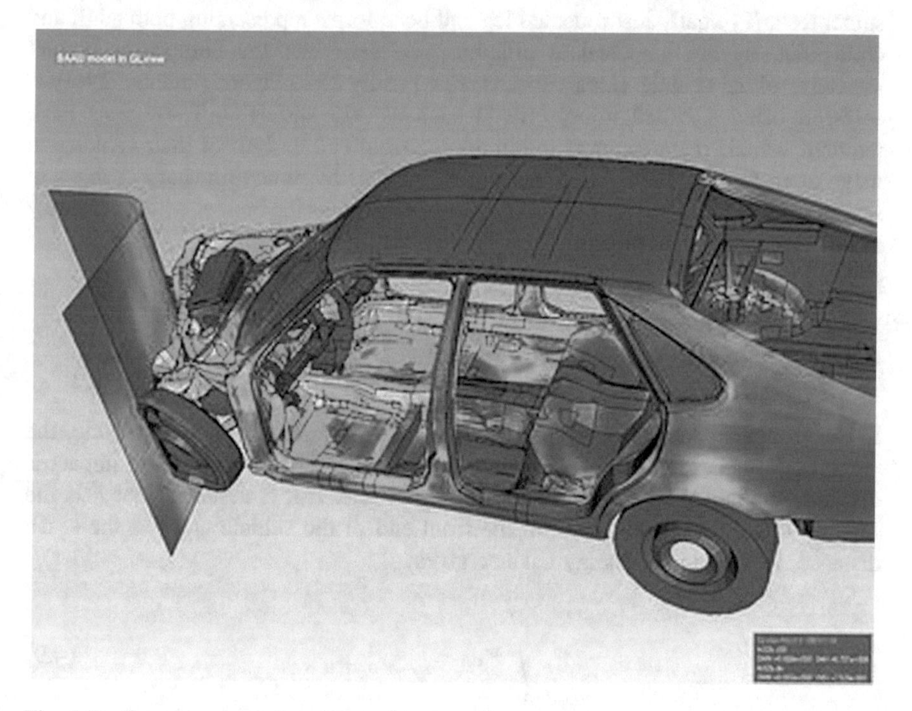

Fig. 4.12 Computer simulation of frontal impact test.
Source Wikipedia Commons. https://commons.wikimedia.org/wiki/File:FAE_visualization.jpg

These offset deformable barrier tests also forms the basis of the current EU standards although the impact speed specified in Europe is 56 km/h. However the requirement for a maximum deceleration of 60 g for any time period longer than 3 ms as measured on instrumented dummies has remained the norm. As well as dummies representing the average male driver, dummy tests are now also carried out for smaller female and child occupants and in the rear as well as the front of the vehicle. Close attention is paid to injury modes such as neck whiplash and lower leg injuries.

In addition to rigorous frontal impact requirements, side impacts are also the subject of legislation. Here worst case tests tend to replicate the side of the vehicle hitting a rigid pole rather than the deformable front end of another vehicle. Since there is very little room for deformable energy absorbing devices in door structures, stiff side impact beams are important in protecting the occupant space from excessive intrusion and close attention is paid to collisions between the dummy (especially the head) and the surrounding interior structures.

Collisions between passenger cars and larger vehicles such as trucks has also been the subject of much scrutiny and legislation. Trucks are now required to have rear and side guards to prevent smaller vehicles "submarining" underneath the truck chassis on impact with obvious dire consequences for the car passengers as mentioned in Sect. 4.1 above.

Finally, protection of pedestrians and other road users (e.g. cyclists) in the event of a collision with a vehicle is now the subject of legislative standards and consumer tests. Typically instrumented leg and head forms representing both adult and child anatomy are launched at different angles against the bumper or bonnet structures of the vehicle. Impact speeds are typically 35 km/h and potential injury is assessed using standard injury criteria such as AIS and HIC. These tests have required vehicle manufacturers to rethink the front end designs of their vehicles to make them far less damaging to pedestrian impacts by removing sharp changes in section and reducing the stiffness of components such as the bonnet which are likely to be involved in the collision.

4.4.2 *Overview of Frontal Impact*

In the frontal impact of a vehicle whether with a barrier or another vehicle, the worst case scenario is that all the initial kinetic energy of the vehicle on impact is dissipated within the structure of that vehicle alone. If it is assumed that F is the force to crush (plastically deform) the front end of the vehicle and s is the crush distance, then a simple energy balance gives:

$$\int_{0}^{Sf} F\,ds = \frac{1}{2} m v_0^2 \tag{4.9}$$

where

m is the mass of the vehicle
v_0 is the velocity on impact
s_f is the crush distance at the end of the impact (total front end deformation).

Now, if it is further assumed that the crush force is constant and known, then the total crush distance is given by:

$$s_f = \frac{mv_0^2}{2F} \tag{4.10}$$

For example, if a vehicle of mass 1000 kg and crush strength 300 kN impacts a rigid barrier at 50 km/h (13.9 m/s), the total amount of front-end crush is:

$$s_f = \frac{1000 \times 13.9^2}{2 \times 300 \times 10^3} = 0.32\,\text{m}$$

The time duration of the impact, t, can be calculated from the momentum equation:

$$F \cdot t = mv_0 \tag{4.11}$$

Giving for the above parameters: $t = \frac{1000 \times 13.9}{300 \times 10^3} = 0.046\,\text{s}$

The average deceleration of the vehicle to the rear of the crush zone can then be calculated from the change in velocity divided by the impact time duration:

$$\begin{aligned} a &= \frac{13.9}{0.046} \\ &= 302\,\text{m/s}^2 \\ &\approx 30\,\text{g} \end{aligned}$$

The above equations clearly illustrate that a very strong front end structure will give a short crush length and impact time duration and therefore a high deceleration.

In practice the situation for any real vehicle is far more complex than suggested by the above equations. In particular, a vehicle absorbs energy by a variety of mechanisms which include crushing, folding, buckling and frictional contact of a number of discrete components within the vehicle structure. The front longitudinal chassis rails in particular, along with other support structures, are designed to absorb energy progressively as the impact load is transferred to the rear of the vehicle (see Fig. 4.13).

Thus the crush strength is not a constant value as assumed above and the deceleration time history (the "crash pulse" as it is known) is complex, as shown by the representative sample in Fig. 4.14. Incidently this pulse only just meets the standard criterion of a maximum deceleration at the occupant restraint position of 30 g for any time period greater than 3 ms.

Fig. 4.13 Load path for frontal impact of vehicle.
Reproduced with kind permission of Audi © Audi AG

Fig. 4.14 Representative frontal impact crash pulse

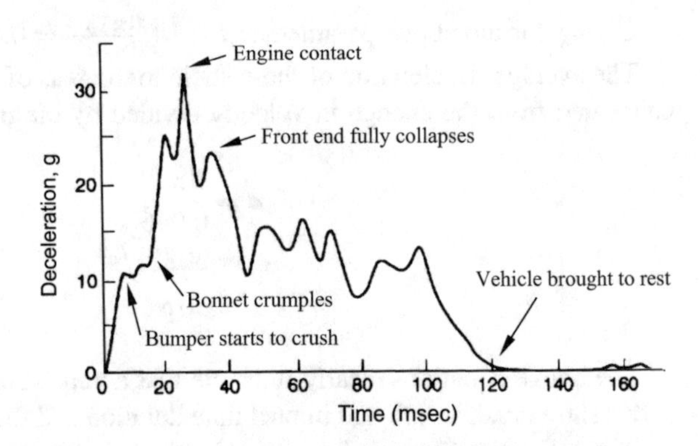

Simple theoretical analysis will not predict such a complex response which requires either a full scale crash test or advanced numerical simulation of the impact event to be carried out. The former is extremely expensive and it is not easy to investigate the complex interactions that occur within a vehicle structure or the effect of changing certain parameters. The latter numerical approach to crashworthiness assessment prior to final impact testing of a vehicle has become the norm as described in the case study presented in Sect. 4.4.4 below.

4.4.3 *Energy Absorbing Devices and Crash Protection Systems*

An ideal energy absorbing (EA) for front-end crash protection has the dynamic force-displacement (P-δ) characteristics shown in Fig. 4.15.

Up to the critical load P_c, the deformation should be elastic, i.e. it should be recovered completely on unloading from a relatively low load (low speed impact). Once P_c is exceeded, the device should deform at constant load, progressively absorbing the energy of impact and restricting the level of force transmitted to other parts of the vehicle. The amount of energy absorbed is of course equal to the area under the force-deflection (P-δ) curve.

The front bumper of a passenger car is normally the first component to be impacted in a frontal collision. Bumper design has evolved from a simple steel construction designed to protect the bodywork from small knocks towards the complex polymer/foam/metal construction of modern bumper systems. Polymer foams, or honeycomb cellular materials, are widely used as impact energy absorbers in devices such as knee bolsters and side impact protection systems as well as for bumpers. Although far from behaving in the ideal manner suggested by Fig. 4.15, polymer foams have many desirable characteristics. They are light, corrosion resistant, responsive to any direction of loading, recover completely when subject to moderate loading below the foam yield stress and can be engineered to meet the particular requirements by varying the density, type and form of the polymer construction. Typical load-deflection curves for polymer foams of varying degrees of porosity (% density of the solid polymer) are shown in Fig. 4.16.

It can be seen from Fig. 4.16 that after the initial elastic response and yield, the load-deflection response becomes approximately linear again (linear strain hardening). The response eventually stiffens as the pores in the cellular material are flattened. However, the energy absorption capabilities of the bumper system can be extended by mounting the bumper on the front longitudinal chassis rails of the vehicle by means of a controlled deformation structure. One such arrangement is the concentric interference-fitted "tube-within-tube" device shown in Fig. 4.17 which is amenable to theoretical analysis as outlined below.

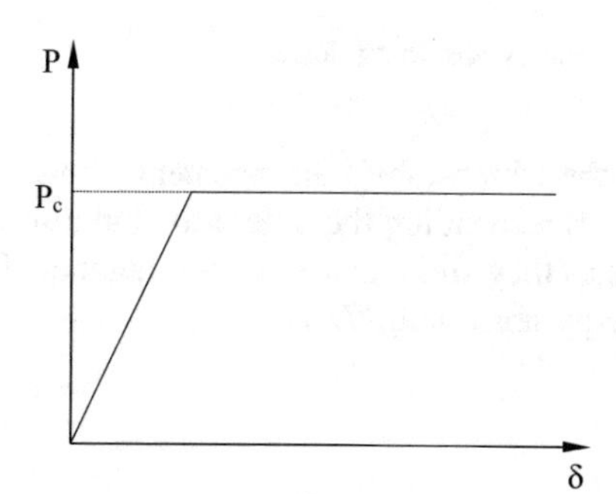

Fig. 4.15 Ideal EA force-deflection response

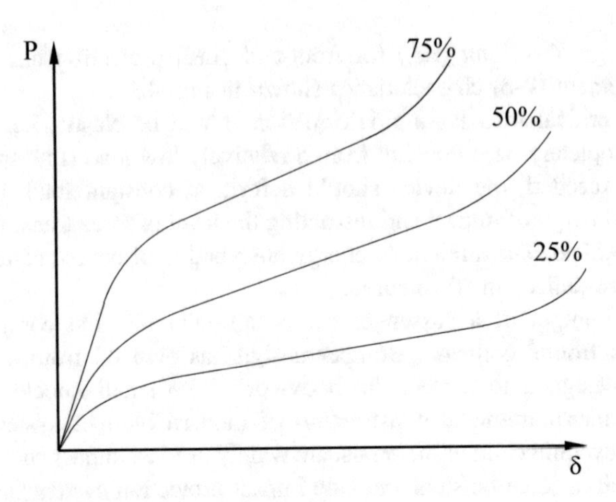

Fig. 4.16 Typical force-deflection plots for cellular polymer foams of different densities (expressed as percentage of fully dense polymer)

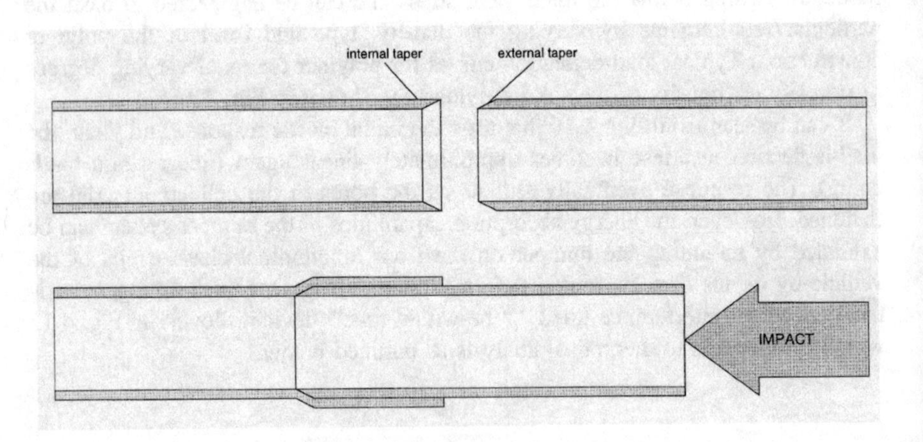

Fig. 4.17 Concentric tube energy absorbing device

The concentric "tube-within-tube" arrangement shown in Fig. 4.17 absorbs energy both in plastically expanding the outer tube but also in overcoming friction between the two tubes as they slide relative to one another. Thus, for small relative movement δx, the energy absorbed, δE, is:

$$\begin{aligned}\delta E &= \delta W_p + \delta W_F \\ &= \delta V \int \sigma d\varepsilon + F \cdot \delta x\end{aligned} \tag{4.12}$$

where

δW_p is the plastic work done
δW_F is the frictional work done
δV is the volume of material undergoing plastic deformation
F is the axial friction force, assumed to remain constant over the small axial movement δx.

Now $\delta V = \pi \, d_m \, t \, \delta x$
where d_m is the mean diameter of the outer tube and t is its thickness.

Assuming perfectly plastic behaviour and that only the outer tube deforms:

$$\int \sigma d\varepsilon = Y \cdot \varepsilon_{tr} = Y \ln \frac{d_o}{d_i} \tag{4.13}$$

where Y is the yield stress (assumed constant), ε_{tr} is the mean true strain and d_o and d_i are the external and internal diameters respectively of the outer tube.

(Note: $d_m = (d_o + d_i)/2, \quad t = (d_o - d_i)/2$)

Furthermore:

$$F = \mu p A \tag{4.14}$$

where

μ is the coefficient of friction
p is the radial pressure between the two tubes
A is the area of interference fit (which increases with length of the tube overlap).

Therefore:

$$\delta E = \pi \, d_m \, t \, Y \ln \frac{d_o}{d_i} \delta x + \mu \cdot p \, \pi \, d_o \, x \, \delta x \tag{4.15}$$

Thus, the total energy absorbed in increasing the length of interference from l_1 to l_2 is given by:

$$\begin{aligned}E &= \int_{l_1}^{l_2} \pi \, d_m \, t Y \ln \frac{d_o}{di} \cdot dx + \int_{l_1}^{l_2} \mu \, p \, \pi \, d_o \, x \cdot dx \\ &= \pi \, d_m \, t \, Y \ln \frac{d_o}{d_i} (l_2 - l_1) + \mu \, p \, \pi \, d_o \, \frac{(l_2^2 - l_1^2)}{2}\end{aligned} \tag{4.16}$$

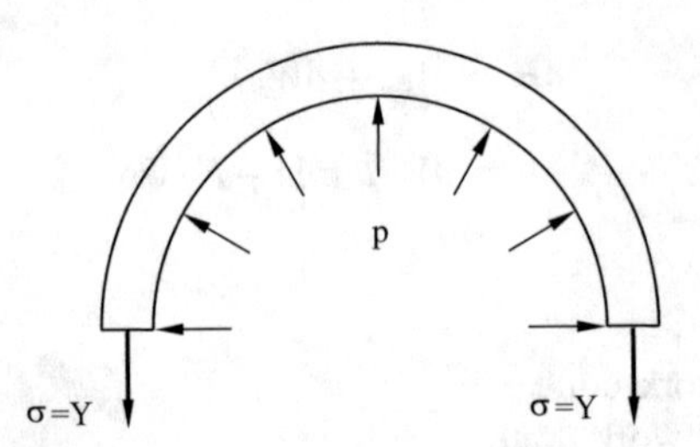

Fig. 4.18 Section though the outer tube of concentric tube EA device

Finally, p can be related simply to Y assuming the entire wall thickness of the outer tube is plastically deformed as shown in Fig. 4.18.

For static equilibrium: $pd_o x = 2Y\,tx$

Therefore:

$$p = \frac{2Yt}{d_o} \tag{4.17}$$

Then, if Y and μ are known or can be estimated, the rate of energy absorption can be calculated for any given set of geometric parameters (d_i, d_o, l_1) and these parameters optimised to give the desired performance.

A typical experimental load-deflection curve for the concentric tube device is shown in Fig. 4.19. Although this is not quite the ideal response as shown in Fig. 4.15, the fact that the device can be reliably tuned, and even re-used if the impact is not too severe, makes it a potentially useful means of limiting the force transmitted from the bumper to the main vehicle structure.

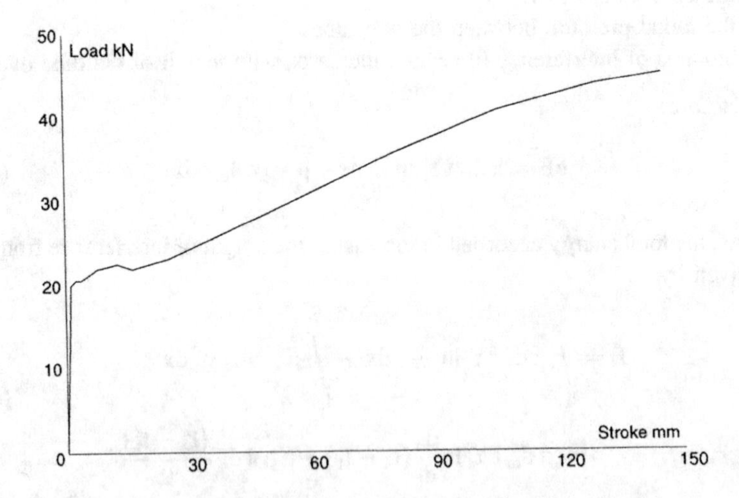

Fig. 4.19 Axial load-deflection curve for concentric tube EA device

4.4.4 Case Study: Crashworthiness of Small Spaceframe Sports Car

As mentioned above, many small sports and racing cars employ a spaceframe in the form of a strong and stiff tubular steel chassis. There is very limited scope within such spaceframes for energy absorption by the large deformation of sheet steel body panels which occurs in mass-produced integral body structures. Furthermore, for styling reasons, it is often not considered possible to mount large bumpers or other forms of external energy absorbing devices at the front of the vehicle. Meeting the crashworthiness legislation is a major challenge for any such vehicle and the research described below aimed to investigate the crashworthiness of a typical small spaceframe vehicle using finite element simulation of the impact of the whole vehicle against a rigid barrier as well as drop weight testing of individual components.

For most small sports/race cars, the wheels are external to the body and, in a frontal impact, the front tyres very quickly came into contact with the barrier, unlike in a normal passenger car where contact between the tyre and front wheel arch occurs much later during the impact event. The overall response therefore depends very much on the impact characteristics of the tyres as well as the wheels and suspension system on which they are mounted. Much effort was devoted in the case study to modelling the tyre and wheel in as realistic yet relatively simplistic manner as possible. This involved developing an air-bag model for the inflated tyre and using a finite element model to simulate the experimental impact of the tyre and wheel against a rigid block using a drop test carriage to house the assembly as indicated in Fig. 4.20.

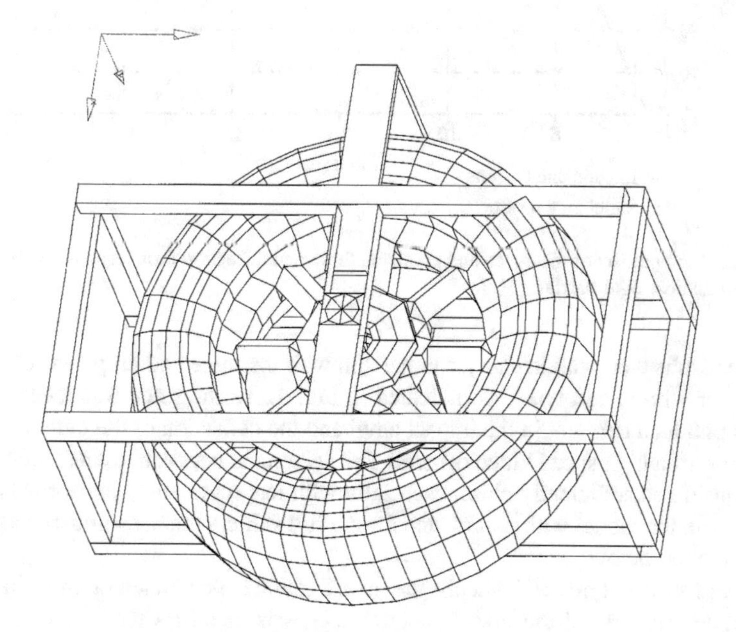

Fig. 4.20 Finite element model of tyre and wheel mounted in drop cage

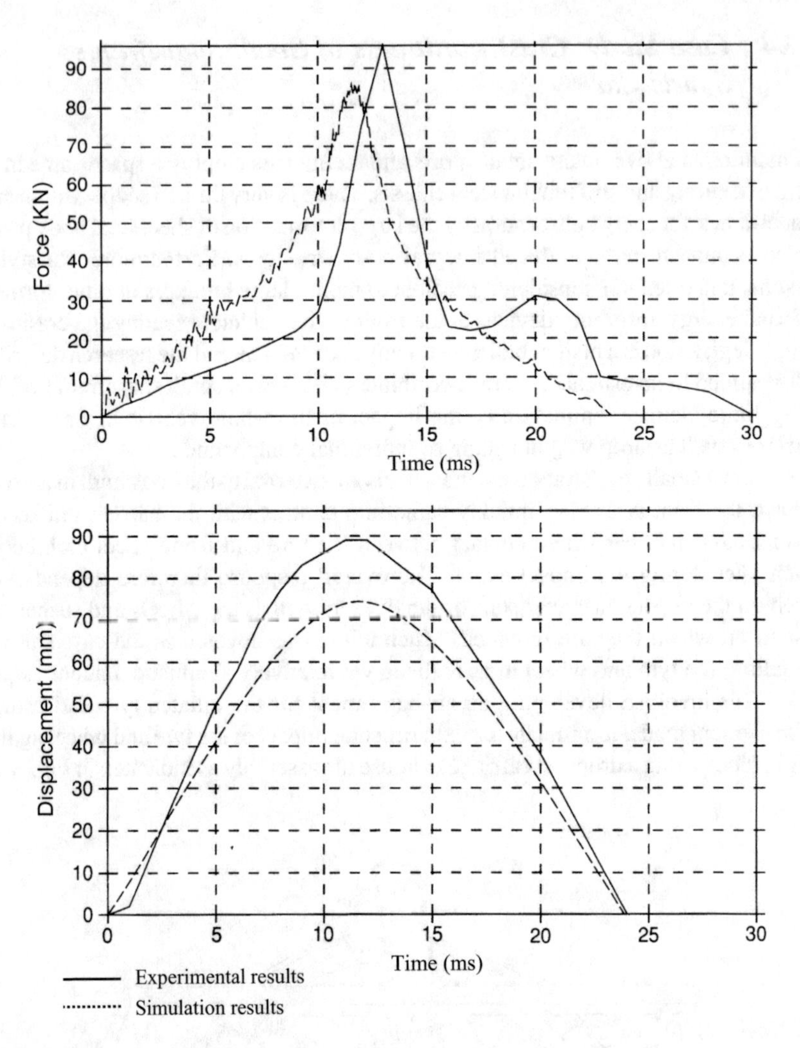

Fig. 4.21 Comparison of measured and predicted force-and deflection-time response of tyre when impacted against rigid barrier

The results were validated by comparison with instrumented drop tests on actual sports car wheels and tyres, as indicated in Fig. 4.21. The force was measured by means of a load cell within the impact anvil and the deflection of the outer tyre wall by means of a high speed video camera. The agreement between test and simulation was considered sufficiently close, especially with respect to energy absorbed by the tyre, to use the model with confidence for the full vehicle impact simulations which are described below.

Apart from the tyres and wheels, the investigation concentrated on modelling the impact deformation of the main spaceframe chassis members with either beam or

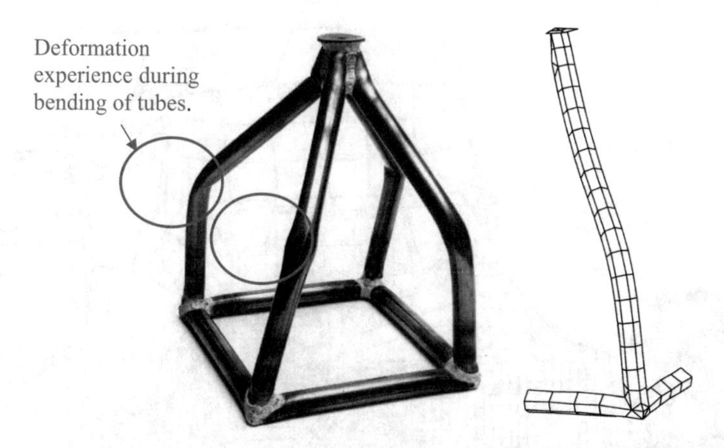

Fig. 4.22 Deformed shape of representative pyramid structure after impact—test piece (left side) and shell element simulation of one quarter of structure (right side)

shell elements. Again the predicted results were validated by testing the dynamic response of representative "pyramid" welded steel structures in a large drop test facility. A comparison between the post-impact experimental and numerical shapes for these simple pyramid structures is shown in Fig. 4.22. It was found that to accurately model the plastic hinge formation and subsequent ovalisation of the longitudinal steel tubes, it was necessary to have a much finer mesh in the local area of the hinge than the mesh indicated in Fig. 4.22.

The complete finite element model of a particular space-framed vehicle shown in Fig. 4.23 makes extensive use of shell-type elements of sufficient mesh density at the front end of the vehicle. Elsewhere simple beam elements are used and the model includes 3D representations of the engine and other heavy components in order to obtain the correct mass distribution. A dynamic analysis using an explicit finite element solver was carried out for this model at an impact velocity of 50 km/h against a rigid barrier. This gave predictions of deformation as shown in Fig. 4.24 which agree well with the limited full car crash test results available. The predicted deceleration time history, monitored at a point equivalent to the driver's chest, has the form shown in Fig. 4.25. The peak deceleration of 87 g is well in excess of the usual limit of 60 g. The incorporation of EA devices such as described above, either within the vehicle nose cone (light plastic structure not included in model) or as an integral part of the chassis structure, was proposed as a potential means of improving the frontal impact characteristics of the vehicle without affecting its overall appearance.

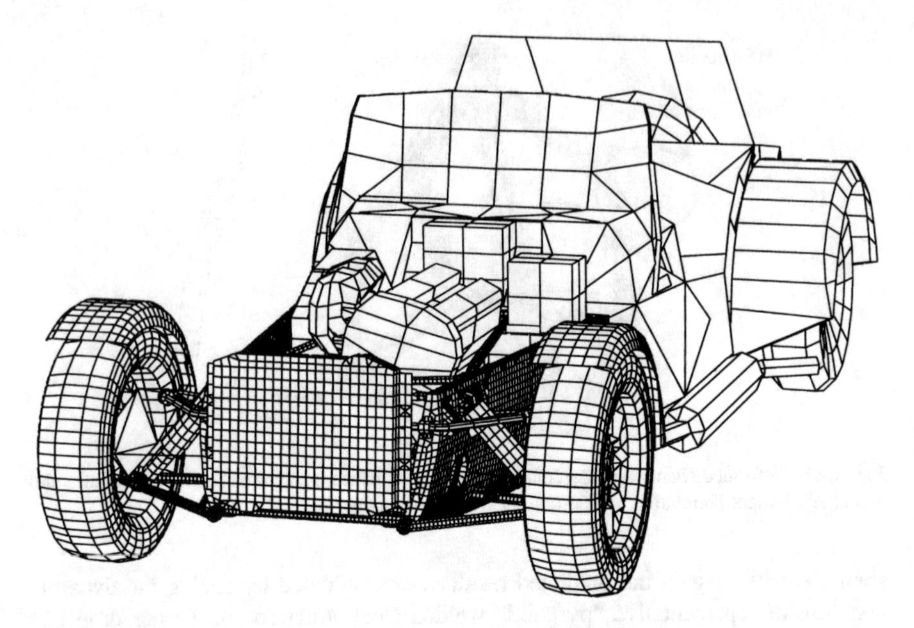

Fig. 4.23 FE model of small space-frame sports car—undeformed

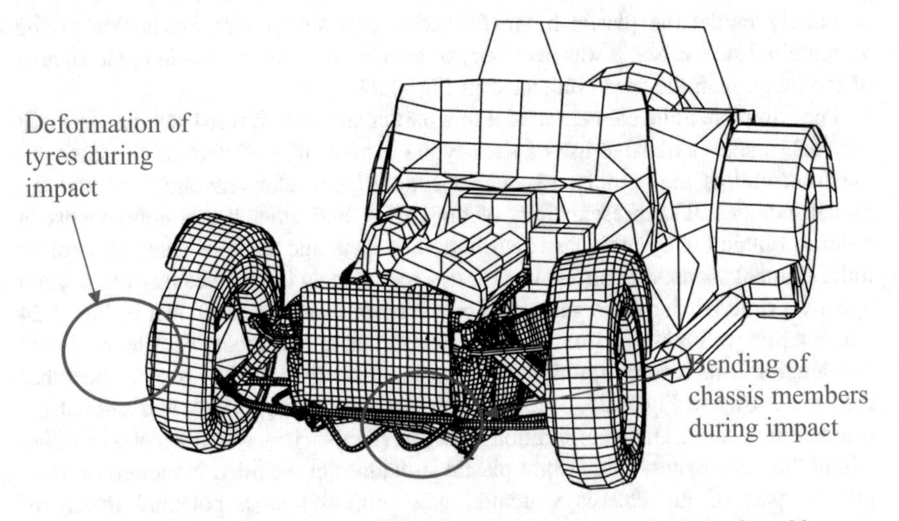

Fig. 4.24 Deformed shape of FE model of small space-frame sports car during frontal impact

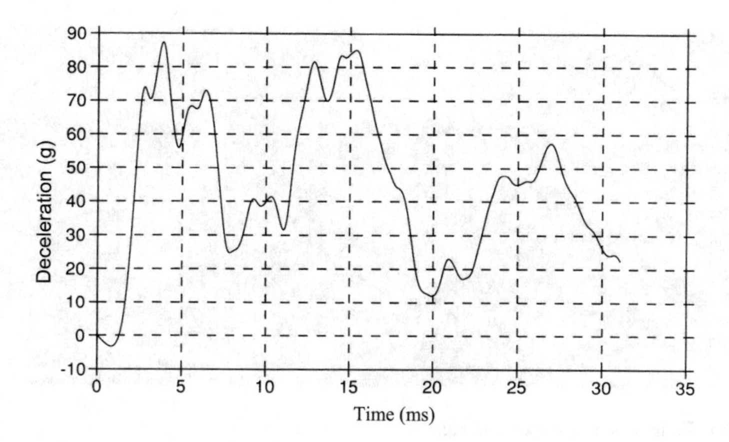

Fig. 4.25 Predicted crash pulse at driver restraint position for frontal impact of small space-frame sports car

4.5 Durability Assessment

4.5.1 Introduction

As discussed above, automotive structures and components rarely fail due to a single application of loading, except of course when they are designed to do so such as in impact crash scenarios. Instead structures and components may ultimately fail over a prolonged service life due to the accumulation of fatigue damage coupled with environmental effects such as corrosion and wear which reduce the load carrying capacity of the material. The study of these long-term loading effects is known as "durability assessment" and is considered an important part of the overall structural reliability of the vehicle.

Experimental testing for durability is carried out in various ways:

- in-service field testing
- accelerated proving ground tests
- laboratory testing.

Field testing with instrumented vehicles is the most realistic form of testing but it is very time-consuming and expensive. In fact, for a well-designed vehicle, no failures should be generated in a reasonable timescale unless the vehicle is subjected to extreme misuse load cases such as hitting kerbs or driving over severe pot-holes. Nevertheless the local acceleration and strain data generated can be useful in other more detailed forms of assessment.

Fig. 4.26 Four poster dynamic test rig

Proving ground tests normally involve driving the vehicle over special tracks incorporating pavé (cobblestone-like) or corrugated surfaces. These are designed to subject the vehicle to more extreme loading than would be the case on normal roads and hence promote failures in a shorter time scale. Again the vehicle will usually be instrumented with accelerometers and/or strain gauges at critical locations to provide useful data on loading conditions.

Finally, laboratory testing is a means of subjecting components and sub-assemblies to realistic loading under carefully controlled conditions. For example, 4 post dynamic test rigs can be used to load all 4 wheels stations of a vehicle independently, as indicated in Fig. 4.26. Since the tests can be run continuously over an extended period, this can be a cost-effective and reasonably rapid means of establishing the durability of automotive structures and components. However the capital investment in test equipment is high and prototype parts and assemblies must be available. This means that testing can only occur quite late in the product design cycle and cannot therefore be used to inform or drive design changes leading to improved solutions.

4.5.2 Virtual Proving Ground Approach

In order to bring more reliable and efficient vehicles to the market place more rapidly, it is highly desirable to substitute or supplement the above traditional durability test methods with predictive software tools. This has led to the concept of a Virtual Proving Ground (VPG) in which computer modelling techniques replace physical tests. Such an approach should enable alternative designs to be assessed far earlier in the product life cycle than previously leading to more efficient and novel solutions being generated. As well as being more cost effective, this could also pave the way for more automated and optimised design methods which may replace some of the reliance on the accumulated experience of previous generations of engineers and/or tried-and-tested solutions.

The heart of any VPG approach to durability assessment is accurate fatigue data for the material or component under consideration, usually in the form of the classic S-N curve where S is the stress (or strain) amplitude and N is the number of load cycles before failure. The relationship should ideally contain design factors to allow for the variability of raw fatigue data and for environmental effects such as the presence of a corrosive environment (e.g. salt water in winter). Also required are robust and accurate algorithms for assessing the equivalent fatigue life under complex dynamic loading conditions and multi-axial stress states. Fortunately sophisticated fatigue analysis software to carry out this complicated assessment is now available as part of established CAE systems or as stand-alone packages.

Assuming that suitable fatigue data and analysis routines are available, the next problem is to obtain appropriately detailed stress distributions for input to the fatigue software. As discussed above, modern finite element methods are capable of analysing stresses and strains in general 3D structures under almost any complexity of loading conditions. The question is how complex does the analysis have to be, as increasing complexity invariably means more expense and time delay. Ideally one would like to make use of linear static stress analysis results but there are limits to the applicability of such data, as discussed in the Case Study below. It may also be possible to make use of a frequency domain approach to account for the dynamic response of the structures considered. The alternative is to carry out a full dynamic transient analysis for a typical time history of loading input to the vehicle. By its very nature, this will be computationally intensive but also by definition will produce the most accurate results.

Whichever form of finite element stress analysis is undertaken, it is necessary to have detailed knowledge of the input loads from the road surface or any other source that give rise to the stresses in the individual components of the assembly or structure. Such data can be generated from in-service or proving ground tests but as discussed above, such tests are time-consuming and require a prototype vehicle to be available. An alternative approach is to use multi-body dynamics (MBD) simulation of the whole or part of the vehicle as it travels over typical road surfaces in order to generate loading data at particularly critical locations. Although MBD software is now readily available, many issues remain concerning its accurate and effective use such as whether a quarter or full car model is required and what type of tyre model to specify for each application.

Assuming that the above software elements are satisfactorily developed and integrated so that accurate estimates of fatigue life distribution throughout the structure/component can be made, there remains the possibility of using these life distributions to automatically modify the designs leading to a more optimal solution in terms of weight and/or efficient use of material. Although structural optimisation routines do exist in some FEA packages, it is fair to say that they are not yet routinely used in the automotive industry due to their limited applicability to real structures subjected to dynamic loading conditions. This is particularly true for the design of components and structures whose primary failure mode is likely to be fatigue due to complex loading rather than from a single load application. The extension of existing optimisation methods and, in particular, the powerful technique of Evolutionary Structural Optimisation (ESO) to this class of problem is presented in the case study that follows.

4.5.3 *Case Study: Durability Assessment and Optimisation of Suspension Component*

The component under consideration is the lower arm from the front suspension of a multi-purpose vehicle as shown in Fig. 4.27. A more detailed view of the existing design of arm is shown in Fig. 4.28. It can be seen that the arm has a number of

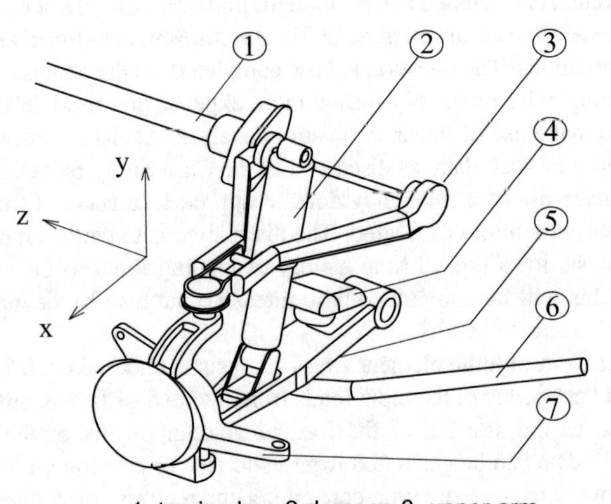

Fig. 4.27 Front suspension system of multi-purpose vehicle

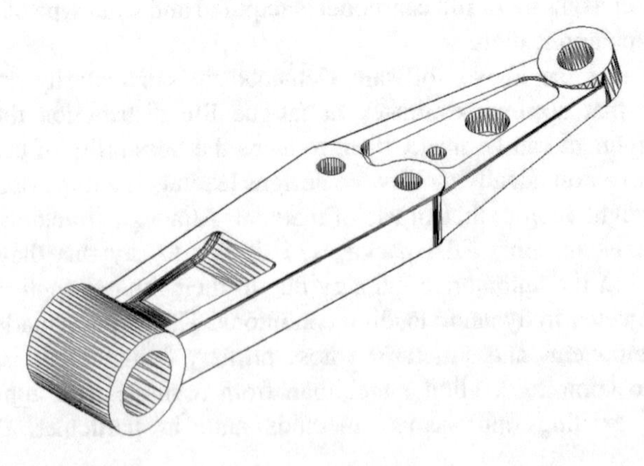

Fig. 4.28 Lower suspension arm of multi-purpose vehicle

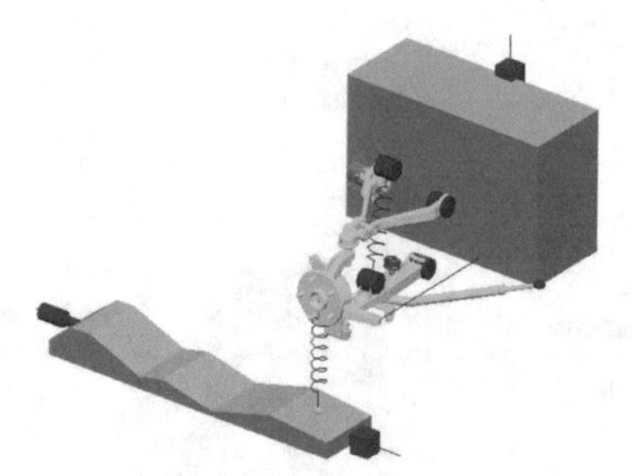

(a) Quarter vehicle model showing simple spring representation of tyre in contact with road

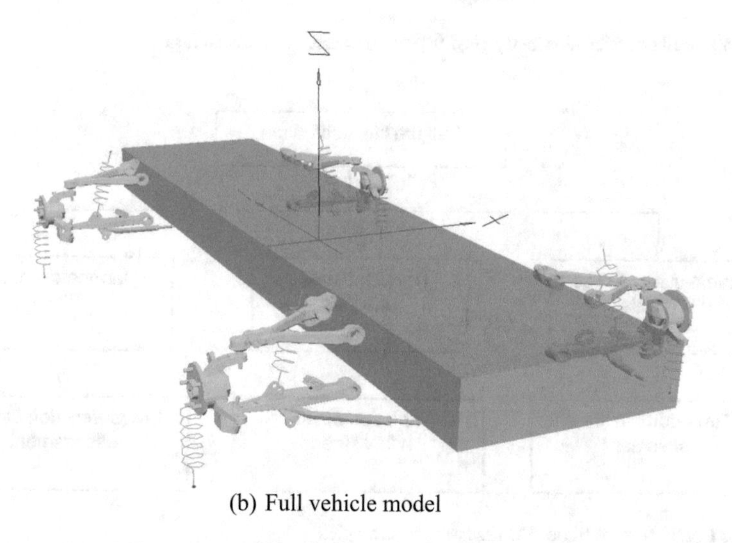

(b) Full vehicle model

Fig. 4.29 Multi-body dynamics (MBD) models of multi-purpose vehicle

interfaces to other components of the suspension. In fact, a total of 19 external load histories can be identified to act in the 3 co-ordinate directions at these interfaces. In addition, there are 9 internal load histories due to the inertia of the arm and its accelerations in 3D space to be considered. Thus the complete loading on the arm is extremely complex and difficult to quantify.

In order to estimate these loads using a VPG approach, MBD models were created and driven over a virtual version of a particular pavé durability track as indicated in Fig. 4.29. Experimental results were also generated by driving an actual vehicle with accelerometers attached at critical locations over the physical pavé test track.

Although quarter vehicle models (QVM) were tried, Fig. 4.29a, it was found necessary to allow for body roll/pitch by moving to a full vehicle model (FVM),

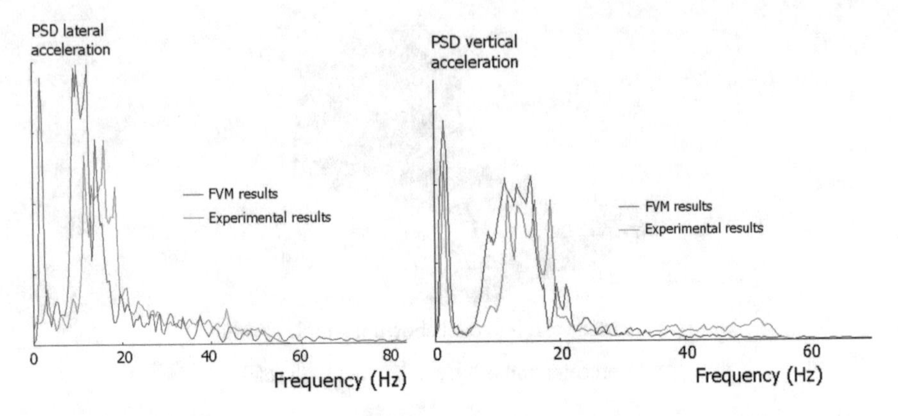

Fig. 4.30 Power spectral density (PSD) plots of wheel hub accelerations

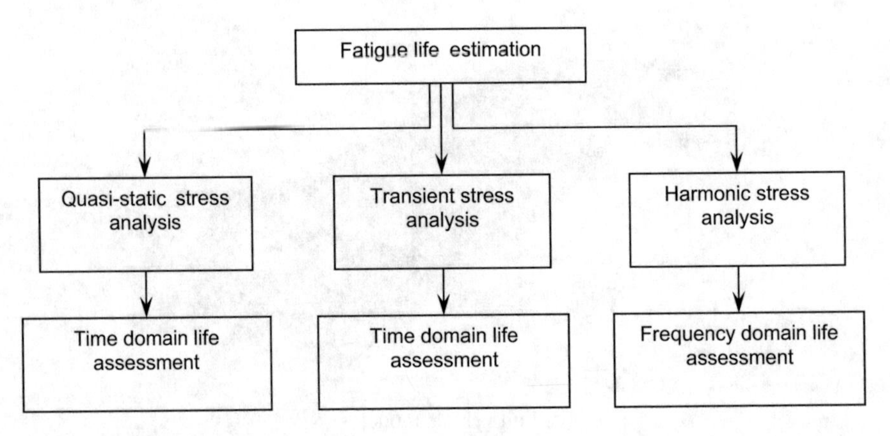

Fig. 4.31 Different fatigue life assessment strategies

Fig. 4.29b, in order to give accurate predictions of measured suspension component loading histories.

Comparisons between the accelerations at the front wheel hub predicted by the full vehicle model and the experimentally measured accelerations are shown in the Power Spectral Density (PSD) plots of Fig. 4.30. It can be seen that both the amplitude and frequency content of the experimental lateral and vertical accelerations of the hub are well characterised by the FVM results which gave confidence that this model could be used for the subsequent design optimisation studies.

Having established the external and internal inertial loading histories applied to the component of interest (the lower suspension arm), the next stage is to carry out a detailed 3D FE analysis of the arm to predict stresses for the subsequent fatigue assessment. Here the user is faced with a choice of at least three different analysis strategies as indicated in Fig. 4.31.

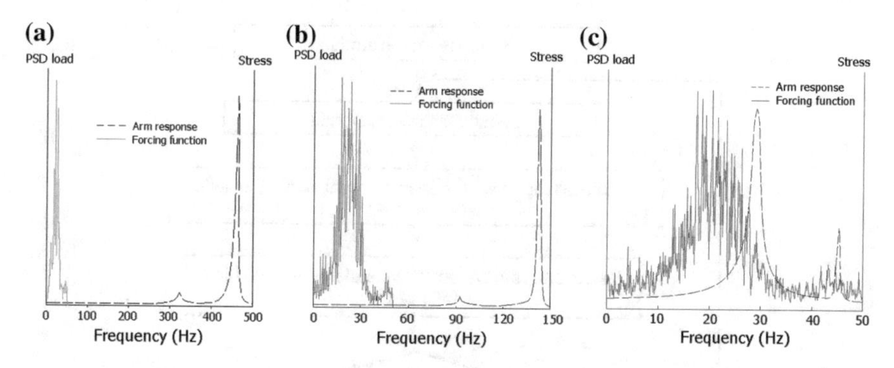

Fig. 4.32 PSD of forcing function compared with arm natural frequency response

The quasi-static analysis strategy requires a single linear static analysis of the arm for unit loading applied in each of the 19 external and 9 internal load directions identified. Using the principle of superposition, the detailed stress history distribution of the arm can then be obtained by factoring and summing the influence coefficients generated by each analysis according to the actual external load histories applied. Although this is the simplest and most economical of the available methods, it was found to be accurate only if the natural frequencies of the arm (which can be readily predicted from the FE model) are well separated from the frequency content of the forcing function input, as indicated in Fig. 4.32a.

If the frequencies are not separated by a factor of at least seven then the quasi-static method was found to be highly erroneous because it does not account for the interactions between the dynamic response of the component and the forcing frequencies, potentially leading to resonance effects.

If the component and loading frequencies are separated by a factor of less than seven as in Fig. 4.32b, then recourse must be made to the computationally accurate but time consuming transient stress analysis method which requires full integration and solution of the dynamic FE model at small time increments for the complete loading time history considered. Such a computationally intensive method is not suitable for the multiple simulations required for design optimisation based on fatigue life even with the power of modern computer systems.

Finally, if the forcing function and component natural frequencies overlap as in Fig. 4.33c, it may be possible to use harmonic stress analysis techniques followed by frequency domain life assessment to estimate the fatigue life distribution in the component. Here, the arm transfer functions which give the stresses per unit load as a function of frequency are firstly generated by applying either full or reduced modal superposition techniques. The transfer functions are then multiplied by the PSD's of the applied loads to give the equivalent PSD's of stress. Finally, the fatigue life can be estimated in the frequency domain using the well-established Dirlik approach.

When using the harmonic stress analysis method, there are a number of parameters that have to be specified such as the number of modes to include in the harmonic analysis, the buffer size used to perform the Fast Fourier Transform

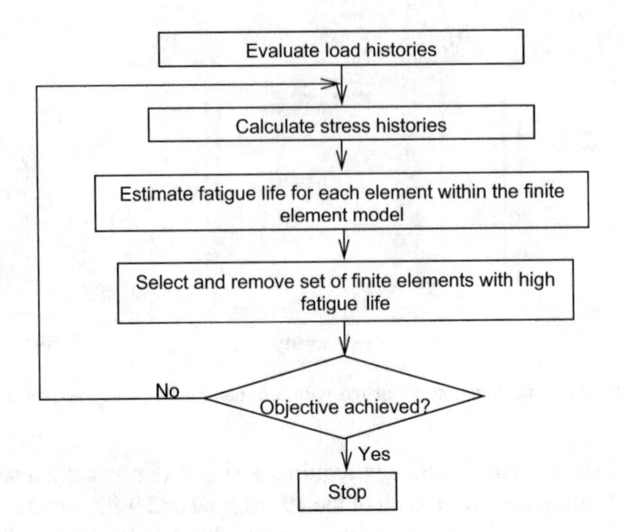

Fig. 4.33 Basic optimisation strategy based on fatigue life

(FFT) to generate the loading PSD's and the frequency step size used to derive the stress PSD's. Although the values of these parameters have been shown to affect the absolute fatigue lives predicted using this method, the distribution of fatigue life within the component is much less sensitive and is in good agreement with the predictions of the most accurate transient stress analysis method. Since optimisation of the design based on fatigue life depends on the distribution rather than absolute values of life, the harmonic stress analysis method can therefore be used for this purpose for cases where the forcing and natural frequencies coincide as in Fig. 4.32c. This is far more computationally efficient than using the full transient analysis method.

A basic optimisation strategy based on fatigue life is shown in Fig. 4.33. An immediate problem arises when attempting to implement such a strategy using standard S-N fatigue data in that a fatigue cut-off limit at around 10^8 cycles is usually specified due to the lack of data for higher numbers of cycles. This means that the majority of material in most automotive components will be predicted to have infinite fatigue life. Since the optimisation algorithm seeks to remove material with maximum life, a distribution which varies throughout the component is really required so that the algorithm can start to remove material at the location with the least damage. A practical solution to this problem is to artificially extend the fatigue life cut-off point much further along the life axis, as indicated in Fig. 4.34. The fact that the fatigue curve beyond the standard cut-off limit may not be completely accurate is irrelevant in terms of the optimisation strategy because material with such very long predicted life will be rapidly removed from the model.

To demonstrate the potential of this optimisation strategy to not only modify existing designs but to generate radical new solutions, the maximum initial domain of the lower suspension arm in question was established as a simple rectangular

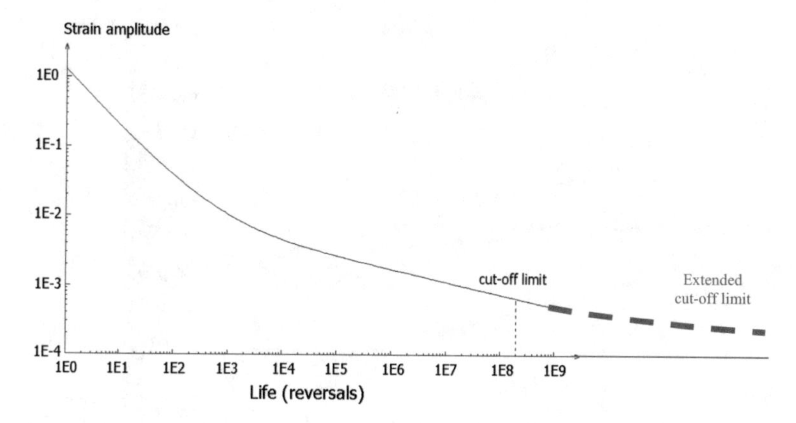

Fig. 4.34 S-N curve with artificially extended fatigue life cut-off limit

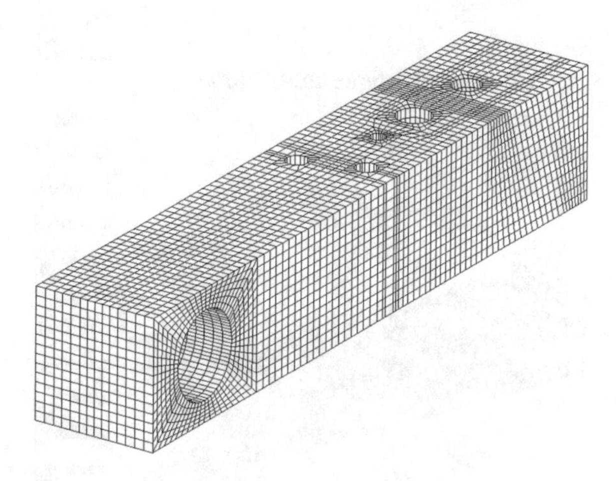

Fig. 4.35 Finite element model of initial domain for optimising lower arm

block, Fig. 4.35. Since it is necessary to have certain features (principally mounting holes) at certain locations in the arm, these features were designated as non-design domains which could not be altered by the optimisation routine.

Figure 4.36a shows that the initial fatigue assessment of the design domain before modifying the fatigue cut-off limit predicts infinite life for most of the material. In contrast, Fig. 4.36b shows the corresponding life contours after the limit was raised to 10^{20} cycles.

Starting from the much more informative life distribution shown in Fig. 4.36b, the optimisation routine was run with the objective of minimising the weight of the arm subject to the constraint that at least 10,000 driving cycles over the pavé

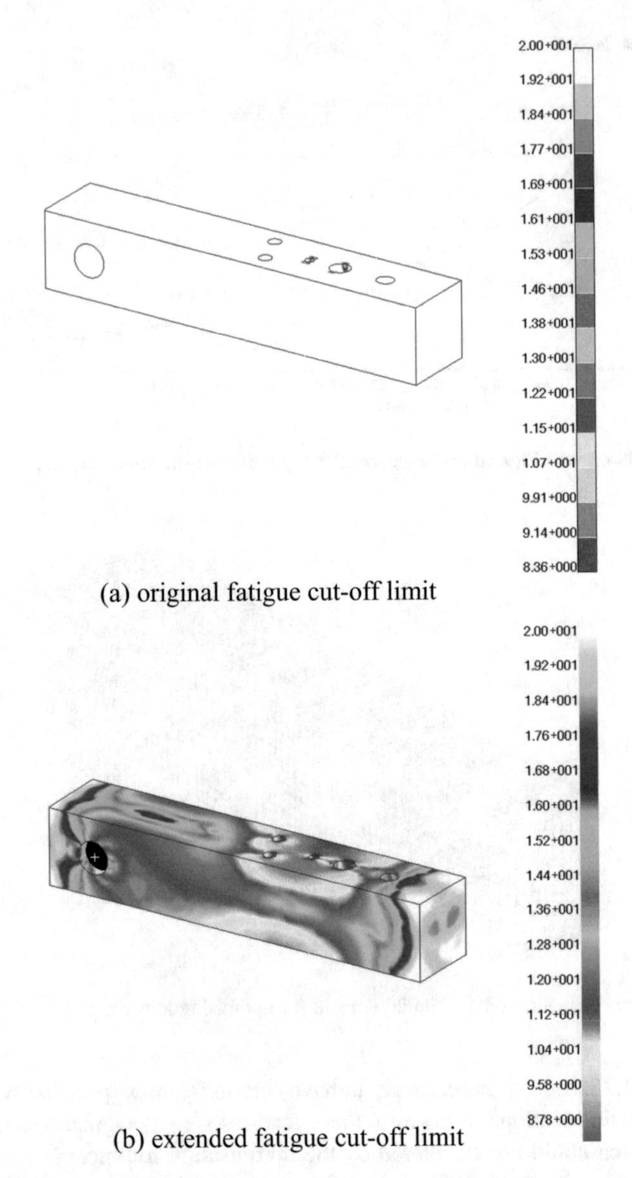

Fig. 4.36 Contour plots of fatigue life distribution in suspension arm

durability road should be possible without fatigue failure. Figure 4.37 shows the history of material removal recorded during this optimisation process.

After about 80 loops of the algorithm shown in Fig. 4.33, the volume of the arm was stabilised at less than 20% of its initial value with the resulting shape and

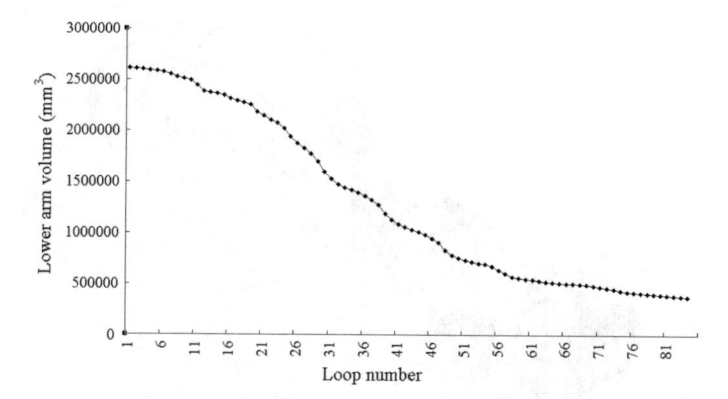

Fig. 4.37 History of material removal during optimisation of arm

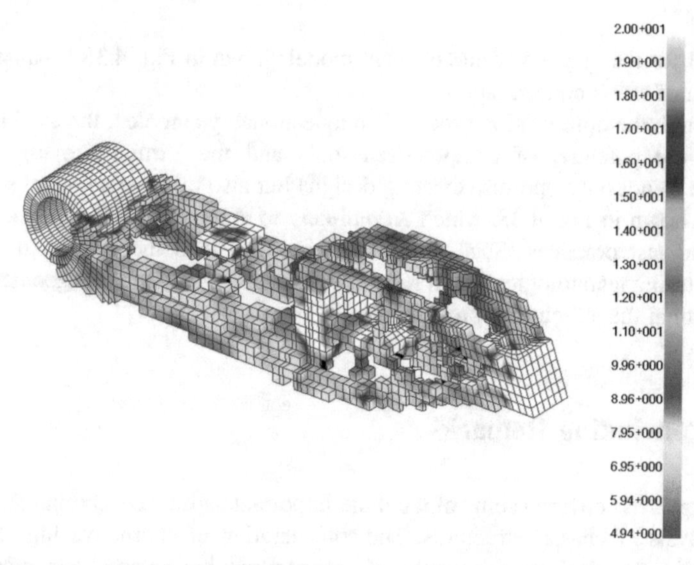

Fig. 4.38 Contour plot of fatigue life distribution for optimised arm

fatigue distribution as shown in Fig. 4.38. The optimised arm has a fatigue life of 36,000 cycles of the pavé durability track compared to only 173 cycles for the existing arm design shown in Fig. 4.27. Simultaneously the weight of the optimised arm has been reduced by 0.5 kg compared with the original design.

Of course, the jagged surfaces created by the particular finite element mesh of the optimised arm shown in Fig. 4.38 are neither desirable nor practical to manufacture. Therefore the final stage of the process was to smooth these surfaces to create the final optimised geometry shown in Fig. 4.39. Although somewhat more complicated to manufacture, this optimised component uses significantly less material than the original design shown in Fig. 4.28 and would have a fatigue life even greater than

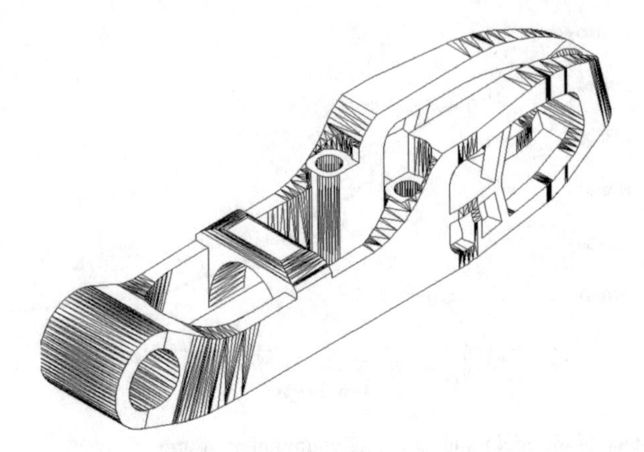

Fig. 4.39 Final optimised design after surface smoothing

predicted for the "jagged" finite element model shown in Fig. 4.38 because of the reduction of stress concentrations.

Although the optimisation process is unquestionably complex, the exercise does indicate the potential of computerised tools and the Virtual Proving Ground approach to not only optimise existing designs but also to generate novel solutions such as shown in Fig. 4.38, which are unlikely to emerge from traditional design, make and test practices. Such novel solutions also lend themselves to additive manufacturing technologies which have fewer constraints on the part geometry than conventional manufacturing processes.

4.6 Concluding Remarks

This chapter has covered some of the more important aspects concerning the design and analysis of chassis structures. The consideration of alternative high strength lightweight materials has been included because they have started to revolutionise the ways that vehicle body structures are designed and fabricated. Although advanced computer analysis methods have become the norm in the industry for both static and dynamic assessment of structures (including their durability and crash safety), it is important that the engineers understand the basic principles and can, from time to time, apply conventional methods of analysis as a reality check on computerised solutions.

Chapter 5
Noise, Vibration and Harshness (NVH)

Abstract This chapter aims to provide chassis engineers with knowledge of the background theory and techniques so that they can make informed judgements on NVH solution strategies at an early stage of vehicle development. The chapter commences with a review of the fundamentals of acoustic theory as this topic is often not covered in detail in Bachelor level engineering programmes. The human response to sound is then outlined followed by a description of general noise measurement and control techniques. The main sources of noise in road vehicles are then reviewed and common assessment and mitigation techniques are outlined for each type of noise. The next section introduces the sources and nature of automotive mechanical vibration as distinct from air-borne noise. There is a focus on vibration arising from the internal combustion engine since this remains the predominant motive power source for the majority of road vehicles. Principles of vibration control are then described with a focus on vibration absorbers and the isolation of engine-induced vibration. The final section of the chapter deals with the particular problems of brake noise and vibration, the latter affecting not only the brake itself but also the entire chassis due to transmission of the vibration through the suspension and steering systems.

5.1 Introduction to NVH

Noise, vibration and harshness (NVH) have become increasingly important factors in vehicle design as a result of the quest for increased refinement. Vibration has always been an important issue closely related to reliability and quality, while noise is of increasing importance to vehicle users and environmentalists. Harshness, which is related to the quality and transient nature of vibration and noise, is also strongly linked to vehicle refinement.

Controlling vibration and noise in vehicles poses a severe challenge to the designer because unlike many machine systems, road vehicles have a number of sources of vibration and noise that are inter-related and speed dependent. In recent years, the trend towards lighter vehicle constructions and higher engine speeds has

© Springer International Publishing AG 2018

D. C. Barton and J. D. Fieldhouse, *Automotive Chassis Engineering*,
https://doi.org/10.1007/978-3-319-72437-9_5

exacerbated NVH problems in conventional internal combustion engine (ICE) vehicles. Moreover the advent of electric and hybrid electric vehicles has tended to increase the potential for noise and vibration from non-engine sources, posing many new problems for automotive engineers. These developments have also coincided with a reduction in the time to market for new vehicles which has created an increased reliance on computer-aided design and analysis with less time spent on prototype testing.

This accelerated development of new and highly refined vehicles is dependent on accurate dynamic analysis of vehicles and their subsystems and calls for refined mathematical modelling and analytical techniques. While NVH analysis has in recent years been aided by developments in finite element and multi-body systems analysis software, there is still an underlying need to apply basic vibration and noise principles in vehicle design.

It is the objective of this chapter to address some important noise and vibration issues arising in vehicle design. It is assumed that the reader has some previous knowledge of noise and vibration theory but not necessarily of acoustics; hence the first part of the chapter presents a brief review of the fundamentals of acoustic theory.

5.2 Fundamentals of Acoustics

An understanding of acoustic fundamentals is essential in controlling noise and interpreting noise criteria. This section outlines some of the basic principles of sound propagation.

5.2.1 *General Sound Propagation*

Sound is transmitted from a source to a receiver by an elastic medium called the *path*. In an automotive context, this is the surrounding air or the vehicle body structure, the latter giving rise to the term *structure-borne sound*.

The simplest form of sound propagation occurs when a small sphere pulsates harmonically in free space (away from any bounding surfaces). The vibrating surface of the sphere causes the air molecules in contact with it to vibrate and this vibration is transmitted radially outwards to adjoining air molecules. This produces a propagating (or travelling) wave having a characteristic velocity c, i.e. the velocity of sound in air. At some arbitrary point on the path, the air undergoes pressure fluctuations that are superimposed on the ambient pressure. A sound source vibrating at a frequency f, produces sound at this frequency. Taking a snapshot of the instantaneous pressure and traversing away from the source, the variation of pressure with distance is also sinusoidal. The distance between pressure peaks is constant and known as the *wavelength* λ. This is related to c and f by the equation:

$$\lambda = \frac{c}{f} \tag{5.1}$$

This equation shows that as f increases, λ decreases (since c is constant for given atmospheric conditions). In the audible range from 20 Hz to 20 kHz, the wavelength correspondingly varies from 17 m to 17 mm.

5.2.2 *Plane Wave Propagation*

The fundamentals of wave motion are most easily understood by considering the propagation of a plane wave (having a flat wavefront perpendicular to the direction of propagation). Denoting the elastic deformation as ξ at some distance x from a fixed datum and combining the continuity and momentum equations for the element with the universal gas law leads to the one dimensional wave equation:

$$\frac{\partial^2 \xi}{\partial t^2} = c^2 \frac{\partial^2 \xi}{\partial x^2} \tag{5.2}$$

The propagation velocity c, is given by:

$$c = \sqrt{\frac{p\gamma}{\rho}} = \sqrt{\gamma RT} \tag{5.3}$$

where p = the ambient pressure, ρ = corresponding density of the medium, γ = ratio of specific heats for air, R = universal gas constant and T = absolute temperature. For air at standard atmospheric pressure and 20 °C, the magnitude of c is 343 m/s.

The general solution to Eq. (5.2) for harmonic waves is:

$$\xi = Ae^{(\omega t - kx)} + Be^{(\omega t + kx)} \tag{5.4}$$

The first term on the right hand side represents the incident wave (travelling away from the source) while the second term represents the reflected wave (travelling in the opposite direction). The *wave number* is given by $k = \frac{\omega}{c}$ or $k = \frac{2\pi f}{c} = \frac{2\pi}{\lambda}$ and hence is defined as the number of acoustic wavelengths in 2π. The values of c vary considerably for fluid and solid materials. Some typical values are shown in Table 5.1. For the solid materials, the speeds shown are for one dimensional stress waves rather than acoustic waves.

Table 5.1 Velocity of wave propagation in various media

Medium	c (m/s)
Air at 1 bar and 20 °C	343
Mild steel	5050
Aluminium	5000
Vulcanised rubber	1269
Water at 15 °C	1440

5.2.3 *Acoustic Impedance, z*

The impedance to the propagation of an acoustic wave is called the *acoustic impedance*. It is defined as the ratio of the acoustic pressure p to the particle velocity u. It can be shown that:

$$z = \frac{p}{u} = \rho c \tag{5.5}$$

For standard air pressure and temperature (101.3 kPa and 20 °C), z is equal to 415 rayls (Ns/m^3).

5.2.4 *Acoustic Intensity, I*

This is defined as the time averaged rate of transport of acoustic energy by a wave per unit area normal to the wavefront. It is given by:

$$I = \frac{p_{rms}^2}{\rho c} \tag{5.6}$$

p_{rms} is the r.m.s. pressure fluctuation. For a harmonic wave, $p_{rms} = \frac{\hat{p}}{\sqrt{2}}$ where $\hat{p}$ is the peak pressure.

Therefore

$$I = \frac{\hat{p}^2}{2\rho c} \tag{5.7}$$

5.2.5 *Spherical Wave Propagation—Acoustic Near- and Far-Fields*

Spherical waves more closely ressemble true source waves, but approximate to plane waves at large distances from a source. It may be shown that the wave equation in spherical coordinates is:

$$\frac{\partial^2(\mathrm{rp})}{\partial \mathrm{t}^2} = \mathrm{c}^2 \frac{\partial^2(\mathrm{rp})}{\partial \mathrm{r}^2} \tag{5.8}$$

where r is the radial distance from the source.

The general solution for an incident wave only (no reflection) is:

$$p = \frac{1}{r} A e^{(\omega t - kr)} \tag{5.9}$$

When Eq. (5.9) is used in conjunction with the definition for acoustic impedance, it may be shown that:

$$\mathrm{z} = \rho\mathrm{c} \frac{(\mathrm{kr})^2}{1 + (\mathrm{kr})^2} + \mathrm{i}(\rho\mathrm{c}) \frac{\mathrm{kr}}{1 + (\mathrm{kr})^2} \tag{5.10}$$

Note $\mathrm{i} = \sqrt{-1}$ and hence z is a complex quantity.

At large distances from the source ($\mathrm{kr} \gg 1$ or $\mathrm{r} \gg \lambda/2\pi$), $\mathrm{z} \to \rho\mathrm{c}$. Here the pressure and particle velocity are in phase and this corresponds to a region called the *acoustic far field* where spherical wavefronts approximate to those of plane waves.

At distances close to the source ($\mathrm{kr} \ll 1$ or $\mathrm{r} \ll \lambda/2\pi$), $\mathrm{z} \to \mathrm{i}\rho\mathrm{ckr}$. Here pressure and velocity are 90° out of phase. This region is called the *acoustic near field.*

The transition from near to far field is in reality a gradual one but the demarcation can be taken to be at kr = 10. Since k = 2 π/λ, the demarcation is given approximately by: r ≈ 1.6 λ. For a harmonic wave in air at 1 kHz (λ ≈ 0.3 m), r = 500 mm; while at 20 Hz (λ ≈ 17 m), r = 27.5 m. The far field/near field transition has important implications for microphone positioning in sound level measurements.

5.2.6 *Reference Quantities*

Certain reference quantities are used for sound emission measurements. For sound transmission in air, the reference r.m.s. pressure is taken to be $\mathrm{p}_{\mathrm{ref}}$ = 20 μPa which corresponds approximately to the threshold of hearing at the reference frequency of 1 kHz. With a reference impedance $z_{ref} = (\rho\ c)_{ref} = 400$ rayls, the reference intensity $I_{ref} = 10^{-12}$ W/m^2 from Eq. (5.6). Since the acoustic power is the intensity times the area, for a reference spherical area $A_{ref} = 1$ m^2, the reference sound power is 10^{-12} W.

5.2.7 Acoustic Quantities Expressed in Decibel Form

The human ear is capable of detecting acoustic quantities over a very wide range, e.g. pressure variations from 20 μPa to 100 Pa. Hence there is a need to represent acoustic data in a convenient form. This is achieved by using the decibel scale. The quantity of interest x is expressed in the form 10 $\log_{10}(x/x_{ref})$, where both x and x_{ref} have units of power. Using the above reference quantities the sound power level (L_W), the sound intensity level (L_I) and the sound pressure level (L_p) in decibels (dB) are as follows:

$$L_W = 10\log_{10}\left(\frac{W}{W_{ref}}\right) \tag{5.11}$$

$$L_I = 10\log_{10}\left(\frac{I}{I_{ref}}\right) \tag{5.12}$$

$$L_p = 10\log_{10}\left(\frac{p}{p_{ref}}\right)^2 = 20\log\left(\frac{p}{p_{ref}}\right) \tag{5.13}$$

When it is noted that the threshold of hearing corresponds to a sound pressure level of 0 dB, it may be shown that for standard temperature and pressure (101.3 kPa and 20 °C), L_W, L_I and Lp in dB are related as follows:

$$L_I = L_p - 0.16 \tag{5.14}$$

$$L_W = L_p + 20\log_{10}\left(\frac{r}{r_{ref}}\right) - 0.16 \tag{5.15}$$

where $r_{ref} = 0.282$ m for the reference spherical area $A_{ref} = 1$ m^2.

From Eq. (5.14), it is seen that L_p and L_I are approximately equal numerically, while Eq. (5.15) is useful for determining the sound power level of an acoustic source from sound pressure level measurements in free field conditions. For these whole field radiation conditions, the reduction in sound pressure level with every doubling of radial distance from the source is 6 dB.

5.2.8 Combined Effects of Sound Sources

It is often necessary to determine the sound pressure level of two or more uncorrelated sound sources when the level for each source is known. This can be achieved by using the equation.

$$L_{p,total} \approx L_{I,total} = 10\log_{10}\left(\sum 10^{0.1\,Lp_i}\right) \tag{5.16}$$

5.2.9 Effects of Reflecting Surfaces on Sound Propagation

When an incident wave strikes a reflecting surface, the wave is reflected back towards the source. In the vicinity of the reflecting surface, the incident and reflected waves interact to produce what is known as a *reverberant field*. The depth of this field is dependent on the absorptive properties of the reflecting surface. A typical interaction between incident and reflected waves is shown in Fig. 5.1 and sound pressure level variations as function of distance "r" from the sound source are as shown in Fig. 5.2.

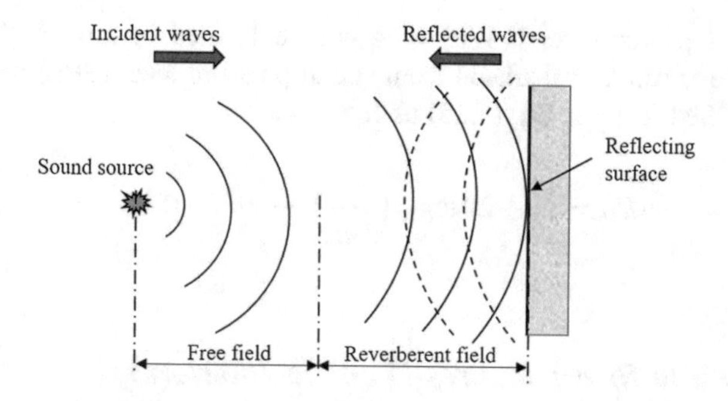

Fig. 5.1 Interaction of incident and reflected waves

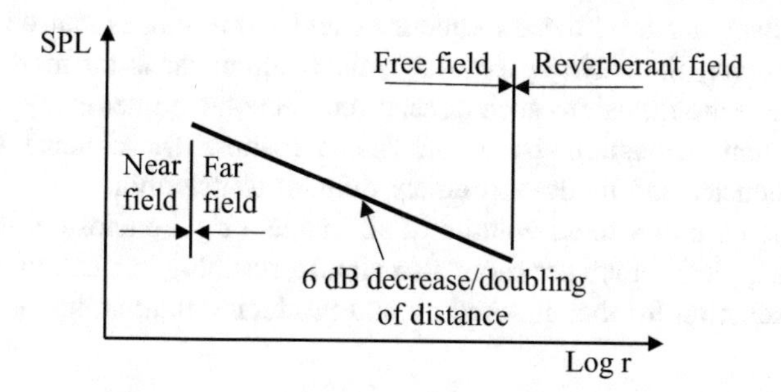

Fig. 5.2 Sound pressure level as a function of distance from a simple spherical source

Various practical situations arise when the sound source is positioned close to a hard reflecting surface. The following are idealised situations:

- Whole-space radiation—when there are no reflecting surfaces, i.e. the source is in free space.
- Half-space radiation—when the source is at the centre of a flat hard (reflecting) surface, e.g. the sound source is on a floor.
- Quarter space radiation—when the source is at the intersection of two flat hard surfaces which are perpendicular to one another, e.g. the sound source is on a floor next to a wall.
- Eighth space radiation—when the source is positioned at the intersection of three flat perpendicular hard surfaces, e.g. the sound source is in a corner.

Increasing the number and proximity of reflecting surfaces to a source increases the acoustic intensity. The situations described above can be represented by the *directivity index* DI expressed in terms of the *directivity factor* Q:

$$DI = 10\log_{10} Q \tag{5.17}$$

- For a whole space Q = 1 (DI = 0 dB)
- For a half space Q = 2 (DI = 3 dB)
- For a quarter space Q = 4 (DI = 6 dB)
- For an eighth space Q = 8 (DI = 9 dB).

The sound power level (PWL) of a source located in any of the positions described above can be calculated from sound pressure level (SPL) measurements using a modified form of Eq. (5.15) as follows:

$$L_W = L_p + 20\log_{10}\left(\frac{r}{r_{ref}}\right) - DI - 0.16 \tag{5.18}$$

5.2.10 Sound in Enclosures (Vehicle Interiors)

The air inside the enclosed volume of a vehicle cabin behaves as an elastic fluid and consequently has a number of natural frequencies and modes (standing waves). For a simple rectangular box, these frequencies and mode shapes can be readily calculated. However, the cabin space of a typical saloon car is far more complex in shape e.g. there are intrusions such as seats and there are numerous types of surface having different (acoustic) absorption characteristics. These make the acoustic natural frequencies and modes extremely difficult to determine.

Unlike the case of a fixed volume of air inside a rigid enclosure, the bounding surfaces of a typical saloon car cabin also vibrate, resulting in a complex distributed source of excitation for the enclosed air and producing fluid-structure interactions.

Nevertheless, it is possible for standing waves to be excited in the cabin under certain conditions. The result is a booming effect. Ideally the driver and passengers should be located at the stationary nodes (points of low SPL) of the standing waves. In practice, boom effects tend to be at low frequencies and can be limited by careful body design and control of structure/airborne noise transmission.

In general the sound field within a cabin space will be diffuse. The central region of the cabin space will tend to be reverberant while regions close to the vibrating surfaces which generate acoustic pressure waves will tend to be in the direct sound field.

In practice there are a number of sound sources emitting noise into vehicle interiors and these can produce discrete components which are superimposed on a lower level of broadband noise. This situation is discussed in more detail in Sect. 5.7.3.

5.3 Subjective Response to Sound

Just as the human body responds differently to vibration excitation at different frequencies, so does the human ear respond differently to sound at different frequencies. Whereas human response to vibration is assessed by weighting the acceleration applied to the body as described in Chap. 3, it is shown below that human response to noise can be assessed by appropriately weighting the SPL. This weighting is derived from an understanding of human response to sound.

5.3.1 The Hearing Mechanism and Human Response Characteristics

The human ear is a delicate and sophisticated device for detecting and amplifying sound. It consists of an outer ear, a middle ear containing an amplifying device (the ossicles) and an inner ear containing the cochlea. This small snail shape element contains lymph nodes and a coiled membrane to which are connected thousands of very sensitive hair endings of varying thickness. These respond to different frequencies, converting the sound stimulus into nerve impulses that are transmitted to the brain. A certain threshold level is required to stimulate the nerve cells, while over-stimulation can lead to temporary or permanent deafness. This latter effect was recognised in the last century as a cause of industrial deafness resulting in a number of regulations to protect workers.

The audible range for a healthy young person lies within the envelope shown in Fig. 5.3. The frequency range extends from 20 Hz to 20 kHz and the SPL extends from the threshold of hearing at the lower boundary to the threshold of feeling (pain) at the upper boundary. The SPL at the upper and lower boundaries vary markedly with frequency. At 1 kHz, the SPL ranges from 0 dB at the lower boundary to 130 dB at the upper boundary. The shapes of the curves for sounds of

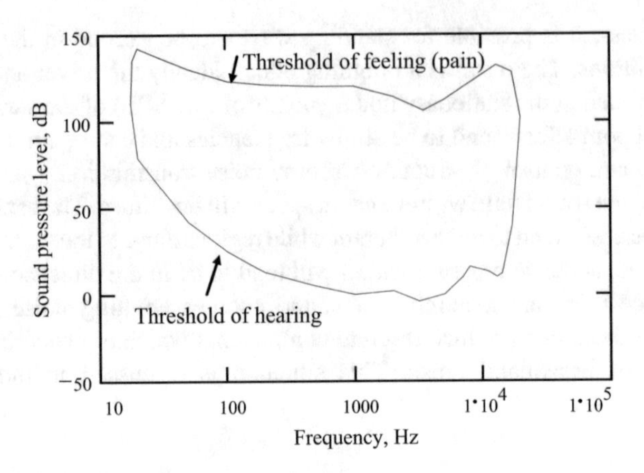

Fig. 5.3 The audible range of human hearing

increasing loudness are generally similar to that for the threshold of hearing. It follows therefore that the human ear is most sensitive between 500 Hz and 5 kHz and is relatively insensitive to sounds below 100 Hz. This characteristic shape dictates the A-weighting characteristic that is used to assess human response to noise (see Fig. 5.4).

5.4 Sound Measurement

Automotive noise measurement is required for a variety of purposes dictating the need for a range of measuring equipment. In development work there is a requirement for measuring continuous noise levels such as that from drive-trains and their ancillaries; there are also requirements for component noise testing to determine sound power output, frequency analysis and source identification. For type approval there are requirements for assessing whole vehicle noise. Controlled test environments are also required to ensure that tests are repeatable and not weather dependent. This requirement is met by special test facilities, such as anechoic chambers, that simulate free-field environments.

5.4.1 Instrumentation for Sound Measurement

5.4.1.1 Sound Pressure Level (SPL) Meter

This is the most basic instrument for sound measurement. It comprises a microphone, r.m.s. SPL calculator with fast and slow time constants, and an A-weighting

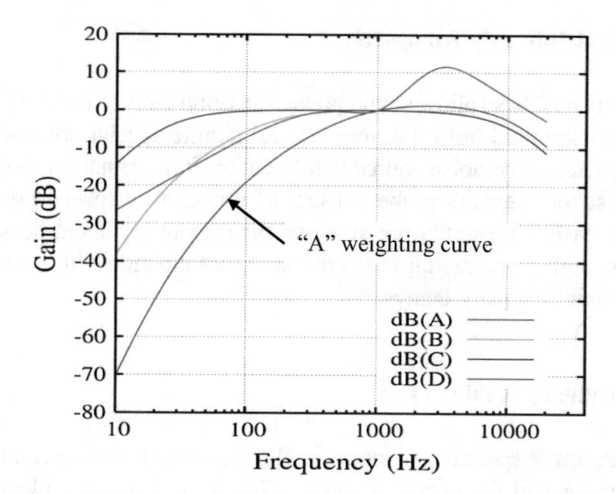

Fig. 5.4 The A-weighting curve

algorithm to relate the measurements made to the human audible response to give the so-called A-weighted noise level L_{pA} expressed in dB(A). Because of the frequency sensitivity of the human ear, the A-weighting network has the form shown in Fig. 5.4. This emphasises the frequencies in the 500 Hz to 5 kHz range and produces increasing levels of attenuation below 100 Hz.

British and other standards specify four quality grades of SPL meter ranging from Type 0 laboratory reference meters to Type 3 industrial grade meters. For development work, Type 1 instruments are generally recommended. If the instrument is required for measurement of transient noise, it should also be equipped with a peak record facility.

Because sound levels are rarely constant (e.g. noise resulting from changes in engine speed), there is a need to average levels over prescribed intervals of time. This type of measurement leads to an equivalent noise level such as $L_{Aeq, T}$, which is related to noise deafness and annoyance criteria. Here the A-weighted noise level is averaged over a measurement period T to give a level having the same energy content as a constant sound of the same numerical SPL. Mathematically this can be written:

$$L_{Aeq,T} = 10\log_{10}\left(\frac{1}{T}\int_{0}^{T}\left(\frac{p_A(t)}{p_{ref}}\right)^2 \mathrm{dt}\right) \tag{5.19}$$

An integrating type of sound level meter is required for this type of measurement.

Another requirement of SPL meters is to determine the level that has been exceeded for a prescribed portion of the measurement time, L_N, e.g. L_{A90} represents the A-weighted level exceeded for 90% of the measurement period and is used in environmental background noise measurements associated with traffic noise.

5.4.1.2 Sound Intensity Analysers

Sound intensity analysers allow sound power measurements to be made *in situ* in the presence of background noise i.e. they do not require special noise testing installations. They also allow noise source identification from sound intensity mapping.

A typical sound intensity probe consists of two closely spaced pressure microphones that measure the sound pressure and the pressure gradient between the two microphones. Signal processing converts these measurements into sound intensity values in a sound intensity analyser.

5.4.1.3 Frequency Analysers

Since the frequency spectrum of noise is closely related to the origins of its production, experimental frequency analysis is a powerful tool for identifying noise sources and assessing the effectiveness of noise control measures.

The simplest frequency analysers split the frequency range into a set of octave bands having the following standardised centre frequencies: 31.5, 63, 125, 250, 500, 1000, 2000, 4000, 8000 and 16,000 Hz. These filters have a constant percentage bandwidth implying that the bandwidth increases with centre frequency giving increasingly poor discrimination at high frequencies. This can be improved with the use of third octave analysis. A number of noise and environmental standards require the use of octave and third octave analysis and much of the performance data for noise control products is expressed in terms of octave band centre frequencies.

For serious noise control investigations, narrow band frequency analysers are a necessity. Instead of switching sequentially through a set of filters, the signal in a narrow band analyser is recorded simultaneously to all the filters in the analysis range.

5.5 General Noise Control Techniques

Sound is produced by air molecules being excited either by a vibrating surface or by changes in pressure in fluid systems. Noise control techniques are based on one or more of the following principles:

- Control the noise at source—this involves a deep understanding of the noise generation process
- Modify the acoustic environment—this relates to sound in enclosures and aims to make the environment less reverberent
- Use sound barriers and enclosures—the aim is to limit the transmission of airborne sound
- Use vibration damping treatments—the aim is to attenuate the amplitude of vibrating surface and hence reduce the level of the noise generated.

5.5.1 Sound Energy Absorption

Absorption is one of the most important factors affecting the acoustic environment in enclosures. Increasing the average absorption of internal surfaces is a relatively inexpensive way of reducing sound levels in enclosures and is effective in vehicle interiors.

Absorption coefficient α, is defined as the ratio of sound energy absorbed by a surface to the sound energy incident on it. Its value is dependent on the angle of incidence and since in an enclosure all angles of incidence are possible, the values of α are averaged for a wide range of angles. Absorption is also dependent on frequency and published data normally quotes values at the standard octave-band centre frequencies.

For an enclosure having a number of different internal surface materials, the average absorption coefficient α can be determined (for n surfaces) from:

$$\alpha = \frac{\sum_{i=1}^{n} S_i \alpha_i}{\sum_{i=1}^{n} S_i} \tag{5.20}$$

where S_i = surface area, α_i = absorption coefficient for the *i*th surface.

In general, materials having high levels of absorption are porous so that movement of air molecules is restricted by the flow resistance of the pores. Absorption coefficients for a typical range of materials can be found in the literature or on company web-sites.

5.5.2 Sound Transmission Through Barriers

One of the principal noise transmission paths in a vehicle is through the bulkhead separating the cabin space from the engine compartment. The bulkhead can be considered to be a sound barrier. The effectiveness of sound barriers is normally quoted in terms of *transmission loss* TL. This is the ratio of the incident sound energy to that of the transmitted sound energy, expressed in dB. For a thin homogeneous barrier and random angle of incidence (in the range from 0° to 72°), it can be shown that the field incidence transmission loss in dB is given by:

$$TL = 20 \log_{10}(f\,m) - 47 \tag{5.21}$$

where f is the frequency of the sound in Hz and m is the mass per unit area of the barrier in kg/m^2.

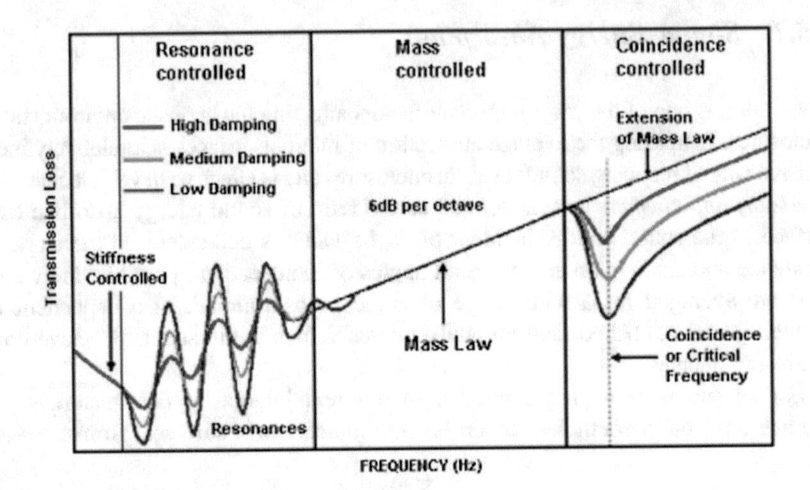

Fig. 5.5 Transmission loss as a function of frequency (http://personal.inet.fi/koti/juhladude/pics/theory/transmission_loss.gif)

Equation (5.21) applies to what is called the mass-controlled frequency region (Fig. 5.5) in which the transmission loss increases by 6 dB per octave increase in frequency. Alternatively doubling the barrier thickness or density increases transmission loss by 6 dB at a given frequency. It is evident from this that an effective means of increasing transmission loss in this frequency range is to use a high-density material for acoustic barriers.

Sound transmission through barriers is governed at low frequencies by panel bending stiffness and panel resonances, see Fig. 5.5. These tend to reduce the low frequency effectiveness by varying amounts dependent on the resonant frequency and amount of damping. Above twice the lowest natural frequency and below a critical frequency f_c, the barrier effectiveness can be considered to be mass-controlled. The critical frequency is related to the ability of the incident sound on a barrier to be transmitted as bending waves. It occurs when the wavelength of the incident wave coincides with the bending mode wavelength of the barrier λ_B. The lowest frequency at which this can occur is when the incident sound grazes the surface of the barrier at almost zero incidence angle in which case the critical or coincidence frequency is given by:

$$f_c = \frac{c}{\lambda_B} \tag{5.22}$$

In practice the range of incidence angles varies from zero to a little less than 90° which means that the decrease in transmission loss associated with coincidence occurs at a frequency somewhat higher than the value given by Eq. (5.22). The effectiveness of barriers is also sharply reduced by even the smallest apertures and can pose problems in noise isolation when electrical trunking and pipework is required to run between the engine and cabin compartments.

In practice layered materials consisting of a dense core and surface layers of absorptive material can perform the dual role of providing sound absorption with a high transmission loss.

5.5.3 *Damping Treatments*

Damping treatments (in the form of high damping polymers) can be used to limit resonant response amplitudes in structures and are particularly effective for flexural vibration of panels and beams. In an automotive context they are used extensively to limit the resonant responses of body panels and bulkheads.

Because of the poor structural strength of high damping polymers, it is necessary to either bond them to the surface of load-bearing elements or to incorporate them into load-bearing elements by sandwich construction. These forms of damping are termed *unconstrained-* and *constrained-layer damping* respectively, with the latter being by far the most effective way of deploying this type of structural damping treatment. Flexing of the load-bearing element produces shearing effects in the damping layer and thus vibrational energy is converted into heat and dissipated. Shear properties of polymer materials are generally temperature and frequency dependent. Furthermore their use in the form of constrained layer damping poses problems with bending and forming in manufacturing processes.

5.6 Automotive Noise—Sources and Control

The frequency composition of a sound is one of its most identifiable features. When the sound occurs at a single frequency it is called a *pure tone*. However, the great majority of sounds are far more complicated than this, having frequency components distributed across the audible range. Because there are number of sources of automotive noise, most of which are cyclic and at different frequencies, the result is what is called *broad band noise*. For internal combustion engine vehicles, the noise usually contains a number of dominant frequency components related to engine speed. The frequency characteristics of noise are represented by its frequency spectrum in a similar way to those for vibration sources.

5.6.1 *Internal Combustion Engine (ICE) Noise*

ICE noise originates from both the combustion process and the mechanical forces associated with the engine dynamics. The combustion process produces large pressure fluctuations in each cylinder and these produce high dynamic gas loads and other mechanical forces such as piston slap. These forces combine with the dynamic forces from inertia and unbalance effects (which are generally dependent on engine configuration and speed) forming the excitations applied to the engine structure. The resulting vibration produces noise radiation from the various surfaces of the engine.

Noise control at source therefore has to be concerned with controlling the extent of cylinder pressure variations (associated with combustion noise) and the choice of engine configuration (associated with mechanical effects). Both of these options tend to conflict with the modern trend for small high-speed fuel-efficient engines. In the case of diesel engines, there is evidence that combustion force reduction can be achieved by controlling the rate of pressure rise in the engine cylinders. This requires careful attention to the design of combustion bowls and the selection of turbocharger and fuel injection options. The mechanical noise associated with piston slap can be reduced by careful choice of gudgeon-pin offset and minimising piston mass.

By ranking the components of engine noise, it has been found that most noise is radiated from the larger more flexible surfaces such as sumps, timing case covers, crankshaft pulleys and induction manifolds. It therefore makes sense to isolate these components from the vibration that is generated in engine blocks. This can be achieved by using specially designed seals and isolating studs. Noise shields can also be effective in attenuating radiated noise from components such as timing case covers and the side-walls of engine blocks. The shields are generally made from laminated steel (see Sect. 5.5.3) or thermoset plastic materials designed to cover the radiating surface from which they are isolated by flexible spacers. The high internal damping of laminated steel can also be used to produce other noise inhibiting components such as cylinder head covers. Noise from crankshaft pulleys can be reduced by using spoked pulleys or fitting a torsional vibration damped pulley. All of these palliatives measures have cost implications.

5.6.2 Transmission Gear Noise

The level of gear noise rises with speed at a rate of 6–8 dB with a doubling of speed, while measurements have shown that gear noise increases at a rate of 2.5–4 dB for a doubling of the power transmitted.

In an ideal pair of gears running at constant speed, power will be transmitted smoothly without vibration and noise. In practice, however, tooth errors occur (both in profile and spacing) and in some cases shaft eccentricities exist. If a single tooth is damaged or incorrectly cut, a fundamental component of vibration is generated at shaft speed f_{ss}. If the shaft is misaligned or a gear or bearing is not concentric, vibration (and noise) is generated at tooth meshing frequency f_{tm} with sidebands f_{s1} and f_{s2} given by:

$$f_{s1}, f_{s2} = f_{tm} \pm f_{ss} \quad (5.23)$$

For a wheel having N teeth rotating at n rev/min, the tooth meshing frequency f_{tm} is given (in Hz) by:

$$f_{tm} = \frac{nN}{60} \quad (5.24)$$

Furthermore gear teeth are elastic and bend slightly under load. This results in the unloaded teeth on the **driving** gear being slightly ahead of their theoretical rigid-body positions and the unloaded teeth on the **driven** gear being slightly behind their theoretical positions. Thus when contact is made between the teeth on driving and driven wheels there is an abrupt transfer of load momentarily accelerating the driven gear and decelerating the driving gear. This leads to torsional vibration in the transmission and consequent noise generation at tooth meshing frequency. Considerable effort has been devoted to correcting standard tooth profiles to account for tooth elasticity effects but, because they are subjected to variable loading, it is impossible to correct for all loading conditions.

5.6.3 Intake and Exhaust Noise

Intake noise is generated by the periodic interruption of airflow through the inlet valves in an ICE, thus creating pressure pulsations in the inlet manifold. This noise is transmitted via the air cleaner and radiates from the intake duct. This form of noise is sensitive to increases in engine load and can result in noise level increases of 10–15 dB from no-load to full-load operation. When a turbocharger is fitted, noise from its compressor is also radiated from the intake duct. Turbocharger noise is characterised by a pure tone at blade passing frequency together with higher harmonics. Typical frequencies are from 2 to 4 kHz.

Exhaust noise is produced by the periodic and sudden release of gases as exhaust valves open and close. Its magnitude and characteristics vary considerably with engine types, valve configurations and valve timing. The fundamental frequency components are related to the engine firing frequency, which for a four-stroke engine is given (in Hz) by:

$$f = \frac{\text{engine speed (rpm)}}{60} \times \frac{\text{number of cylinders}}{2} \quad (5.25)$$

Levels of exhaust noise vary significantly with engine loading. From no-load to full-load operation, these variations are typically 15 dB. Turbo-charging not only reduces engine-radiated noise by smoothing combustion, but also reduces exhaust noise. However the turbo-charger itself might be a source of noise.

Attenuation of noise at engine intakes and exhausts calls for devices that minimise the pressure fluctuations in the gases while allowing them to flow relatively unimpeded. Such devices are effectively acoustic filters. The operational principles of intake and exhaust silencers ('mufflers' as they are called in the USA) can be divided into two types, dissipative and reactive. In practice, silencers are often a combination of both types.

Intake noise attenuation is generally incorporated into the air filter and is achieved by designing the filter to act as a reactive silencer based on the Helmholtz resonator principle. For an intake system (Fig. 5.6) comprising an intake venturi pipe of mean cross-sectional area A and length L together with a filter volume V, the resonance frequency is given by:

$$f = \frac{c}{2\pi}\sqrt{\frac{A}{LV}} \tag{5.26}$$

where c = the velocity of sound in air.

This type of design produces a low frequency resonance (a negative attenuation) but an increasing attenuation at higher frequencies. This can however be offset by high frequency resonances within the intake venturi.

Dissipative silencers contain absorptive material that physically absorbs acoustic energy from the gas flow. In construction, this type of silencer is a single chamber device through which passes a perforated pipe carrying the gas flow. The chamber surrounding the pipe is filled with sound absorbing material (normally long-fibre mineral wool) which produces attenuation across a very broad band of frequencies above approximately 500 Hz. The degree of attenuation is generally dependent on the thickness and grade of the absorbing material, the length of the silencer and its wall thickness. Figure 5.7 shows the cross-section through a typical dissipative exhaust silencer with a tuned venturi tube.

Reactive silencers operate on the principle that, when the sound in a pipe or duct encounters a discontinuity in the cross-section, some of the acoustic energy is reflected back towards the sound source thereby creating destructive interference. This is an effective means of attenuating low frequency noise over a limited range of frequencies. The effectiveness of the technique can be extended by having

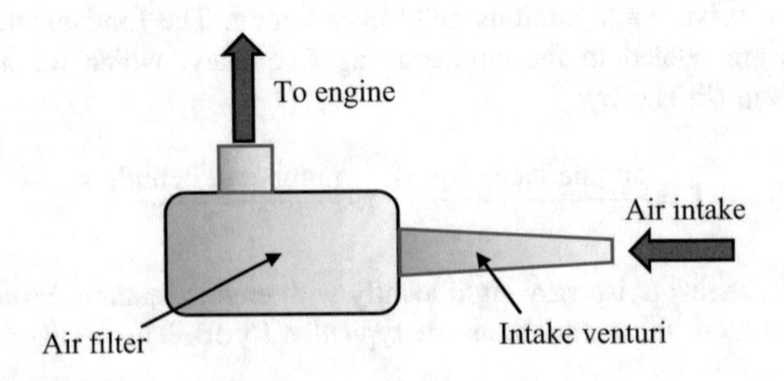

Fig. 5.6 Air intake with venturi tube

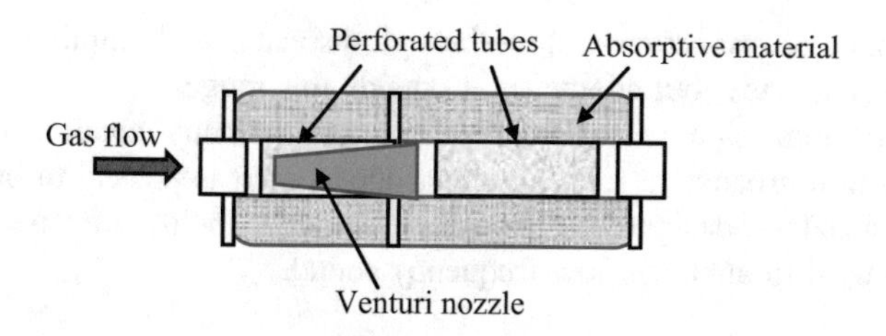

Fig. 5.7 Two chamber dissipative exhaust silencer with venturi

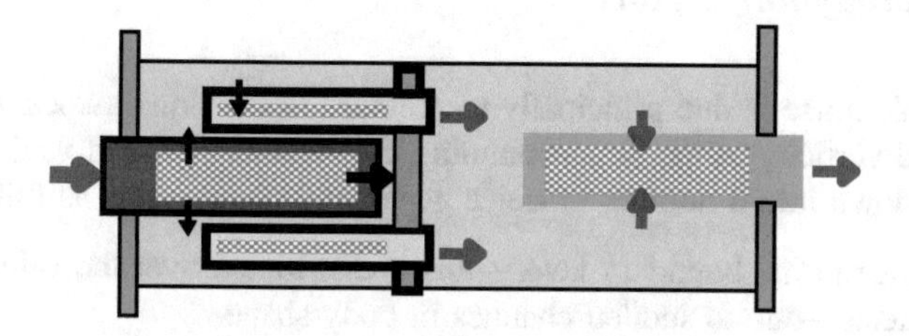

Fig. 5.8 Two-chamber reactive exhaust silencer

several expansion chambers within the same casing connected together by pipes of varying lengths and diameters (Fig. 5.8). Silencers of this type increase the exhaust back-pressure and result in some power loss from the engine.

Exhaust systems on modern vehicles are required to perform the dual task of reducing both the exhaust gas pollutants and exhaust noise. Catalytic converters are fitted immediately downstream from exhaust manifolds to ensure that they quickly achieve operating temperature and thus become quickly effective in urban driving. In addition to acting as exhaust gas scrubbers, catalytic converters also have an acoustic attenuation effect resulting from the gas flow through narrow ceramic pipes. This produces attenuation by both acoustic interference and dissipation effects.

Silencers are positioned in the exhaust system downstream from the catalytic converter. Together with their pipework, they form an acoustically resonant system that needs to be carefully tuned. Furthermore, there is a need to prevent the transmission of airborne noise from the shell of the exhaust to the body structure of the vehicle. For this reason it is common for silencers to have a double skin and insulating layer. This also provides thermal insulation.

Exhaust systems also need to be structurally isolated from vehicle bodywork to prevent the transmission of structure-borne sound and for this reason are suspended from the underbody of the vehicle by flexible suspension elements. There is also a risk that the noise emitted from the tailpipe can cause body resonances if the exhaust is not properly tuned.

The following are some of the devices used to overcome specific silencer tuning problems:

- The Helmholtz resonator—a flow-through resonator that amplifies sound at its resonant frequency, but attenuates it outside this range.
- Circumferential pipe perforations—these create many small sound sources resulting in a broadband filtering effect due to increased local turbulence.
- Venturi nozzles—designed to have flow velocities below the speed of sound, they are used to attenuate low frequency sound.

5.6.4 Aerodynamic Noise

Aerodynamic noise is due principally to pressure fluctuations associated with turbulence and vorticity of the air surrounding the vehicle. For road vehicles this can be broken down into a number of noise generating components as follows:

- Turbulence in the boundary layer—this is distributed over the vehicle body
- Edge effects—due to sudden changes in body shape
- Vortex shedding—occurs at various locations on the vehicle body
- Cooling fans.

5.6.4.1 Boundary Layer Noise

This tends to be random in character and is spread over a broad band of frequencies. Even though a whole vehicle is covered by a boundary layer, the noise that it produces is not normally troublesome. The higher order frequency components in the spectrum can easily by attenuated with absorbent materials inside body panels.

5.6.4.2 Edge Noise

This is produced by flow separating from sharp corners and edges on the body structure. As the flow separates from an edge, it rolls up into large vortices which also break up into smaller ones. It is the intermittent formation and collapse of these vortices that leads to the narrow band characteristics associated with edge separation. Noise levels associated with edge noise are generally higher than those for boundary layer noise and have a more defined band of frequencies. This band of frequencies is a function of vehicle speed to the extent that changes in the noise signature can be detected as vehicle speed changes.

It is possible to reduce edge noise by minimising protrusions from the body surface, making the body surface smooth and ensuring that gaps around apertures, such as doors, are well sealed. There is also a strong tendency for vortices to be produced at the A-pillars supporting the front windscreen. The vortices produced here extend rearwards along the sides of the vehicle enveloping the front sidelights.

These tend to have a low resistance to sound transmission and there is generally very little which can be done to improve this problem. Changing the profile of the A-pillars to a well-rounded contour will improving the aerodynamics, but may be unacceptable from a vision point of view. Edge noise also arises at protuberances such as door mirrors and at wheel trims. In these cases there is generally scope for improvements to the profiles without impairing their function.

5.6.4.3 Vortex Shedding

This occurs when airflow strikes a bluff-body producing a periodic stream of vortices downstream. This results in the production of pure tones (subjectively the most annoying of all sounds) at the vortex shedding frequency. The frequency f of the vortices are related to the relative air speed U and depth d of the bluff body by the equation:

$$f = \frac{SU}{d} \tag{5.27}$$

where S is the Strouhal number.

Typically S = 0.2 for a long thin rod. This means that for a vehicle fitted with a roof-rack having 10 mm diameter bars (fitted across the air flow) and travelling at 113 km/h (70 mph), the vortices are shed at a frequency of 640 Hz. This is in the frequency range where the human ear is most sensitive. A much quieter form of roof carrier is the enclosed pod-shaped design which tends to avoid the production of vortices.

5.6.4.4 Engine Cooling Fan

In this case the fan blades shed helical trailing vortices which result in periodic pressure fluctuations when they strike downstream obstacles. To overcome this problem, fan rotors are made with unevenly spaced blades and with odd numbers of blades. This spreads the noise over a band of frequencies—subjectively this is better than a pure tone. The thermostatically-controlled electrically-driven fans used in modern vehicles ensure that fan noise does not increase with engine speed, unlike the former belt-driven designs.

5.6.4.5 Noise from Internal Airflows

Internal airflow into and out of the passenger compartment is designed for ventilation and occupant comfort. The resulting noise is becoming more of a problem as overall cabin noise is reduced. It is essential for modern vehicles that inlet and outlet apertures are carefully sited and designed to ensure that they do not themselves generate noise and also that noise from outside the compartment is not carried into the cabin space by ventilating air.

5.6.5 Tyre Noise

As engine noise is progressively reduced and with the introduction of electric vehicles, tyre noise is emerging as a serious problem. Tests have shown that tyre noise can be broken down into two components, tread patterns and road surfaces being the main culprits. While the problem of road surface excited noise is the province of highway engineers that of tread pattern excitation noise clearly belongs to the automotive engineer.

Tyre designers are concerned with reducing tyre noise at source while chassis engineers are concerned with reducing the transmission of noise from the tyre contact patch to the vehicle interior. The mechanism of tyre tread noise generation is due to an energy release when a small block of tread is released from the trailing edge of the tyre footprint and returns to its undeformed position. There is also a contribution from the opposite effect at the leading edge of the footprint.

With a uniform tyre block pattern, tonal noise (at a single frequency, with harmonics, when the wheel rotates at a constant speed) is generated. To overcome this problem tyre designers have produced block pitch sequences that are designed to reduce this effect by distributing the acoustic energy over a wide band of frequencies. When tread patterns are taken into account, there is a need to analyse the effect of the individual impulses produced across the width of the tyre. Computer software has been developed to aid this aspect of tread evaluation at the design stage. Models of tyres that take account of their structural dynamic characteristics and the air contained within them are also used at the design stage. Road surface design plays a significant role in road noise generation, concrete roads tending to be noisier than asphalt roads because of their different surface characteristics.

5.6.6 Brake Noise

Despite sustained theoretical and experimental efforts over many years, the mechanism of noise generation in disc and drum brakes is still not fully understood. The problem of brake noise is one of the most common reasons for warranty claims on new vehicles with evidence from one manufacturer suggesting that over 25% of owners of one year old medium sized cars complain of brake noise problems.

The intractable nature of the problem arises from the complex assemblage of components in which shoes or pads are held in contact with either a drum or disc under hydraulic and friction loading. A dynamically unstable brake system results in vibration of the brake components and this is transmitted to those components having significant surface area such as brake drums and discs. These are particularly effective radiators of airborne noise.

Progress towards a better understanding of the noise generating mechanisms has been aided by sophisticated experimental investigations and various mathematical models have been put forward to aid the design of quieter brakes. See Sect. 5.11 for a more detailed description of brake NVH problems and solutions.

5.7 Automotive Noise Assessment

As a result of the ever-increasing number of vehicles on our roads, the level of road traffic noise has continued to grow relentlessly. This is in spite of regulations imposed by governments and the significant reductions in noise levels that have already been achieved by new vehicles, especially electric vehicles. The quest for quieter vehicles coupled with good design of new roads is set to continue in an attempt to drive down overall noise levels.

Vehicle manufacturers have been faced with increasingly stringent noise regulations, e.g. in the period from 1976 to 1996 the EEC drive-by noise requirements for new cars (with less than 9 seats) has been reduced from 82 to 74 dBA and for buses/trucks from 91 to 80 dBA. Given that only 3 dBA implies a 50% reduction in acoustic power emissions, these are very significant reductions. The limits for drive-by noise of new vehicles is part of the process of vehicle homologation and is currently determined in the EEC by directive 92/97/EEC. The test procedure is contained in ISO 362-1 (2007) and is outlined in the next section.

5.7.1 *Drive-by Noise Tests (ISO 362)*

The procedure is to accelerate the vehicle in a prescribed way and in a prescribed gear past a precision sound level meter set up at a height of 1.2 m above a hard reflecting surface and 7.5 m from the path of the vehicle.

The test site should be a flat open area of minimum radius 50 m containing a road of at least 3 m width constructed in accordance with ISO 10844 which defines the texture and porosity of the road surface. In the noise measurement area, the road widens to 20 m by 20 m as shown in Fig. 5.9.

In a typical test, the vehicle is driven along the centre of the road at a fixed speed and in a certain gear until the measurement area is reached. As the front of the vehicle enters the area, the driver fully depresses the accelerator and maintains it in this position until the rear of the vehicle leaves the area. For passenger cars with manual 5-speed transmissions, the approach speed should be 50 km/h with measurements taken for 2nd and 3rd gear operation. The microphone of the sound level meter should be positioned at A and B (Fig. 5.9).

The maximum A-weighted noise level (sound level meter on the fast setting) is recorded as the vehicle passes through the measurement area. At least 4 valid measurements are made in positions A and B for each of the two gear settings. Each

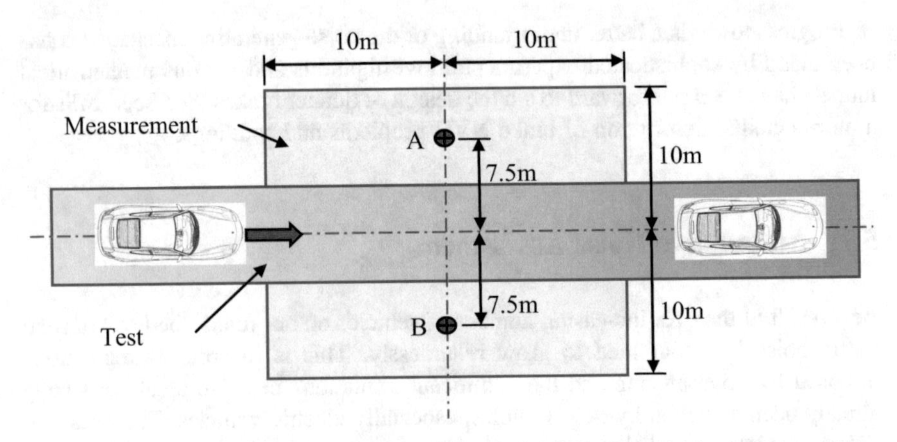

Fig. 5.9 Drive-by test site dimensions and measurement locations

of the four sets of readings are logarithmically averaged and for a given gear setting the higher of the two averages represents the intermediate value. The overall result of the test is the logarithmic average of the intermediate values for each gear setting.

In addition to the results of these measurements, vehicle details such as loading, rating, capacity and engine speeds should be reported. Note that the most recent versions of the standard take account of more realistic acceleration conditions in urban environments and include constant velocity as well as constant acceleration drive-by tests.

5.7.2 *Noise from Stationary Vehicles*

Since exhaust noise is one of the major sources of vehicle noise and vehicles spend a significant amount of time stationary in traffic queues, noise measurements are often taken from stationary vehicles in the vicinity of the exhaust silencer. The EEC directive 92/97/EEC requires measurements to be carried out with the engine running at 75% of the speed at which it develops maximum power. Maximum A-weighted sound pressure level is recorded with the microphone of the sound level meter located 0.5 m from the exhaust tailpipe of the stationary vehicle, 0.2 m above the ground and at 45° angle of incidence to the tailpipe axis. The results from three such tests should be reported as part of the vehicle certification process.

5.7.3 *Interior Noise in Vehicles*

The interior noise within a vehicle is produced by the sources discussed above. In most cases they are periodic, i.e. they have a fundamental component with higher harmonics. These higher harmonics are called “orders”. In the frequency domain they can be considered to be sets of excitation components spread over a range of frequencies. Furthermore these orders are speed dependent—they vary with engine speed (for an ICE vehicle) and vehicle speed. Their strength depends on power output.

These excitation components produce periodic vibration responses of the vehicle structure (chassis, sub-frames and body shell) causing resonance when orders coincide with natural frequencies of the structure. Amplitude and phase variation at points on the body shell can be attributed to path dynamics and source characteristics. This interaction between a noise source and the vehicle structure leads to *structure-borne noise* in the vehicle interior. In this case the dynamic characteristics of the vehicle structure modifies the noise spectrum that is transmitted into the vehicle.

Interior noise within a vehicle can also be produced by *airborne noise* leaking into a cabin space through apertures in the firewall and poor door seals.

At a given location in the cabin space, e.g. at the driver’s ear, the sound field may comprise both reverberant and direct sound field components. Changes in operating conditions modify the noise signature and sound fields.

A procedure for measuring vehicle interior noise is specified in BS 6086 (ISO 5128). The results of such measurements can be used in the assessment of interior noise by playing back recordings of the noise to an experienced panel of judges in a laboratory. Alternatively, the subjective rating of cabin noise can be conducted by having an experienced team of assessors riding in vehicles on a test route. This allows various noise and vibration attributes to be determined.

The above techniques require prototype vehicles for testing. In order to reduce the development time of vehicles, methods have been developed for predicting noise levels in vehicle interiors at the design stage. Noise path analysis can be used to investigate the effect of body structure and vibration isolators on structure-borne noise into a cabin space. One such approached is the so-called **Transfer Path Analysis** (TPA) which is outlined here in the context of noise emitted from an IC engine.

Figure 5.10 illustrates the connection of an engine to a chassis by 3 mounts. Engine excitation forces/moments produce dynamic forces in the mounts and chassis (see Sect. 5.10). These in turn are transmitted to the vehicle structure producing vibration and hence noise within the passenger compartment. Each mount has its own local x, y and z-axes, with different characteristics along each axis. Hence there are 9 different transmission paths between the mount system and the receiver in the passenger compartment.

In TPA, the path characteristics are determined in terms of mechanical-acoustic **frequency response functions** $H_{ix}(f)$, $H_{iy}(f)$ and $H_{iz}(f)$ where x, y and z are excitation directions at the *i*th mount (f = frequency).

TPA can be conducted experimentally in two different ways. One way to determine the $H_i(f)$’s is to use a shaker to excite the chassis at each mount position in turn in the x, y and z-directions over a range of discrete frequencies and measure the sound pressure in the passenger compartment using a microphone. The second method is a reciprocal one in which a loudspeaker replaces the microphone and the

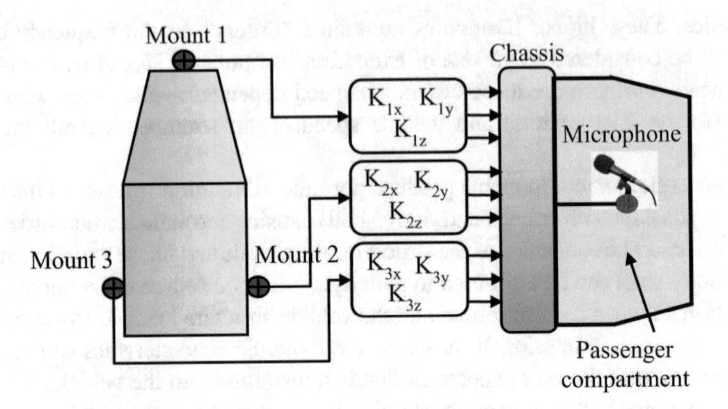

Fig. 5.10 Structure-borne noise from an engine-mount system

effect of loudspeaker output produces vibration at the engine mounts. This is measured by a set of accelerometers mounted on the body side of the engine mounts.

In order to determine the noise level produced via all transmission paths, there is a need to know the complex stiffness of each mount in the x, y and z-directions. This can be determined by separate bench tests. The total sound pressure P produced via all 9 paths is then given by:

$$P = \sum_{i=1}^{3} (K_{ix}\, X_i\, H_{ix} + K_{iy}\, Y_i\, H_{iy} + K_{iz}\, Z_i\, H_{iz}) \tag{5.28}$$

where K_{ix}, K_{iy}, K_{iz} are the stiffnesses of the *i*th mount and X_i, Y_i, Z_i are the dynamic amplitudes of the vibration at the *i*th mount in the x, y and z-directions.

Changes in mount stiffness and/or chassis structures can be investigated using TPA to control the noise level at various locations in the cabin space.

5.8 The Sources and Nature of Automotive Vibration

Vibration arises from a disturbance applied to a flexible structure or component. Common sources of vibration in all vehicles are ground inputs to suspensions, gear manufacturing errors and tooth loading effects in transmissions, and generation of fluctuating dynamic forces in constant velocity joints. ICE vehicles have additional sources of vibration due to engine rotating and reciprocating unbalance, fluctuating combustion loads on the crankshaft, and inertia and elasto-dynamic effects in the engine valve train.

Vibration sources are characterised by their time and frequency domain characteristics. In automotive engineering, most vibration sources produce continuous periodic disturbances. The only exception of note is that of ground inputs to wheels in which the excitation is random. This is the dominant source of vibration in a vehicle (responsible for *primary ride*) and its effect on chassis vibration has been described in Chap. 3.

The simplest form of periodic disturbance is harmonic and might typically be produced by rotor unbalance. In the time domain, this disturbance is represented by a sinusoid and in the frequency domain by a single line spectrum. It might be noted that a full representation in the frequency domain requires both amplitude and phase information. This is important when the disturbance includes several frequency components, each of which may be phased differently to one another. Typical are the general periodic disturbances produced by reciprocating engine unbalance and crankshaft torque. These are responsible for *secondary ride*.

All mass-elastic systems have natural frequencies, i.e. frequencies at which the system naturally wants to vibrate. For a given (linear) system these frequencies are determined only by the mass and stiffness distribution. Corresponding to each natural frequency there is an associated *mode of vibration*. See Appendix A for an outline of basic vibration theory.

Lightly damped structures can produce high levels of vibration from low-level sources if the frequency components in the disturbance are close to one of the system's natural frequencies. This means that well designed and manufactured sub-systems, that produce low level disturbing forces, can still create problems when assembled on a vehicle. In order to avoid these problems at the design stage, it is necessary to model the system accurately and analyse its response to anticipated disturbances.

5.9 The Principles of Vibration Control

5.9.1 Control at Source

While it is recognised that the ideal form of vibration control is "control at source", there is a limit to which this can be carried out. For example the most dominant on-board source of vibration in ICE vehicles is the engine. Here, engine combustion loading and reciprocating inertia combine to produce a complex source of vibration that varies with engine operating conditions as discussed in Sect. 5.10. Even though careful arrangement of crank angles can lead to cancellation of some force and moment components, they can never be completely removed.

Another important source of on-board vibration in all vehicles is due to the imbalance of rotating parts. While these may meet balancing standards, it should be appreciated that there is no such thing as "perfect balance". Hence, small amounts of allowable residual unbalance remain leading to unwanted but acceptable levels of vibration.

It follows that even with the best attempts to eliminate vibration at source, there will always be some unwanted sources of vibration present in all vehicles. It is then necessary to minimise the effects of these on driver and passengers. In this section we review some of the ways in which this can be achieved.

5.9.2 Vibration Isolation

This is a way of localising vibration to the vicinity of the source, thereby preventing its transmission to other parts of a (vehicle body) structure where it may result in the generation of noise or other forms of discomfort to the driver and passengers. Isolation can be achieved by the use of either passive or controllable vibration isolators. Passive isolators range from simple rubber components in shear, or combinations of shear and compression components, to quite sophisticated hydro-elastic elements.

The basic principles for selecting the appropriate isolator can be illustrated with reference to the SDOF model shown in Fig. 5.11. This represents a machine of mass m, subjected to a harmonic excitation arising from rotating unbalance $m_e r$, supported on elastomeric mounts having a complex stiffness k* described by $k^* = k(1 + \eta i)$, where k is the dynamic stiffness and η the loss factor.

The effectiveness of the isolation (a function of frequency ω) can be defined by the transmissibility:

$$T(\omega) = \frac{P}{F_0} \tag{5.29}$$

where P = the amplitude of the force transmitted to the foundation and F_0 is the amplitude of the excitation force due to unbalance.

Applying Newton's second law to the FBD in Fig. 5.11b gives:

$$m\ddot{x} + k(1 + \eta i)x = m_e r\omega^2 \sin \omega t = F_0 f(t) \tag{5.30}$$

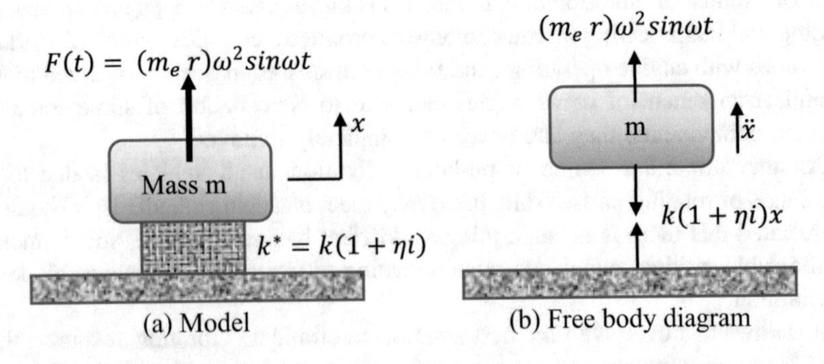

Fig. 5.11 SDOF vibration isolation model and free-body diagram

The force transmitted to the foundation is:

$$P(t) = k(1 + \eta i)\,x(t) \tag{5.31}$$

Employing the standard approach for harmonic excitation outlined in the Appendix, it can be shown that:

$$T(\omega) = k\sqrt{\frac{1+\eta^2}{\left(k - m\omega^2\right)^2 + (k\eta)^2}} \tag{5.32}$$

To appreciate how T(ω) varies with frequency, it is helpful to show Eq. (5.32) in dimensionless form by dividing numerator and denominator by k to give:

$$T(\omega) = \sqrt{\frac{1+\eta^2}{\left[1-\left(\frac{\omega}{\omega_n}\right)^2\right]^2 + \eta^2}} \tag{5.33}$$

where $\omega_n = \sqrt{\frac{k}{m}}$ is the undamped natural frequency of the mass on its isolators.

For low damping elastomers η is of the order of 0.05. The variation of transmissibility with frequency ratio $r = \frac{\omega}{\omega_n}$ for η = 0.05 is shown in Fig. 5.12.

For an isolator to be effective, the transmissibility must be less than unity, i.e. P must be less than F_0. From Fig. 5.12, it is clear that for frequency ratios from 0 to approximately 1.4, P is greater than F_0 and therefore the isolator **magnifies** the force to the foundation in this range. Importantly around resonance (r = 1), the isolator magnifies the force to the foundation approximately 20 times for η = 0.05 emphasising the danger during the run up to operating speed (at which r > 1). As the value of r increases beyond 1.4, the isolator becomes increasingly effective. However, for very large values of r, it is possible to induce wave-effects in isolators. These are due to local resonances in the distributed mass and elasticity of the isolator material and produce additional resonance peaks in the transmissibility curve (and reduced isolator performance) at certain frequencies of excitation.

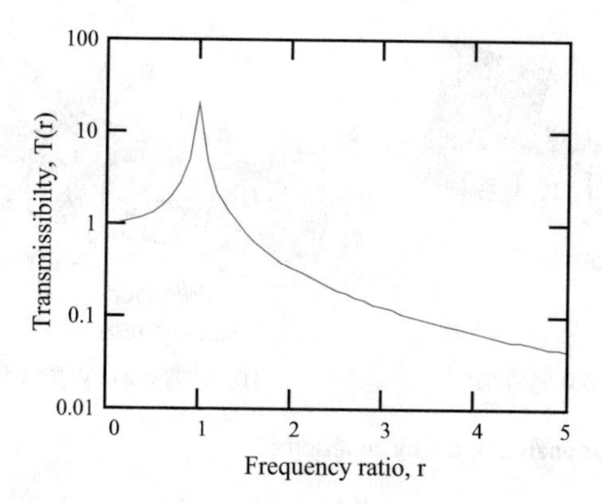

Fig. 5.12 Transmissibility of an elastomeric isolator (η = 0.05)

An alternative way of describing isolator effectiveness is to calculate the isolation efficiency defined as:

$$E_{iso} = [1 - T(\omega)] \tag{5.34}$$

A typical engine isolation system could be expected to have an isolation efficiency of 90% at idle speed.

5.9.3 Tuned Vibration Absorbers

Vibration absorbers are useful for attenuating a resonant response in a system. This can arise in an automotive context when a harmonic component in the excitation coincides with a natural frequency of the system.

Vibration absorbers consist of a spring-mass sub-system that is added to the original system. In effect, energy is transferred from the original system to the absorber mass which can vibrate with significant amplitude depending on the amount of damping contained in the absorber sub-system. These devices are effective attenuators, but add another degree of freedom to the overall system, producing new natural frequencies above and below the original natural frequency. Resulting resonant amplitudes can be controlled with an appropriate choice of damping in the absorber. This enables absorbers to be used for variable speed applications.

The principles of undamped and damped tuned absorbers can be analysed by examining the damped absorber and then treating the undamped absorber as a special case of this. The analysis requires that the original system can be represented by a single-DOF system (normally based on its fundamental mode). Because the applications are often of a torsional nature due to rotational out-of-balance, we will base our analysis on torsion systems. The analysis is of course equally applicable to translation systems.

Consider the original SDOF system shown in Fig. 5.13a. By adding an absorber, the two-DOF system shown in Fig. 5.13b is formed.

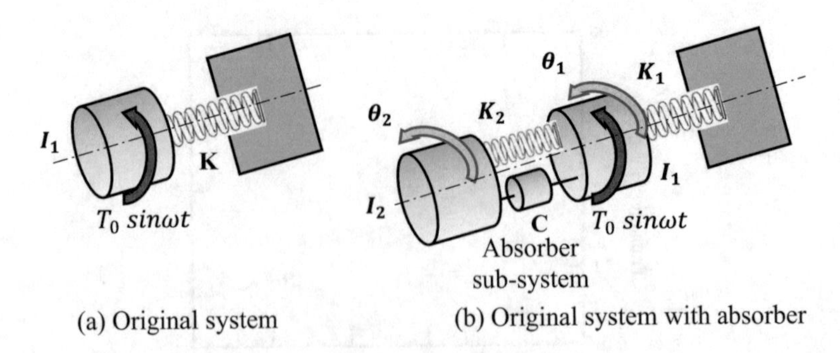

Fig. 5.13 Models for analysing the tuned absorber

In Fig. 5.13, I_1 and I_2 are moments of inertia, K_1 and K_2 are torsional stiffnesses, C is the torsional damping coefficient of the absorber, $T_0 \sin \omega t$ is the excitation and θ_1, θ_2 are the coordinates describing the rotational position of the two masses.

The natural frequency of the original system is $\omega_1 = \sqrt{\frac{K_1}{I_1}}$ and at resonance $\omega = \omega_1$. By drawing the FBDs for the two-mass system, it can be shown that the equations of motion in matrix form are:

$$\begin{bmatrix} I_1 & 0 \\ 0 & I_2 \end{bmatrix} \begin{Bmatrix} \ddot{\theta}_1 \\ \ddot{\theta}_2 \end{Bmatrix} + \begin{bmatrix} C & -C \\ -C & C \end{bmatrix} \begin{Bmatrix} \dot{\theta}_1 \\ \dot{\theta}_2 \end{Bmatrix} + \begin{bmatrix} (K_1 + K_2) & -K_2 \\ -K_2 & K_2 \end{bmatrix} \begin{Bmatrix} \theta_1 \\ \theta_2 \end{Bmatrix} = \begin{Bmatrix} T_0 \\ 0 \end{Bmatrix} \sin \omega t \tag{5.35}$$

It can also be shown that the (complex) steady state responses $\bar{\Theta}_1$ and $\bar{\Theta}_2$ of the two masses are given by:

$$\begin{Bmatrix} \overline{\Theta}_1 \\ \overline{\Theta}_2 \end{Bmatrix} = \begin{bmatrix} ((K_1 + K_2 - I_1 \omega^2) + C \omega i) & K_2 + iC\omega \\ K_2 + iC\omega & ((K_2 - I_2 \omega^2) + C \omega i) \end{bmatrix}^{-1} \begin{Bmatrix} T_0 \\ 0 \end{Bmatrix}$$

This produces the amplitudes of vibration given by:

$$\Theta_1 = T_0 \sqrt{\frac{(K_2 - I_2 \omega^2)^2 + (C\omega)^2}{\left[(K_2 - I_2 \omega^2)(K_1 + K_2 - I_1 \omega^2) - K_2^2\right]^2 + (C\omega)^2 (K_1 - I_1 \omega^2 - I_2 \omega^2)^2}} \tag{5.36}$$

and

$$\Theta_2 = \frac{T_0 \sqrt{K_2^2 + (C\omega)^2}}{\sqrt{\left[(K_2 - I_2 \omega^2)(K_1 + K_2 - I_1 \omega^2) - K_2^2\right]^2 + (C\omega)^2 (K_1 - I_1 \omega^2 - I_2 \omega^2)^2}} \tag{5.37}$$

5.9.3.1 The Undamped Tuned Absorber (C = 0)

In this case the amplitudes are given by:

$$\Theta_1 = \frac{T_0 (K_2 - I_2 \omega^2)}{\Delta(\omega)} \tag{5.38}$$

$$\Theta_2 = \frac{T_0 K_2}{\Delta(\omega)} \tag{5.39}$$

where

$$\Delta(\omega) = \left(K_2 - I_2\,\omega^2\right)\left(K_1 + K_2 - I_1\,\omega^2\right) - K_2^2 \tag{5.40}$$

In order to make Θ_1 zero, $K_2 - I_2\,\omega^2$ must be zero and hence $\omega = \sqrt{\frac{K_2}{I_2}} = \omega_2$ the natural frequency of the absorber sub-system. It follows that, for this case, $\omega_1^2 = \omega_2^2 = \frac{K_1}{I_1} = \frac{K_2}{I_2}$ and hence the natural frequencies of the two sub-systems must be the same.

Resonance of the complete 2-DOF system (i.e. when Θ_1 and Θ_2 tend to infinity) occurs when ω coincides with the system's natural frequencies Ω_1 and Ω_2. This occurs when $\Delta(\omega) = 0$, representing the characteristic equation for the 2-DOF system.

In designing an undamped absorber it is necessary to consider the magnitude of the absorber mass in relation the original system mass m_1. In general the larger the inertia ratio $\mu = I_2/I_1$, the more widely separated are the frequencies Ω_1 and Ω_2 and the wider is the range of frequencies at which the system can operate without exciting resonance.

The general response of the complete system is best described in terms of dimensionless amplitudes and frequencies ratios. Denoting the dimensionless amplitudes of I_1 and I_2 as $A_1 = \frac{K_1}{T_0}\Theta_1$ and $A_2 = \frac{K_1}{T_0}\Theta_2$ together with the frequency ratio as $r = \frac{\omega}{\omega_1}$ enables Eqs. (5.38) and (5.39) to be re-written as:

$$A_1 = \frac{1 - r^2}{(1 - r^2)(1 + \mu - r^2) - \mu} \tag{5.41}$$

$$A_2 = \frac{1}{(1 - r^2)(1 + \mu - r^2) - \mu} \tag{5.42}$$

These equations enable the amplitude responses to be plotted for various values of μ. Figures 5.14 and 5.15 show plots for of the response of the original mass and the absorber mass respectively for $\mu = 0.2$.

5.9.3.2 The Damped Tuned Absorber (C > 0)

The undamped absorber transfers energy from the original system to the absorber sub-system resulting in large amplitudes of vibration of the absorber mass. This can lead to the possibility of fatigue failure in the absorber spring. To overcome this problem it is necessary in practice to add some damping to the absorber. This also allows a wider operating range and limits the resonant amplitudes in the region of the two new system natural frequencies.

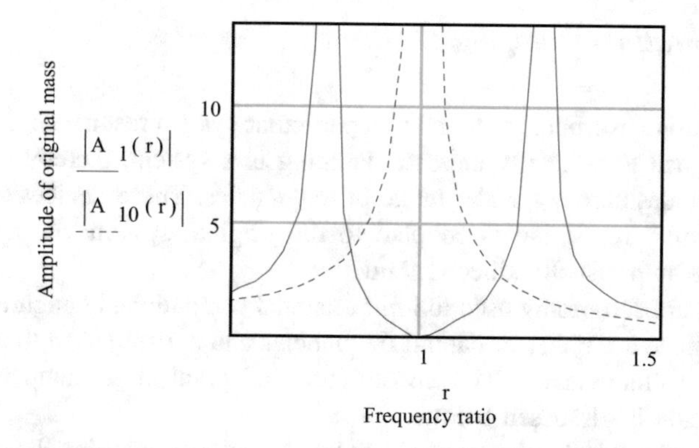

Fig. 5.14 Response of original mass before (A_{10}) and after (A_1) addition of undamped absorber ($\mu = 0.2$)

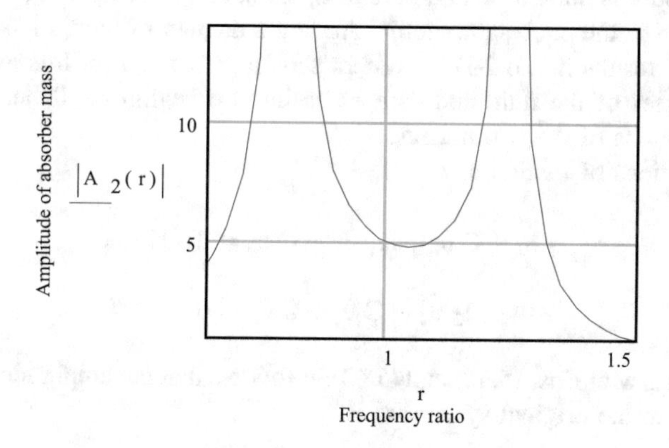

Fig. 5.15 Response of absorber mass (A_2) for undamped system ($\mu = 0.2$)

In this case, the system is represented by the model shown in Fig. 5.13b. The amplitude responses are given by Eqs. (5.43) and (5.44). The damping ratio is given by $\zeta = \frac{C}{2\sqrt{I_1 K}}$. The two new resonant peaks created by the addition of the absorber are reduced with the addition of damping but it is not possible to reduce the amplitude of the main inertia I_1 to zero at the original natural frequency of the system (unlike for the undamped absorber as shown in Fig. 5.14). Thus some of the effectiveness of the absorber is lost at this resonant frequency when damping is introduced. However, by careful optimisation of the absorber's parameters, it is possible to minimise the response of the main inertia across a frequency range. For example, for an inertia ratio $\mu = 0.2$, $\zeta = 0.32$ can be shown to be the optimum damping ratio. This is the underlying principle of the crankshaft absorber as a device designed to attenuate undesirable vibration effects across a certain range of speeds.

5.9.4 Vibration Dampers

While vibration absorbers are tuned to a particular system resonance, a damper is a device designed to generally increase damping in a system, thereby reducing resonant amplitudes across a wider range of frequencies. These devices consist of an inertia (seismic) mass that is coupled to the original system via some form of damping medium, usually silicone fluid.

Dampers are commonly used to limit torsional oscillations in engine crankshafts since these have a number of natural frequencies and are subjected to a wide range of excitation frequencies. The construction of crankshaft dampers and their implementation is discussed below.

The principles of vibration control can be studied by modelling the crankshaft as a single disc inertia mounted at the end of a torsional spring fixed at the other end. This model is based on the first mode of torsional vibration of the crankshaft. The resulting model is shown in Fig. 5.16a. $T_0 \sin\omega t$ is the component of crankshaft toque applied to the equivalent inertia. Adding a damper of inertia I_2 and damping coefficient C results in the 2-DOF system shown in Fig. 5.16b. It is assumed here that the masses of the fluid and damper casing are negligible. θ_1 and θ_2 are the angular positions of the two masses.

The equations of motion are:

$$I_1\,\ddot{\theta}_1 + C\,\dot{\theta}_1 + K_1\,\dot{\theta}_1 - C\,\dot{\theta}_2 = T_0\,\sin\omega t \tag{5.43}$$

$$I_2\,\ddot{\theta}_2 - C\,\dot{\theta}_1 + C\,\dot{\theta}_2 = 0 \tag{5.44}$$

Comparing with Eqs. (5.42) and (5.43), it follows that the amplitude of vibration of the mass in the original system is:

$$\theta_1 = T_0\sqrt{\frac{(I_2\,\omega^2)^2 + (C\omega)^2}{[(I_2\,\omega^2)(K_1 - I_1\,\omega^2)]^2 + (C\omega)^2(K_1 - I_1\,\omega^2 - I_2\,\omega^2)^2}} \tag{5.45}$$

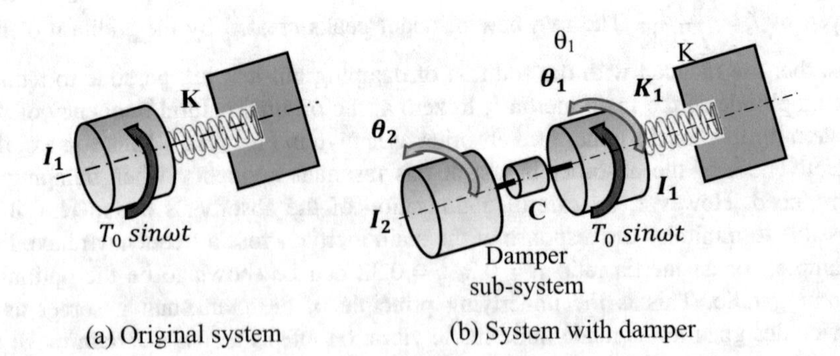

Fig. 5.16 Models for analysing a purely viscous damper

Using the following notation:

undamped natural frequency of the original system, $\omega_n = \sqrt{\frac{K}{I_1}}$,

damping ratio, $\zeta = \frac{C}{2\sqrt{I_1 K}}$,

inertia ratio, $\mu = \frac{I_2}{I_1}$,

dimensionless amplitude of I_1, $A_1 = \frac{K\theta_1}{T_0}$,

and frequency ratio, $r = \frac{\omega}{\omega_n}$,

it can be shown that:

$$A_1 = \sqrt{\frac{(\mu r)^2 + 4\zeta^2}{(\mu r)^2(1 - r^2)^2 + 4\zeta^2[\mu r^2 - (1 - r^2)]^2}} \quad (5.46)$$

A_1 is thus a function of r, μ and ζ. For a given value of ζ the response will exhibit a single peak similar to that for a damped SDOF system. The extreme values of damping are ζ = 0 and ∞. When ζ = 0 the system response is that of the original SDOF system having a natural frequency ω_n. and when ζ = ∞, both masses move together as one and the undamped natural frequency is $\sqrt{K/(I_1 + I_2)}$. When A_1 is plotted on the same axes for these two extreme cases and for a given μ, the curves intersect at a point P. It can be shown that the curves for other values of damping also pass through P. These features are illustrated in Fig. 5.17.

Clearly the optimum value of damping is the one which gives the maximum attenuation at P. It can be shown that this is given by:

$$\zeta_{opt} = \frac{\mu}{\sqrt{2(1+\mu)(2+\mu)}} \quad (5.47)$$

Thus, for μ = 1.0 as in Fig. 5.17, ζ_{opt} = 0.288.

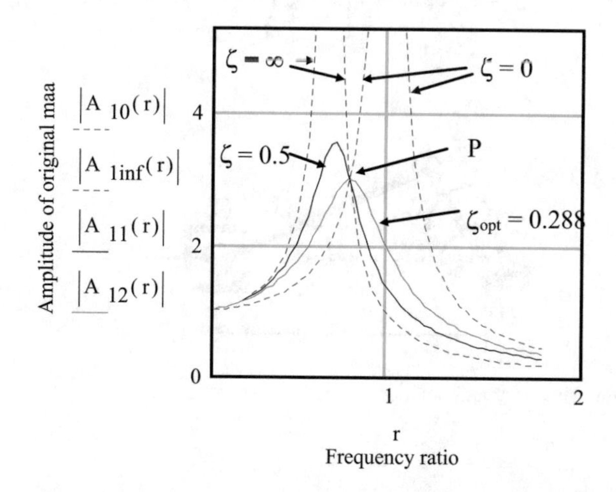

Fig. 5.17 Response of original mass (A_1) for purely viscous damper for different levels of damping (μ = 1.0)

5.10 Engine-Induced Vibration

5.10.1 Single Cylinder Engines

The engine is the most dominant onboard source of vibration for an ICE vehicle (still by far the majority of vehicles on our roads) and this section of the chapter is devoted to analysing this source of vibration. Since a multi-cylinder inline engine can be considered to be a set of single cylinder engines connected to the same crankshaft, we begin by analysing the dynamics of a single cylinder engine.

5.10.1.1 Kinematic Analysis of a Single Cylinder Engine

Consider the engine mechanism shown in Fig. 5.18, where the crankshaft is assumed to rotate at a constant speed ω. The component parts are number 1 to 4, where 1 is assigned to the engine frame. From this geometry:

$$x_B = r\cos\omega t + l\cos\phi \tag{5.48}$$

From the sine rule: $\frac{r}{\sin\phi} = \frac{l}{\sin\omega t}$ and denoting $\frac{l}{r}$ with n, $\sin\phi = \frac{\sin\omega t}{n}$

Then:

$$x_B = r\cos\omega t + l\left(1 - \sin^2\phi\right)^{1/2} = r\cos\omega t + l\left[1 - \left(\frac{\sin\omega t}{n}\right)^2\right]^{1/2} \tag{5.49}$$

Expanding the last term using the Binomial Theorem gives the infinite series:

$$x_B = r\cos\omega t + l\left[1 - \frac{1}{2}\left(\frac{\sin\omega t}{n}\right)^2 + \cdots\right] \tag{5.50}$$

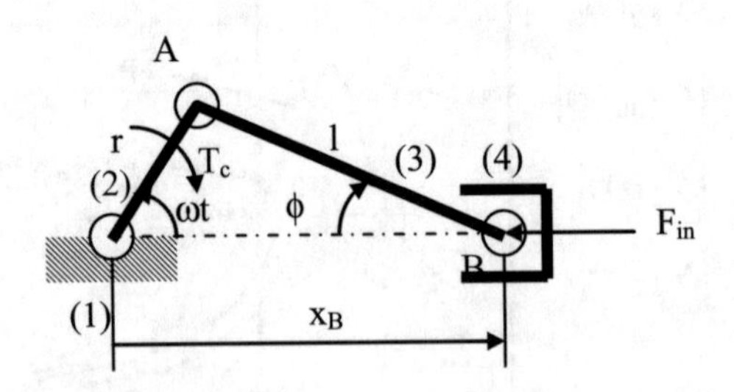

Fig. 5.18 Engine geometry and notation

Piston velocity $\dot{x}_B$ is obtained by differentiating Eq. 5.50 with respect to time:

$$\dot{x}_B = -r\omega \sin \omega t - l\omega \left[\frac{2 \sin \omega t \cos \omega t}{2n^2} + \cdots \right]$$

or (5.51)

$$\dot{x}_B = -r\omega \sin \omega t - r\omega \frac{\sin 2\omega t}{2n} + \cdots$$

A second differentiation produces an expression for piston acceleration $\ddot{x}_B$:

$$\ddot{x}_B = -r\omega^2 \left(\cos \omega t + \frac{\cos 2\omega t}{n} + \cdots \right) \quad (5.52)$$

Equation (5.52) can be written more generally as:

$$\ddot{x}_B = -r\omega^2 (A \cos \omega t + B \cos 2\omega t + C \cos 4\omega t + D \cos 6\omega t + \cdots) \quad (5.53)$$

where A = 1, B = 1/n and C, D etc. are functions of n.

Thus $\ddot{x}_B$ is described by a component at fundamental frequency ω plus a series of even harmonics (orders), whose amplitudes (which are functions of n) decrease with increasing order. x_B and $\dot{x}_B$ also contain fundamental components at frequency ω plus higher orders.

5.10.1.2 Total Force and Moment on a Single Cylinder Engine Frame

These comprise a *shaking force* attributed solely to the inertia effects and *a rocking moment* equal and opposite to the crankshaft torque. An understanding of these is important in engine mounting, the objective being to minimise the transmission of forces to the vehicle chassis.

The *shaking force* F_{in} acting on the engine frame along the line of stroke of the engine is solely due to inertia effects. This is given from Eq. (5.53) by:

$$F_{in} = m_B \ddot{x}_B = -m_B r\omega^2 \left(\cos \omega t + \frac{\cos 2\omega t}{n} + \cdots \right) \quad (5.54)$$

where m_B is the equivalent mass acting at the piston end of the connecting rod.

Thus the *shaking force* comprises a fundamental component at engine speed, together with a set of even harmonics (orders)—see Eq. (5.53).

The *rocking moment* acts about an axis parallel to the crankshaft in an opposite direction to the crankshaft torque T_c shown in Fig. 5.19.

The combined force and moment produce a translational and rotation excitation of the engine block, Fig. 5.19. The transmission of the force and moment to the chassis is controlled by a set of engine mounts supporting the engine.

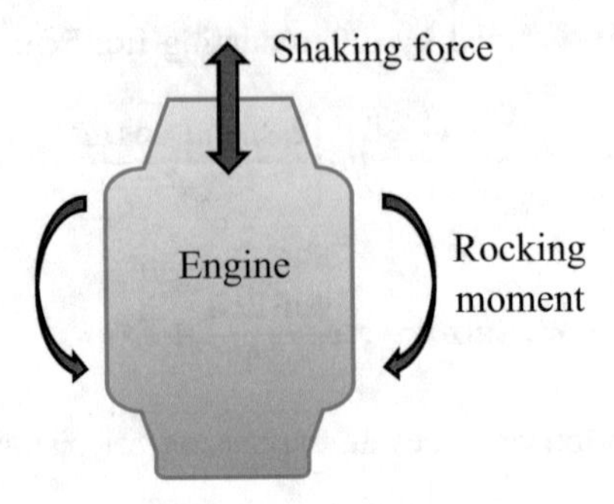

Fig. 5.19 Excitation force and moment on a single cylinder engine

5.10.2 Multi-cylinder Engines

Since the combustion torque from each cylinder is basically a pulse occurring during the expansion stroke, it is essential that for a multi-cylinder engine, the torque contribution from each cylinder produces a regular train of pulses. This in turn dictates that the crank throws, in the order of firing, must have a regular angular spacing. For two-stroke and four-stroke engines, this spacing is 360/n and 720/n degrees respectively, where n is the number of cylinders. It follows that a 4-cylinder, 4-stroke engine must have crank throws of 180° in the order of firing. Possible firing orders are 1-3-4-2 and 1-2-4-3. The crank throws for these two firing orders (viewed along the crankshaft) are shown in Fig. 5.20.

The spacing of the cylinders along the crankshaft means that the combined effect of the shaking force along each cylinder results in a shaking moment about an axis perpendicular to the plane containing the cylinders, i.e. about the z-axis. The resultant crankshaft torque due to the 180° crank throws will contain ½ order components (4-stroke engines). Hence there is an equal and opposite rocking moment exerted on the engine block.

The overall effects of inertia forces and combustion loading are thus:

- A shaking force due to inertia forces along the x-axis
- A shaking moment about the z-axis due to inertia forces acting at the offset of each cylinder from the centre plane
- A rocking moment about the y-axis due to the reaction to the crankshaft torque.

These effects are shown in Fig. 5.21.

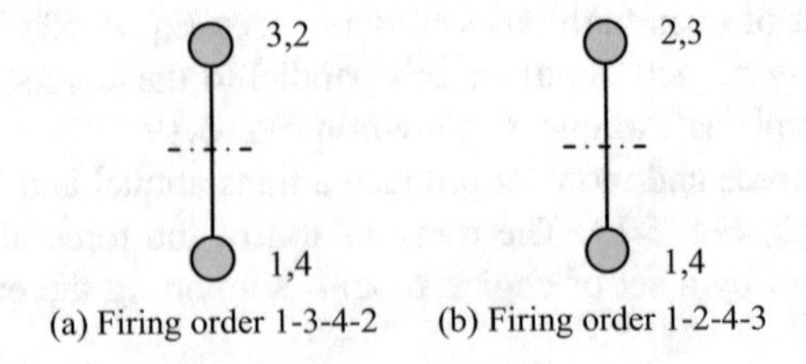

Fig. 5.20 Crank throws for a 4-cylinder, 4-stroke engine

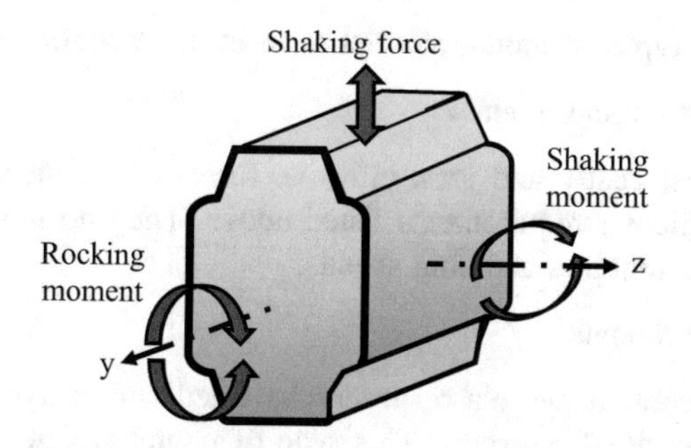

Fig. 5.21 General excitation of a multi-cylinder engine block

With certain crank configuration, it can be shown that some components of these forces and moments are eliminated. For example, it can be shown that with a firing order 1-2-4-3 for a four stroke engine, the 1st and 3rd order inertia force components are balanced. Thus, there are only 2nd, 4th and 6th order shaking force components and only a 2nd order shaking moment component (the 1st, 4th and 6th shaking moments being balanced).

5.10.3 The Isolation of Engine-Induced Vibration

Engine mounting is an example of vibration isolation. The objective here is prevent the excitation forces on the engine block from being transmitted to the chassis. This requires that the engine block be mounted on a set of vibration isolators. The simpler forms of isolator are a cost-effective solution for sources with a limited range of operating conditions (amplitudes and frequencies). For situations where the source of vibration produces a range of operating conditions (as in ICE vehicles), it is often necessary to consider the use of hydro-elastic or controllable isolators. Some of these devices are discussed below. In all cases it is essential to understand the basic principles of vibration isolation (Sect. 5.9.2) to achieve the best results.

The vibration isolation theory discussed in Sect. 5.9.2 suggests that the natural frequency of the engine on its mounts should be much less than the excitation forces arising in the engine. This suggests a low stiffness for the engine mounts. However, there are other considerations. If the engine is supported on very soft springs there will be significant movement of the engine when the vehicle traverses over rough terrain. Furthermore the excitation forces on the engine will vary with engine speed.

The requirements for satisfactory mounting of engines are therefore:

- A low spring rate and high damping during idling
- A high spring rate and low damping for high speeds, manoeuvring and when traversing rough terrain.

The following types of mount attempt to meet these conflicting requirements:

(a) Simple rubber engine mounts

These are the least costly and least effective forms of mount but clearly do not meet all the conflicting requirements listed above. They do not provide the high levels of damping required at idling speeds.

(b) Hydro-elastic mounts

These generally contain two elastic reservoirs filled with a hydraulic fluid. Some also contain a gas filled reservoir. This type of mount which is a form of tuned vibration absorber exploits the feature of mass-augmented dynamic damping. In operation there is relative motion across the damper which produces flexure of the rubber component and transfer of fluid between chambers, thereby inducing a change in mount transmissibility.

(c) Semi-active mounts

These are a form of controllable mount. Their operation is dependent on modifying the magnitude of the forces that they transmit. They may be implemented via low-bandwidth, low-powered actuators that are suited to open-loop control. Some forms of hydraulic semi-active (adaptive) mount use low powered actuators to induce changes in mount properties by modifying the hydraulic parameters within the mount. The actuators may then operate in on-off (adaptive) or continuously variable (semi-active) mode. Considerable effort has been devoted to this type of technology in recent years.

(d) Active mounts

This type of mount controls both the magnitude and direction of the actuator force. Active vibration control is typically implemented by closed-loop control. The actuators need to operate at high speed and together with the sensors require an operating bandwidth to match the frequency spectrum of the excitation. Power consumption is generally high in order to satisfy the response criteria.

5.11 Braking Systems NVH

5.11.1 Introduction

Dynamic oscillations in friction brakes which cause noise, vibration and harshness may be categorised into 2 mechanisms, firstly a dynamic instability which results in a constant resonant frequency independent of rotor speed and secondly a mechanical vibration with a frequency related directly to rotor speed. The former makes use of the generalised term of "brake noise" and the second is often referred

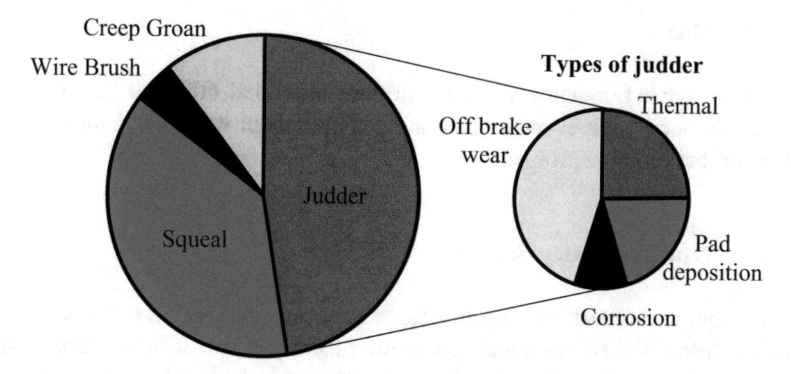

Fig. 5.22 Typical distribution of warranty issues reported by high performance vehicle manufacturer

to as "judder" or "drone". These will be discussed separately in this section which will focus mainly on disc brake systems. A typical estimate of the proportion of warranty claims based on the different categories of brake NVH is shown in Fig. 5.22.

5.11.2 Brake Noise and Vibration Terminology

5.11.2.1 Groan

Groan is a semi-resonant vibration, typically with a frequency of less than 100 Hz. The mechanism appears to be of a stick-slip nature due to a decreasing friction/velocity characteristic. Groan is most prevalent at lower speeds where the friction/velocity gradient is steepest. It may be experienced when a car is allowed to creep forward with the foundation brake or parking brake slightly activated, hence the term "creep groan".

5.11.2.2 Hum

Hum is a resonant sinusoidal vibration and is characterised by a frequency of between 100 and 400 Hz. The vibration tends to be about the radial axis of the brake system and is associated with those installations which have a low calliper mounting torsional stiffness. Hum generally occurs under off brake conditions and unlike groan the vibration is independent of speed. It is also observed in drum brakes where the back-plate is caused to vibrate.

5.11.2.3 Moan

Moan is a higher frequency vibration to hum at around 600–700 Hz and results from whole body calliper movement and bending about its spine. Again observed with drum brake noise problems.

5.11.2.4 Low Frequency Squeal

In low frequency squeal, the unstable vibration occurs between 2 and 4 kHz and is therefore below the fundamental frequency of the pad and disc assembly. Of all instances of geometrically induced instability, which involve transverse disc vibration, low frequency squeal is the mechanism which has the lowest frequency. The disc modes are of a low order diametrical nature and typically have between 2 and 4 full nodal diameters. The nodal spacing, in this case, is far greater than in the case of a traditional brake pad as per the two-opposed-piston calliper design. Low frequency squeal induces a flexural mode that causes the resultant wavelength of the vibrating disc to be long compared to the length of the brake pad. Such an excitation of the disc allows the contact patch to be considered practically flat. This means that the section can be represented by rigid beams with two degrees of freedom, transverse and rotational. This model also allows for direct application of "binary flutter" analysis (see below).

5.11.2.5 High Frequency Squeal

Although high frequency squeal was originally characterised as "squeak" it is now generally classified as high frequency squeal. High frequency squeal is characterised by the involvement of higher order disc modes and this is what sets it apart from low frequency squeal. The modes have, typically, 5 to 10 nodal diameters and the frequency range is between 4 and 15 kHz. In high frequency squeal the nodal spacing is less than the pad length and disc natural frequencies are now greater than the fundamental pad bending mode. High frequency squeal is often present with the installation of single piston sliding callipers which have high aspect ratio (length to width) pads.

5.11.2.6 Squelch

This is a quite complicated noise and has been likened both to the sound of wash leather on glass and also a washing machine filling up with water. Squelch is, in fact, an amplitude modulated version of squeak noise, with the modulation depth being 100%, indicative of the fact that the frequency sources are related. This modulation of the vibration causes a beating effect, caused by general asymmetry of the disc.

5.11.2.7 Wire Brush

This mechanism is observed immediately prior to development of an unstable squeak. The vibration frequency is very high, up to 20 kHz, and considered to be non-resonant, although recent studies have shown it to be periodic, at a squeak frequency, but with a random amplitude modulation.

5.11.2.8 Judder

This is a low speed non-resonant vibration whose frequency decreases as velocity decreases. The judder frequency is found to be typically less than 10 Hz and is propagated by the non-uniformity of the disc rubbing path, caused either by disc thickness non-uniformity or variable frictional characteristics. Low speed judder is of a stick-slip nature and causes wheel and brake vibration about the suspension or chassis.

There is also the phenomenon of high speed judder, which again is non-resonant and has a typical frequency of about 200 Hz. This occurs when the vibration frequency coincides with road wheel rotational velocity (or a multiple thereof). High speed judder is often accompanied by an event called "blue (or hot) spotting" where the disc rubbing path has thermally deformed. High speed judder is often referred to as ***drone***.

5.11.2.9 Disc Thickness Variation (DTV)

Most brake judder complaints are caused by DTV where the rotor varies in thickness about its circumferential rubbing path. DTV can be a permanent geometric condition leading to "cold judder" or a transient condition occurring at elevated temperatures due to non-uniform thermal distortions ("hot judder").

5.11.3 Disc Brake Noise—Squeal

Above around 2000 Hz noise frequency, it should not be necessary to include the suspension within any dynamic considerations. Disc brake callipers tend to fall into 2 categories, opposed piston and sliding "fist", see Fig. 5.23. Above 3000 Hz, the calliper tends not to play a part and only the disc, pads and the calliper carrier bracket (with sliding "fist") need be considered, see Fig. 5.24. The disc has predominantly two types of vibration as described below.

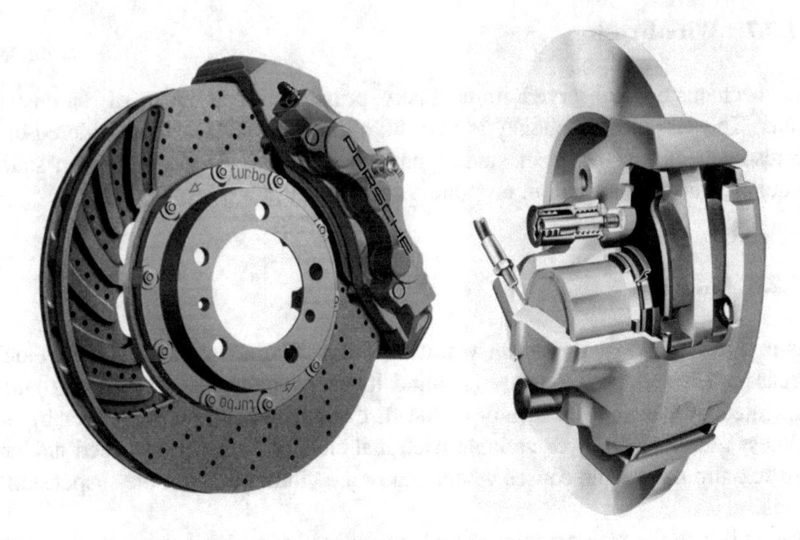

Fig. 5.23 Opposed piston (left) and sliding "fist" (right)

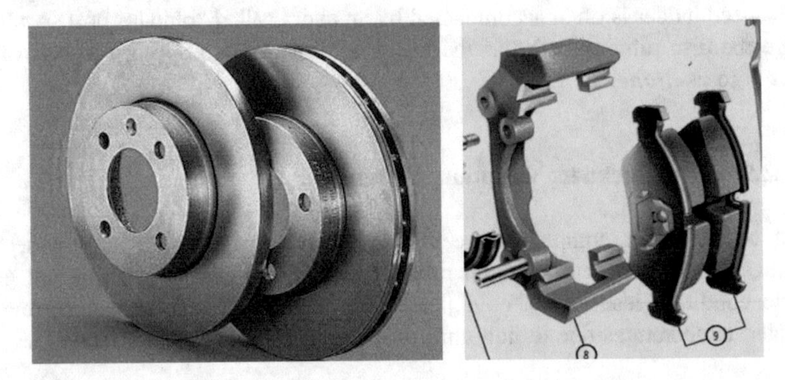

Fig. 5.24 Only the disc (left) and calliper carrier bracket and brake pads (right) need to be considered above 3000 Hz

5.11.3.1 Disc Circumferential Modes of Vibration

This occurs when the disc vibrates axially, and in phase, circumferentially. This causes the nodes of vibration (points of zero displacement) to be circumferential in their nature. The action of these nodes causes the disc to take the shape of a cup then a cone due to the action of each cycle. Although a common mode of vibration, it is rare to experience brake noise associated with such modes. A "Belleville" (cone) washer deformed shape would be typical of a circumferential mode order.

5.11.3.2 Disc Diametrical Modes of Vibration

This occurs when the disc tends to vibrate about one or more diameters at one time, see Fig. 5.25. The action of the nodes in this case causes the disc to take the shape of a "wave" washer. The number of diametrical nodes gives the so-called mode number of the vibration.

Disc brake noise results usually from the disc being excited in one or more of its diametrical mode orders (Fig. 5.26) and as such the noise may be related back to the disc diametrical mode natural frequencies. The mode order may be plotted against frequency which should show a smooth curve as in Fig. 5.27. When disc/pad interface

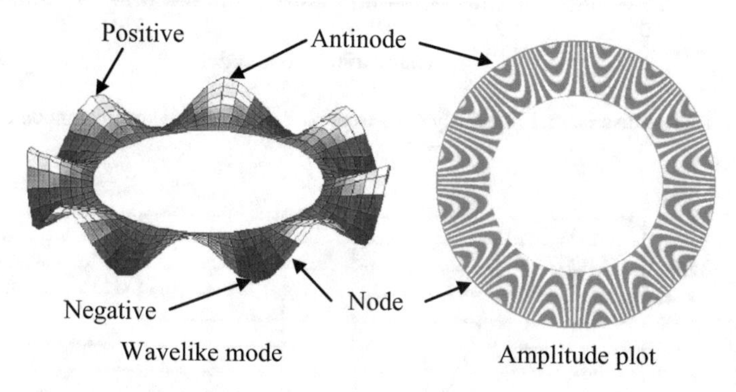

Fig. 5.25 Typical 8 diametrical mode of vibration—16 antinodes

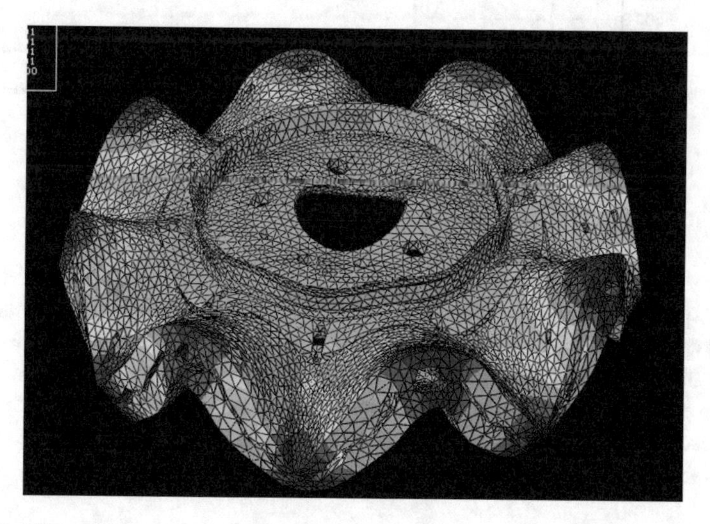

Fig. 5.26 Disc excited with an 8 diametrical mode order

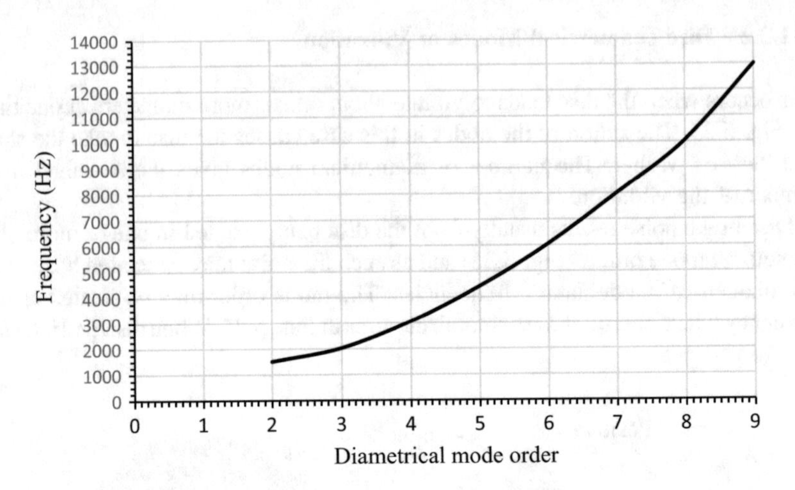

Fig. 5.27 Typical disc natural frequency (free-free) plotted against diametrical mode order

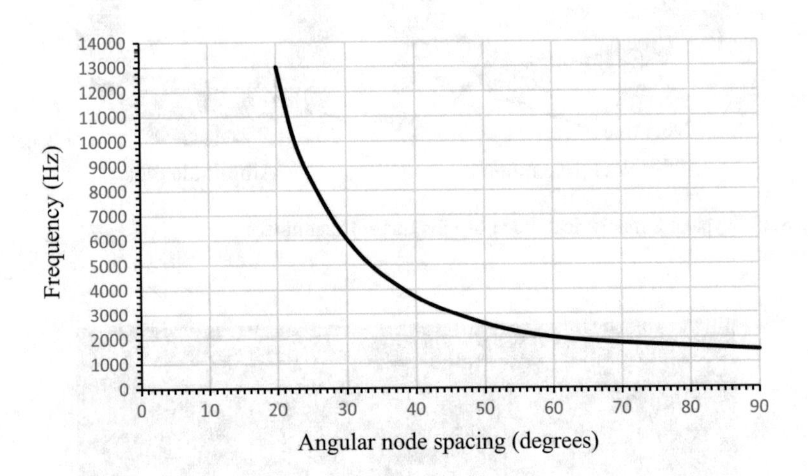

Fig. 5.28 Frequency plotted against antinode angular spacing

geometry is considered, it can be more useful to plot frequency against antinode angular spacing (Fig. 5.28)—the third diametrical mode is equivalent to 60° antinode angular spacing, the fifth diametrical mode equivalent to 36°, and so on.

The disc may also exhibit an in-plane mode of vibration and some research findings indicate that both the in-plane and out-of-plane (diametrical modes) need to coalesce in order for noise to be generated.

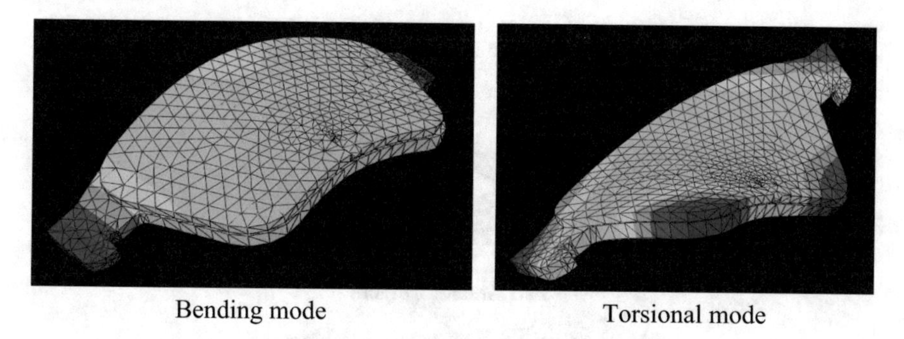

Fig. 5.29 Pad modes of vibration—bending mode (left) and torsional mode (right)

5.11.3.3 Pad Modes of Vibration

The pads may vibrate with two possible modes of vibration—bending or torsional—as shown in Fig. 5.29. The modes do not appear to follow any general trend, such as in the disc, and their occurrence will vary as the pad natural frequency changes due to wear. As such the modes will change and noise may emanate at any time during the wearing process. It is suspected that there is a link between the disc mode and the pad mode for noise to be generated. This may be evaluated during the design stage by considering the interface geometry and contact pressure distribution.

5.11.3.4 Calliper Carrier Bracket Modes of Vibration

The calliper carrier bracket vibrates in a complex manner which is difficult to predict because of the friction variables involved, see Fig. 5.30.

It should be noted that the pad anchor fingers and the transfer beam are connected when a combined trailing and leading pad abutment is used (often referred to as "hammer head"). This connection tend to promote the fingers to be in-phase (but not exclusively). The beam undergoes a "tortuous" mode of vibration that is difficult to confirm unless a vibration visual technique is employed. To remove the beam, and so simplify the system, the anchor fingers need to be stronger as the trailing fingers (forward & reverse) now have to withstand the total braking force at the pads. The in-plane vibration of the pad may now become significant and may need to be considered when the μ/velocity profile is considered (see stick-slip model below).

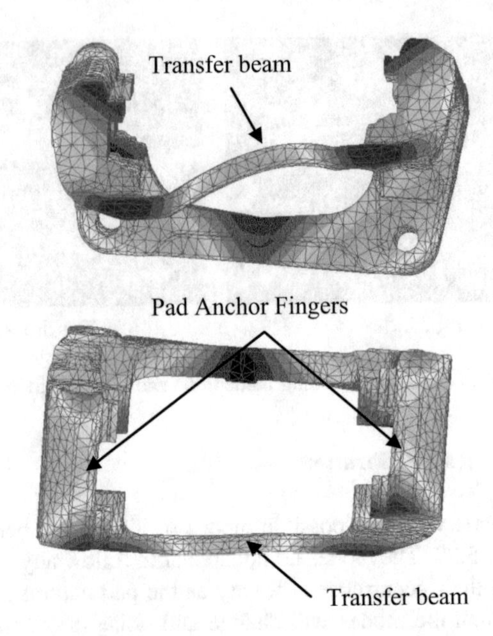

Fig. 5.30 Typical vibration characteristic of a calliper carrier bracket

5.11.4 Brake Noise Theories and Models

The mechanisms that cause brake noise to be generated are still poorly understood, even after many years of research. It has been stated that the USA automotive industry spend $2 billion a year on brake NVH research, the global warranty claims per year nearing $1 billion. It is a quality issue and if the problem becomes large enough to damage the vehicle manufacturers reputation then the whole of the supply chain becomes involved in seeking a solution.

The theories are numerous—the principal ones being stick-slip (negative μ/velocity profile of the friction material), sprag-slip and binary flutter (8 degree of freedom model) as outlined below. Early experimental techniques used to investigate the phenomena were based on pin-on-disc and cantilever/disc models. Experimental techniques used to investigate vibrations of a brake system include Electronic Speckle Pattern Interferometry (ESPI), Double Pulsed Laser Holography and Laser Doppler Velocimetry. Computational methods include FEA and Complex Eigenvalue Analysis (CEA).

5.11.4.1 Slip-Stick Model

For the simple sliding spring-mass system shown in Fig. 5.31, the resistive force of the mass to sliding is given by mgμ where μ = static friction coefficient (stiction). The spring force is given by kx where x is the displacement from static equilibrium position.

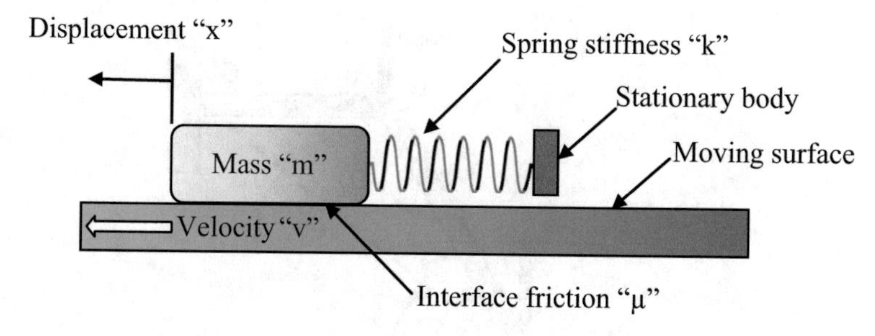

Fig. 5.31 Stick-slip model

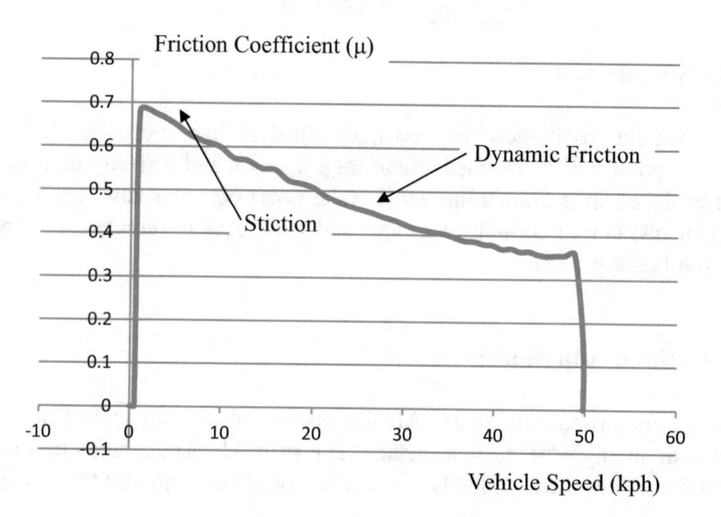

Fig. 5.32 Typical μ/velocity curve for a typical friction material

If mgμ > kx then the mass will move with the surface. As the relative velocity between the mass and moving surface is zero, μ is then the static (stiction) value. As the spring extends, the spring force increases until eventually the spring force kx = mgμ. At this point the mass ceases to move with the surface. As the relative velocity between the mass and moving surface increases, the friction falls to a value according to the interface μ/velocity curve (see Fig. 5.32 for typical example) and the spring force kx > mgμ_{dyn} where μ_{dyn} is now the dynamic friction coefficient. This is the "slip" phase.

The mass is then pulled back against the moving surface and in doing so the relative interface speed increases so the dynamic friction reduces still further. At the same time the spring extension (x) reduces and so does the spring force kx. A point is reached where again mgμ_{dyn} > kx and the mass stops moving against the surface. The friction level then increases to the static value. This is the "stick" phase.

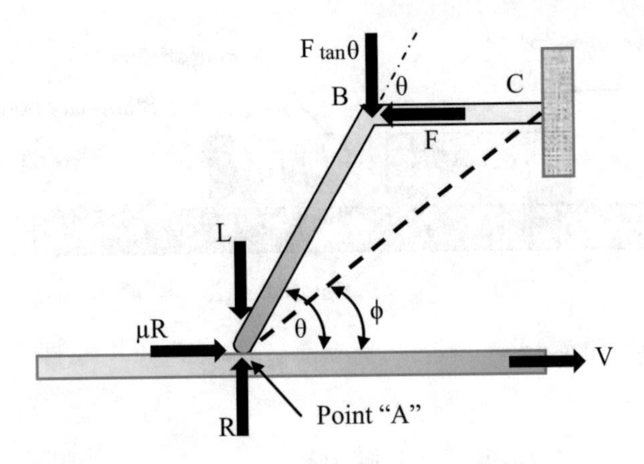

Fig. 5.33 Sprag-slip model

The stick-slip oscillation may be maintained at low speed but if the speed increases a point will be reached where $mg\mu_{dyn} = kx$ and a steady non-oscillatory state is achieved. It is known that most brake noise occurs at low speed (downhill braking or just before coming to a stop) and will cease if vehicle speed increases during the braking event.

5.11.4.2 Sprag-Slip Model

The model comprises a rigid strut AB that contacts a moving surface at point "A" and tilted at an angle "θ" to that surface. The strut AB is connected to a relatively rigid cantilever beam BC which is encased in a rigid body at point "C", as shown in Fig. 5.33.

Resolving forces vertically gives:

$$R = L + Ftan\theta \tag{5.55}$$

and taking moments about "B" gives:

$$L \times AB\ cos\theta + \mu R \times AB\ sin\theta - R \times AB\ cos\theta = 0 \tag{5.56}$$

which reduces to:

$$L + \mu R tan\theta - R = 0 \tag{5.57}$$

Substituting for R in equation above gives:

$$L + \mu(L + Ftan\theta)tan\theta - (L + Ftan\theta) = 0 \tag{5.58}$$

resulting in:

$$\mu L + \mu F tan\theta - F = 0 \tag{5.60}$$

and rearranging gives:

$$\mu L = F(1 - \mu tan\theta) \tag{5.61}$$

so that:

$$F = \frac{\mu L}{(1 - \mu tan\theta)} \tag{5.62}$$

Note that F approaches infinity as μ tends towards cot θ, which is when spragging will occur. When the inclination angle is set at the "sprag" angle of $\tan^{-1}\mu$, or greater, the strut will "dig-in". The normal force to the rubbing surface then increases until flexure of the system allows a secondary strut arrangement "AC" (with angle ϕ to the surface) to be established whereby the contact angle reduces below the sprag angle. The forces then reduce and the strut continues to slide. The sprag angle is often referred to as the "locking angle" in general engineering applications (e.g. Morse tapers and the metal inserts in CV belts).

The calliper represents a more complex suppport system with a multiple of "sprag angles" and it is this sprag-slip mechanism that may be used to address a limited number of noise issues relating to the coefficient of friction at the disc/pad interface.

5.11.4.3 Binary Flutter

The proposal is that the disc may be considered as a rigid body over the length of the pad. The disc can therefore exhibit both axial and rotational motion as shown in Fig. 5.34. This is similar to an aircraft wing which is able to exhibit both bending and torsion modes of vibration. The principal node for bending (zero displacement) is at the aircraft fuselage whilst that for torsion is along the length of the wing to the tip. If the two modes coalesce at the same frequency then the vertical movement of the wing (flapping) may become dangerously large. This can only occur if the natural frequencies are close and the phase difference between the modes is 90° i.e. the wing twist (or the angle of attack) is at a maximum when the wing is level (zero bending). The angle of attack then promotes wing lift (or dip) to exacerbate the forced bending of the wing.

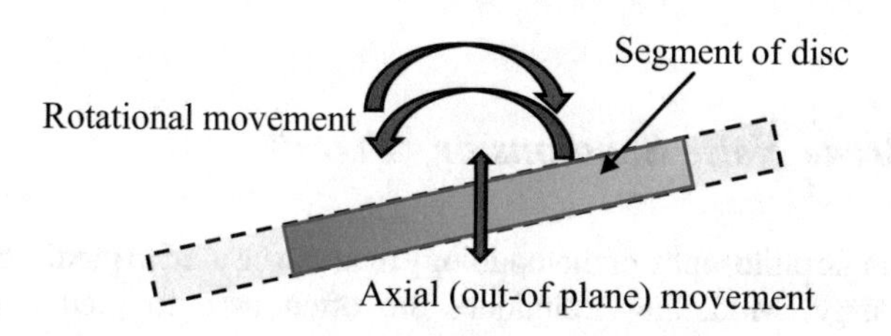

Fig. 5.34 Binary flutter model showing disc segment exhibiting both out-of-plane and rotational movement

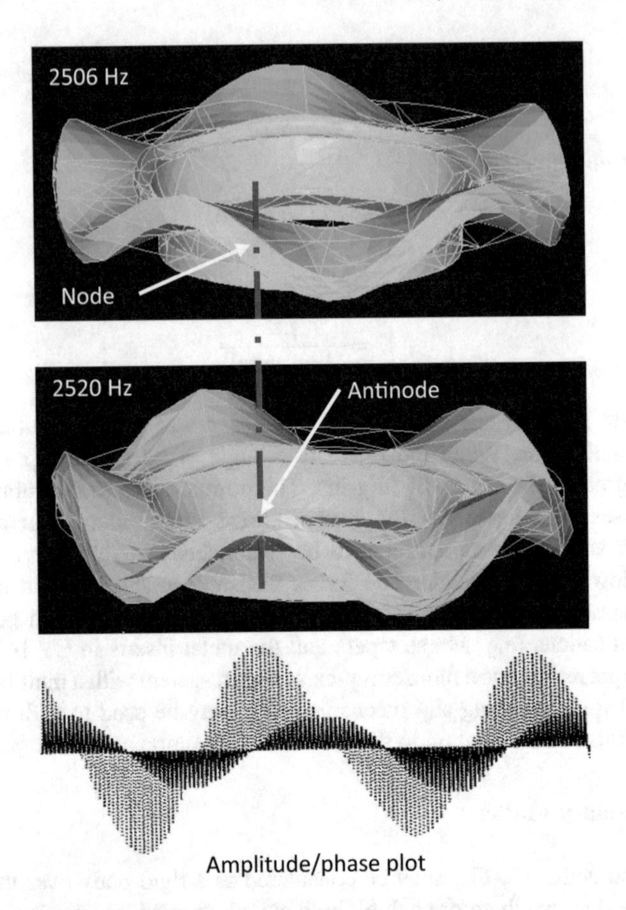

Fig. 5.35 Potential "binary flutter" modes for brake disc

To prevent such mode coalescence, the frequency difference between modes needs to be greater and in the case of aircraft wings a mass is often added to the tip of the wing. This changes the bending frequency but as the mass lays on the torsional nodal line, the torsional frequency remains unchanged.

Figure 5.35 shows potentially coupled modes for a brake disc. A disc may "hold" 2 normal modes of vibration simultaneously with very close natural frequencies. The two modes are displaced circumferentially by half a node pitch such that one node is positioned at the node of the other as indicated by the amplitude/phase plot in the Figure. Coalescence of these two modes may result in brake squeal.

5.11.5 Brake Noise Solutions or "Fixes"

There is no single philosophy or methodology than can be used to predict brake noise at the design stage. Modelling techniques are often used to predict noise but if unreasonable assumptions are made within the model it generally becomes invalid. Such methods include the analysis of the Centre of Pressure (CoP) at the disc/pad interface to assess the potential for "spragging". Another method is to consider the disc/pad interface geometry to determine if an integer number of disc antinodes may "fit" below the pads and whether there is a potential pad mode at that frequency. More advanced techniques such as Complex Eigenvalue Analysis (CEA) based on sophisticated finite element models of the brake assembly have come to the fore in recent years. However such methods tend to overpredict the number of unstable modes of vibration (that may give rise to squeal) compared to experimental measurements.

Thus brake noise remains unpredictable and, if noise becomes a problem during the early months of vehicle release, then the solution (referred to as a "fix") is often a retrofit device or design modification to the brake, as outlined below. The problem then arises that once a noise "fix" is found to work, no individual will have the courage to remove it and so it becomes a permanent feature of the brake design.

5.11.5.1 Noise "Fix" Shims

Visco-elastic or constrained layer shims are constructed from a thin metallic plate (typically 0.75 mm), with rubber bonded to one or both sides and adhesive on both sides. They are in effect a damping medium and are attached by adhesive (hot or cold) between the pad back-plate and the calliper pistons(s). If the pad begins to vibrate (or oscillate linearly), the rubber deforms and energy is lost through hysteretic damping. The principal concern is delamination of the shim during operation which is especially likely to occur at elevated temperatures. In addition, if dust penetrates the bond interface then delamination and corrosion may progress and the shim becomes ineffective. Typical shims are shown in Fig. 5.36.

In some cases the shim is purely of stainless steel construction to which a thin layer of grease is applied. The principle is that the grease acts rather like the elastomeric shim and induces damping into the system. In addition it reduces the in-plane friction forces between the back-plate and the pistons so increasing the pad

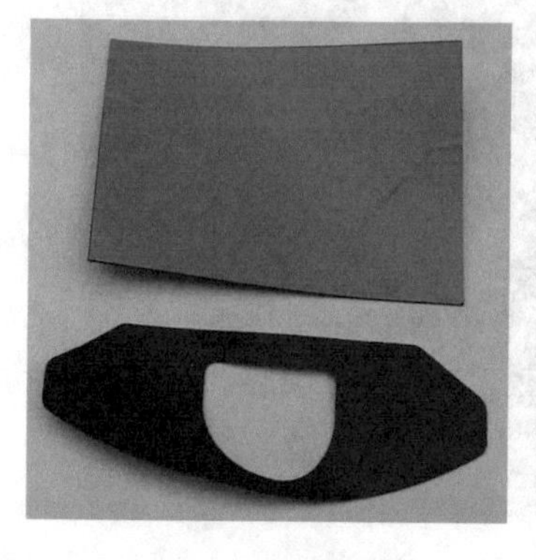

Fig. 5.36 Flexible noise "fix" shims (left) and constrained layer shims (right)

abutment/anchor bracket force and inhibiting pad vibration. Molybdenum disulphide is one such high temperature grease.

5.11.5.2 Added Mass

Adding mass to specific parts of the braking system is a common method of alleviating noise. The aim is to shift the frequency of one part so avoiding a "coupling" of frequencies or modes of vibration. Carrier brackets are first tested with bolted masses and then the casting is modified to include the mass (Fig. 5.37). Small masses are sometimes fitted to the ends of the pad in order to move the frequency of the pad away from a disc frequency (Fig. 5.38).

The excitation of the back-plate of a drum brake may result in "hum". To alleviate the noise in a typical drum brake, a 100 g mass may be added to the back-plate antinode (Fig. 5.39 top). In other more severe noise conditions, inertial

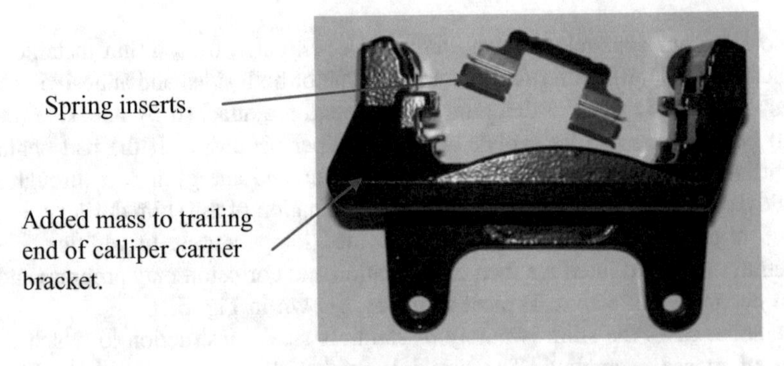

Fig. 5.37 Mass added to one side of calliper carrier bracket—normally trailing end

Fig. 5.38 Mass added to back-plate of pad shifts frequency and changes mode of vibration

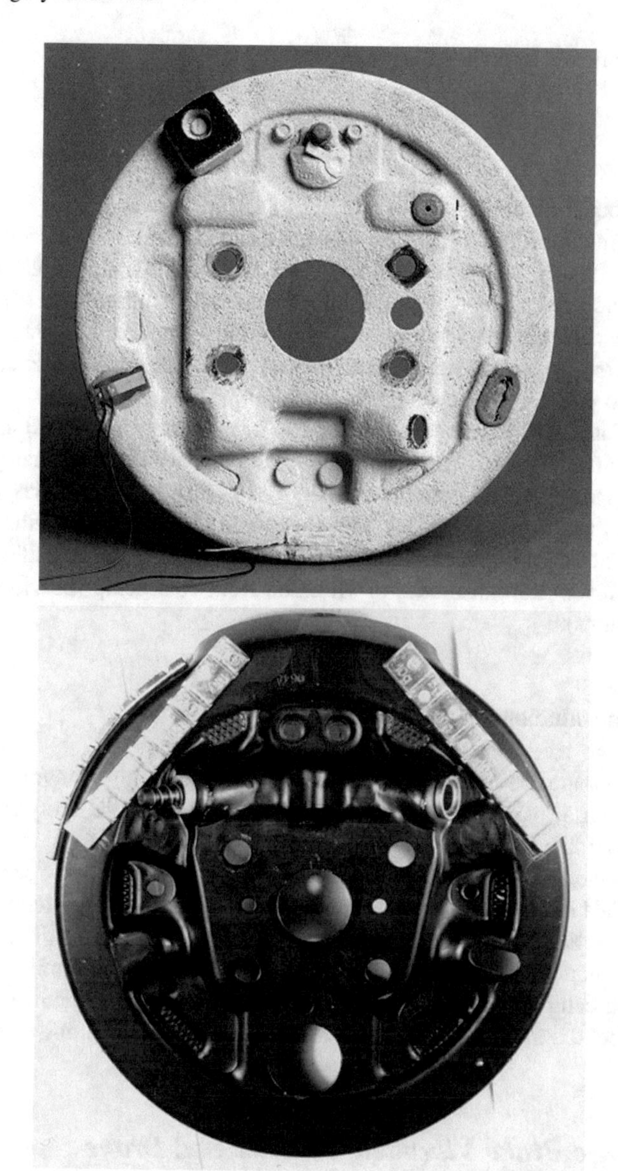

Fig. 5.39 100 g mass added to antinode of back-plate (top) and inertial dampers added to back-plate (bottom)

dampers can be used (Fig. 5.39 bottom). Note that the latter are not only adhered across the antinode positions but also on the rim of the back-plate.

With both discs and drums, a noise may be eradicated by adding or removing mass to or from the rotor. This must be done such that the mass is positioned at the node of one mode which will then be at the antinode of the other (potentially coupled) mode. This introduces asymmetry to the rotor.

The general rule is:

$$\frac{2n}{z} = Integer$$

where n = squeal mode order, z = number of added masses.

So if the mode order n = 3, the number of added masses z = 2, 3 or 6.

Although this is very effective and does work, it is only applicable to that particular mode order. Also the addition or removal of mass from the rotor may result in unacceptable temperature gradients around the rotor and thus lead to the onset of judder. Some masses are constructed to include elastomeric interfaces so damping also plays a part in reducing noise propensity.

For ventilated disc brakes, the distribution of mass may be altered to promote a frequency shift). The ventilated rotor design shown in Fig. 5.40 changed the frequency difference between two potentially-coalescing normal modes from 20 to 769 Hz. This was for a heavy but low speed vehicle so the problems of ensuing judder were not as prominent. Such modifications address the condition of binary flutter and prevent normal modes of vibration from coalescing (possible with both discs and drums).

5.11.5.3 Introducing Asymmetry

Brake systems are invariably symmetrical, the reason being to avoid parts being fitted the wrong way round. Such symmetric systems are more likely to experience instability and the introduction of asymmetry may often result in better noise alleviation. One technique is to offset the pistons to give a trailing “centre of brake pressure”. This is often done to prevent pad taper wear but it also tends to stop the “spragging” action where the pad effectively “digs-in” at its leading edge. Such “spragging” induces a stick-slip mechanism, resulting in noise. Similarly, it may be seen that the calliper “fingers” may be offset or differ in size (width). Such small changes affect the symmetry of the system and tend to reduce noise propensity.

5.11.6 Disc Brake Vibration—Judder and Drone

Brake judder is a mechanically induced vibration with a frequency related to the rotational speed of the wheel. There are two forms of judder, cold and hot. All judder will be detected as a brake torque variation but not all will demonstrate a pressure variation. Judder is felt as a vibration at the steering wheel and brake pedal. It may also be heard as a drumming sound within the cabin.

Fig. 5.40 Disc test design for a 5 diametrical mode order noise "fix". Drilling the periphery of the disc preceded this final test design

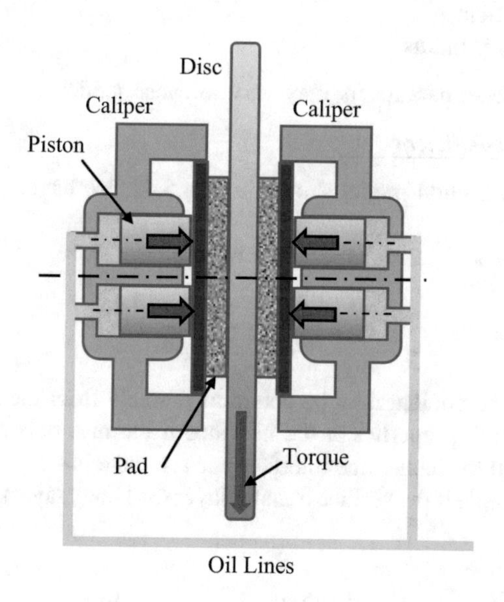

Fig. 5.41 Typical arrangement of a disc brake

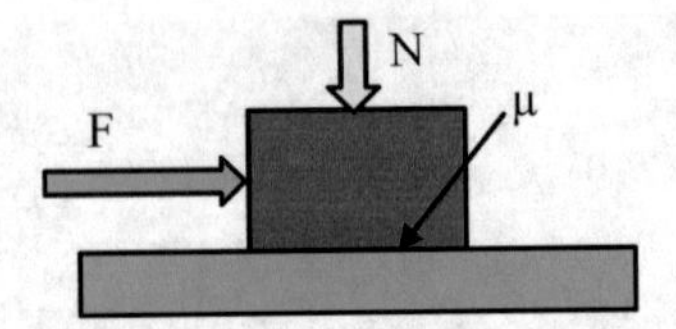

Fig. 5.42 Basic friction model

5.11.6.1 Cold Judder

Generally the causes of cold judder are well understood and may be listed as brake wear, corrosion, and surface film transfer (or pad deposition), often referred to as the third body layer. These effects are considered below with reference to the basic equation for braking torque. For the typical disc brake arrangement shown in Fig. 5.41, the braking torque (T) is given by:

$$T = F\mu r$$

where

F Total piston force
μ Friction coefficient
r Effective friction radius

How each of these parameters may vary to cause judder is now considered.

Variable friction coefficient "μ"

From the basic friction model shown in Fig. 5.42, we have:

$$F = \mu N \tag{5.63}$$

$$\mu = \frac{F}{N} = \frac{\tau A}{\sigma A} = \frac{\tau}{\sigma} \tag{5.64}$$

So the friction coefficient may be determined solely from the relative shear and compressive strength properties at the interface if the materials remain the same.

Therefore, for μ to change, the interface surface materials or characteristics must change as for example if the surface transfer layer (3rd body layer) is incomplete, see

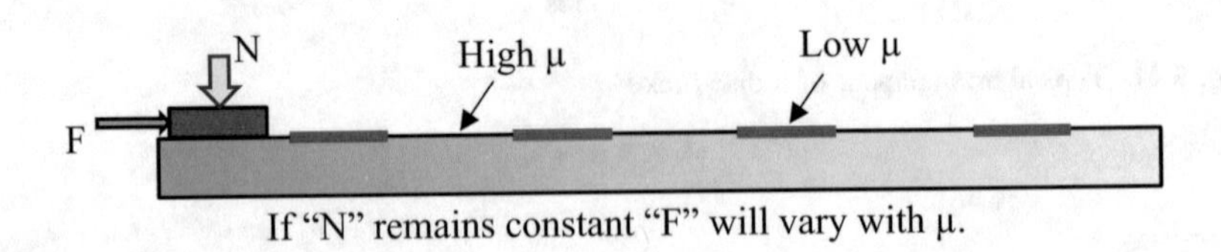

Fig. 5.43 Effects of variable friction surface

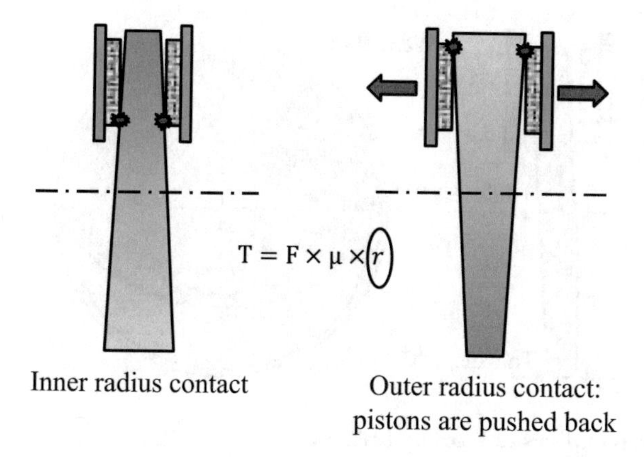

Fig. 5.44 Disc machined but not parallel

Fig. 5.43. This will produce a variable friction coefficient which will not be seen as a pressure variation but as a torque variation. To avoid such variable conditions, the brake must be properly "bedded" or "run-in" to ensure the film is properly transferred from pad to disc early after a service change. It is known that if both disc and pad are not properly "bedded", the material (surface) transfer between the two materials can be deposited in a random fashion leading to a variable surface transfer film.

If, after a high speed stop, the pad remains in contact with the disc then the disc cannot readily cool in this region. This results in a visible impression on the disc known as "pad imprinting", the shape being that of the pad. This condition also results in variable friction levels and judder.

Variable effective disc radius "r"

Such a situation may occur due to a manufacturing error such as a non-parallel disc as indicated in Fig. 5.44. This will produce a variable contact radius and a variable torque. It will also be seen as a pressure variation because the pistons will be pushed back by the thicker regions of the disc.

Even if the piston pressure were to be considered as constant, it is clear that the effective radius of the disc changes considerably and so will the torque.

Variable Force "F"

$$\text{Piston Force F} = \text{Pressure} \times \text{Piston Area} = \text{pA}$$

As the piston areas remain constant it is clear that a variable force may only vary because of a pressure change. With this consideration it is necessary to understand why the pressure could change.

Manufacturing errors and tolerance allowances may cause the disc to "wobble" when rotated, a condition known as "swash" or "runout". If this effect is significant

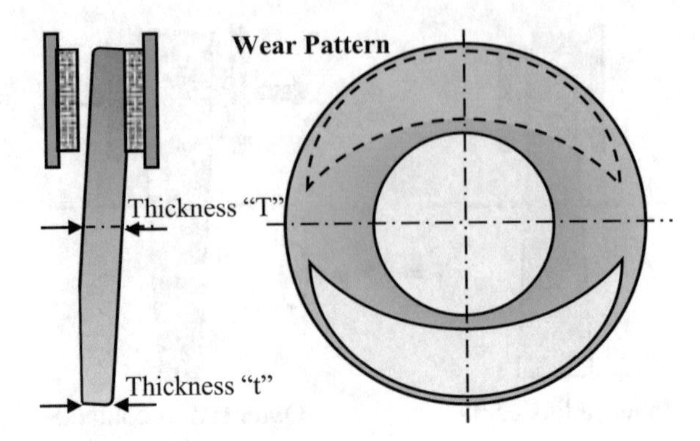

Fig. 5.45 Effects of disc swash and off brake wear

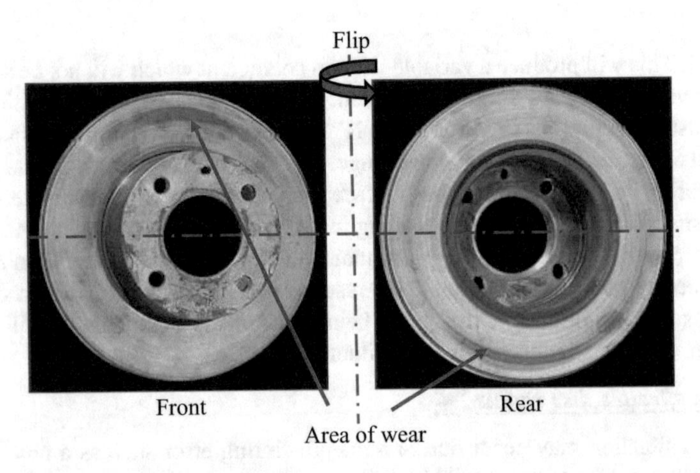

Fig. 5.46 Front and rear view of a disc showing signs of "opposed" wear which will eventually result in disc thickness variation (DTV)

then the disc may in fact touch the brake pads during disc rotation even if the brakes are not applied. This can result in "off brake wear" and lead to disc thickness variations (DTV) as shown in Fig. 5.45. Figure 5.46 shows signs of initial wear on opposite sides of a disc which will progress to DTV. Both swash and DTV can cause cyclic variations in piston force leading to judder.

To prevent off brake wear, techniques are employed such as "seal rollback" when the seal tends to "pull" the piston away from the pads so allowing the pads to be "pushed back" from the disc, known as "piston knockback". In some instances the designer may account for "swash" and indeed tolerance the brake assembly accordingly. This may lead to a "soft" pedal feel and excessive pedal travel. As such "seal rollback" is a favoured approach.

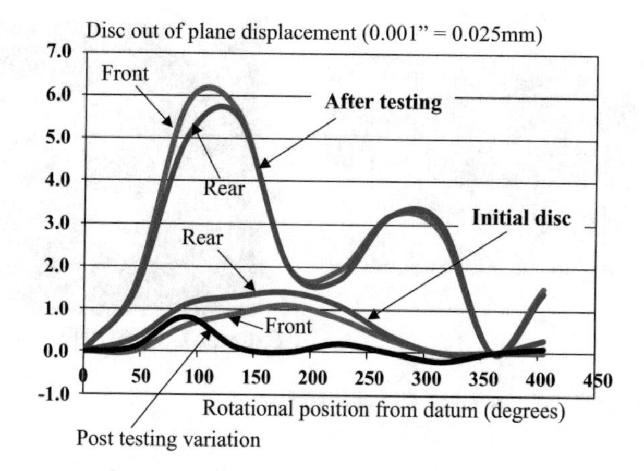

Fig. 5.47 Profile of a disc before testing and after testing at elevated temperatures

5.11.6.2 Hot Judder

Hot judder is the least understood of all mechanisms. In general "hot judder" is experienced after a heavy braking application followed by light braking, typically when exiting a motorway. Research has shown that during a heavy braking application, the kinetic energy absorbed by the brake causes the disc to "warp"; the disc experiences a thermal deformation equivalent to a waveform often exacerbated by the presence of vanes in a ventilated disc. As this "wave" passes through the caliper/pads, it pushes the pistons backwards so inducing a pressure variation leading to judder. In some instances this "wave" is temporary and will reduce as the disc cools. This makes it difficult for test engineers to identify the underlying cause. In more severe braking, the "wave" becomes permanent so judder is then experienced with every brake application. The only solution then is regrinding or replacement of the disc.

Figure 5.47 shows the face profile of a brake disc before and after testing at elevated temperatures. It should be noted that the initial disc did demonstrate a degree of runout (swash) in addition to DTV (similar to Fig. 5.45). In addition the post-tested disc showed a second order wave in addition to DTV (post-testing variation curve). Such an out of plane displacement did result in judder. The "wave" was permanent and did not show further deformation. After regrinding the disc was retested and the second order "wave" was re-established.

5.11.6.3 Disc Brake Drone

Disc brake drone is the result of a high frequency judder mechanism. It emanates from raised areas within the disc generated during braking which are seen as "blue" or "hot" spots after the disc cools as shown typically in Fig. 5.48. This generation

Extreme Case of "Hot spotting"

Fig. 5.48 Blue (hot) spots seen on a disc after cooling

of blue spots is often referred to as "hot spotting". On cooling the spots become recesses so subsequent "hot spotting" occurs elsewhere aound the disc (Fig. 5.48).

Cast iron is an iron alloy comprising carbon, silicon and possibly other alloying elements. If any areas of the disc are slightly prouder than the surrounding areas (possibly due to pad imprinting) then they become hotter than the surrounding material. At high temperatures exceeding 650–700 °C, areas of martensite can be formed on the disc surface. Martensite is an abrasive hard material which will expand more than the surrounding material because of the elevated temperatures. As such the problem is exacerbated with temperatures ever increasing. The process appears to be random in nature (Fig. 5.48) but early signs of "hot spotting" indicate that there may be an element of symmetry (Fig. 5.49). It was noticed that these "hot spots" were antisymmetric, on the rear face being between the visible "spots" as seen in Fig. 5.49, on the front, a wavelike shape being apparent.

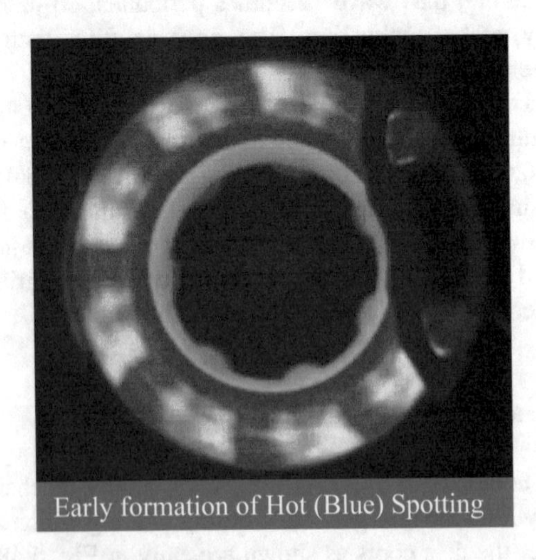

Early formation of Hot (Blue) Spotting

Fig. 5.49 Early signs of hot (blue) spotting

The spots are often related to the disc vent design and are diametrically opposed. If these "vent" effects do not merge to form hot spots then they may form a ripple on the disc surface and cause very high frequency drone. An attempt to resolve this effect would be to make the number of vents equal to a prime number, so making merging less likely. Experiments show these spots to "hunt" about the disc profile, wearing away and then re-establishing themselves somewhere else. The radial position also changes over time. The high number of spots gives a higher frequency than "judder" and vibration is not felt at the steering wheel or brake pedal. The most prominent feature of this form of instability is an annoying "droning" sound that is heard in the cabin. Hence the term "brake drone" is used to define this problem.

5.11.6.4 Judder Summary

Regardless of the above general comments, judder is not simply a brake problem and the suspension system plays a significant role in exacerbating the problem. If the vehicle suspension system is over-sensitive to brake torque variations then modifications to the elastomeric bushes can often alleviate the problem. The issues regarding hot judder are often a result of excessive thermal gradients, uneven heating and dissipation, which can be addressed by careful design of the disc venting system. It is also necessary to stress relieve the discs before machining, otherwise "in-service" stress relieving will occur if elevated temperatures are experienced.

5.12 Concluding Remarks

This chapter has provided a foundation for understanding the important NVH issues in vehicles and the processes needed to mitigate these issues leading, to improved vehicle refinement. It includes reviews of basic acoustic and vibration theory and considers all the major sources of automotive noise and vibration, with a particular focus on internal combustion engines (still the predominant source of motive power in most road vehicles) and on friction brakes (still required even on fully electric vehicles). The fact that NVH is covered in the final chapter of this book is no coincidence. It is often the final part of the vehicle design process to be considered and often continues once series production has commenced with the need to find "fixes" to urgent NVH problems. Yet vehicle refinement (i.e. lack of NVH issues) is one of the key selling points of any vehicle and helps to distinguish one vehicle manufacturer from their rivals, especially at the quality end of the market.

Appendix
Summary of Vibration Fundamentals

The general approach to vibration analysis is to:

- Develop a mathematical model of the system and formulate the equations of motion
- Analyse the free vibration characteristics (natural frequencies and modes)
- Analyse the forced vibration response to prescribed disturbances
- Investigate methods for controlling undesirable vibration levels if they arise.

This Appendix outlines approaches to the first 3 of these topics. Methods for controlling vibration are outlined in the relevant chapters (principally Chaps. 3 and 5).

Mathematical Models

These provide the basis of all vibration studies at the design stage. The aim is to represent the dynamics of a system by one or more differential equations. The most common approach is to represent the distribution of mass, elasticity and damping in a system by a set of discrete elements and assign a set of coordinates to the masses.

- Elasticity and damping elements are assumed massless—there is a need to know the constituent equation describing the character of these elements
- The number of degrees-of-freedom (DOFs) of the system is determined by the number of coordinates, e.g. the lumped-model of a simply-supported beam in Fig. A.1
- Each coordinate z_1, z_2... etc. is a function of time t
- The number of DOFs chosen dictates accuracy. The aim to have just sufficient DOFs to ensure that an adequate number of natural vibration modes and frequencies can be determined while avoiding unnecessary computing effort
- An n-DOF system will be described by n-second order differential equations and have n-natural frequencies and modes
- The simplest model has only one DOF!

© Springer International Publishing AG 2018

D. C. Barton and J. D. Fieldhouse, *Automotive Chassis Engineering*,
https://doi.org/10.1007/978-3-319-72437-9

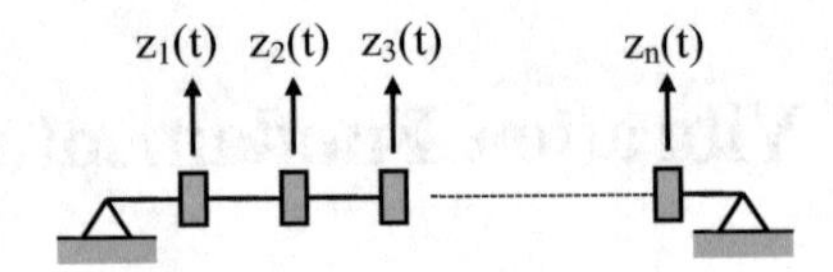

Fig. A.1 Lumped-parameter model of a simply-supported beam

Formulating Equations of Motion

- Equation(s) of motion can be determined by drawing a free-body diagram (FBD) of each mass including all relevant forces and applying Newton's 2nd law
- Each equation of motion is a second order differential equation
- For a multi-DOF system, the equations can be assembled in matrix form
- Figure A.2 shows the FBD's of the quarter vehicle model discussed in Chap. 3
- $z_1(t)$ and $z_2(t)$ are the generalised coordinates (measured from the static equilibrium position). $x_0(t)$ represents the dynamic displacement input from the road surface.

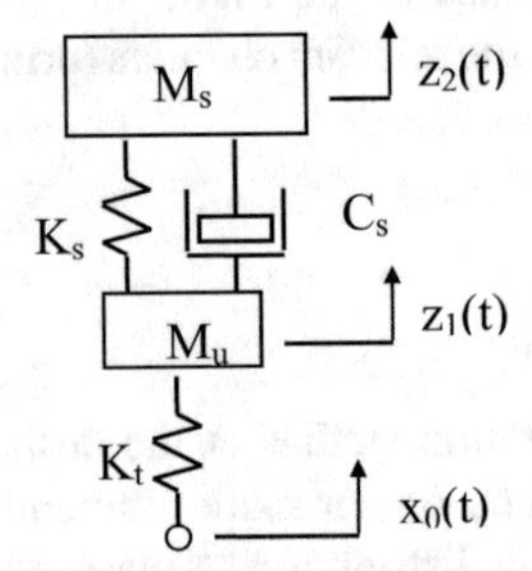

(a) Quarter vehicle model

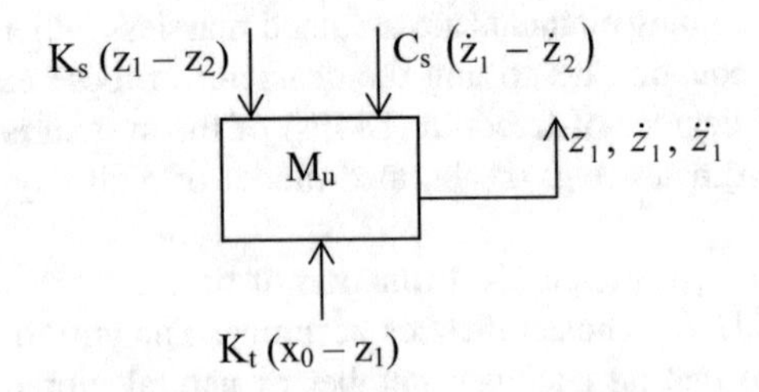

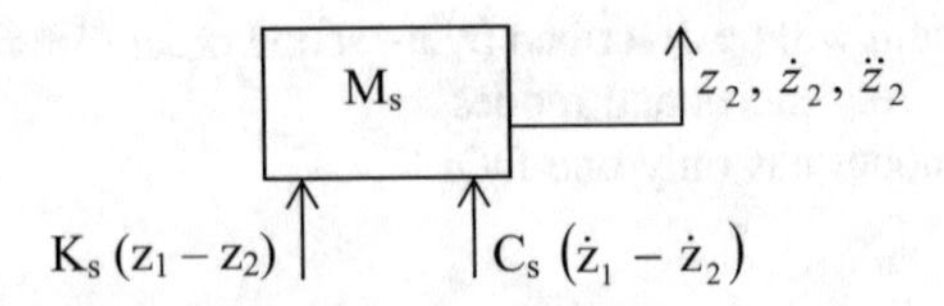

(b) Associated free-body diagrams

Fig. A.2 **a** Quarter vehicle model and **b** Associated free-body diagrams

- Applying Newton's second law to the unsprung mass M_u in Fig. A.2b gives:

$$K_t(x_0 - z_1) - K_s(z_1 - z_2) - C_s(\dot{z}_1 - \dot{z}_2) = M_u\ddot{z}_1$$

$$M_u\ddot{z}_1 + C_s(\dot{z}_1 - \dot{z}_2) + K_s(z_1 - z_2) + K_t z_1 = K_t x_0$$

- Applying Newton's second law to the sprung mass M_s gives:

$$K_s(z_1 - z_2) + C_s(\dot{z}_1 - \dot{z}_2) = M_s\ddot{z}_2$$

$$M_s\ddot{z}_2 - C_s(\dot{z}_1 - \dot{z}_2) - K_s(z_1 - z_2) = 0$$

- These equations can be re-expressed in matrix form as:

$$\begin{bmatrix} M_u & 0 \\ 0 & M_s \end{bmatrix}\begin{Bmatrix} \ddot{z}_1 \\ \ddot{z}_2 \end{Bmatrix} + \begin{bmatrix} C_s & -C_s \\ -C_s & C_s \end{bmatrix}\begin{Bmatrix} \dot{z}_1 \\ \dot{z}_2 \end{Bmatrix} + \begin{bmatrix} (K_t + K_s) & -K_s \\ -K_s & K_s \end{bmatrix}\begin{Bmatrix} z_1 \\ z_2 \end{Bmatrix} = \begin{Bmatrix} K_t\, x_0 \\ 0 \end{Bmatrix}$$

The equations can be written more concisely as:

$$[M]\{\ddot{z}\} + [C]\{\dot{z}\} + [K]\{z\} = \{F(t)\}$$

where [M], [C] and [K] are the mass/inertia, damping and stiffness matrices $\{z\}$, $\{\dot{z}\}$ and $\{\ddot{z}\}$ are displacement, velocity and acceleration vectors $\{F(t)\}$ is the excitation force vector.

Single Degree of Freedom (SDOF) Systems

- Knowledge of SDOF behaviour provides an understanding of more complex systems.
- For the SDOF system shown in Fig. A.3 the equation of motion is:

$$m\ddot{z} + c\underline{z} + kz = F(t)$$

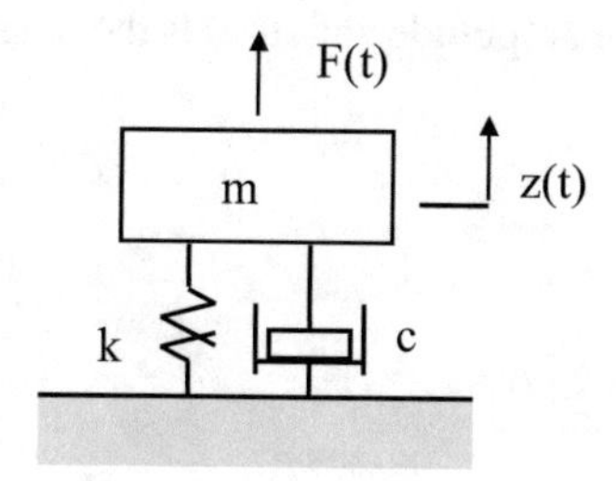

Fig. A.3 Notation for SDOF system

- The system characteristics are obtained from the free-vibration behaviour ($F(t) = 0$)
- With $c = 0$, there is simple harmonic motion:

$$z = Z \cos(\omega_n t - \phi),$$

where Z (the amplitude) and ϕ (the phase change) are constants (determined from the conditions at $t = 0$) and ω_n is the undamped natural frequency given by:

$$\omega_n = \sqrt{\frac{k}{m}}$$

- When $c \neq 0$, the level of damping can be described in terms the damping ratio ζ defined as:

$$\zeta = \frac{c}{c_c} \quad \text{where } c_c = 2\sqrt{mk}$$

- When disturbed, the mass slowly returns to its equilibrium position either with or without oscillation.
- Two possible characteristics:

  $\zeta < 1$, the system is under damped—oscillation with decaying amplitude
  $\zeta \geq 1$, the system is over damped—the mass returns to its equilibrium without oscillation

- When $\zeta = 1$ the system is critically damped ($c = c_c$).
- For an under damped system: $z = Z e^{-\zeta \omega_n t} \cos(\omega_d t - \phi)$, where Z and ϕ are constants representing the amplitude and phase shift respectively and ω_d is the damped natural frequency given by: $\omega_d = \omega_n \sqrt{1 - \zeta^2}$
- If $F(t) \neq 0$ the solution to the equation of motion has two components: the Complementary Function (CF) and the Particular Integral (PI).
- The CF is identical to the free vibration solution and quickly dies away with realistic levels of damping to leave z = PI. In practice ζ ranges from approximately 0.02 for low damping elastomers to 0.5 for vehicle suspensions.
- If $F(t) = F_0 \sin \omega t$, the steady-state response of the mass (after the CF has become zero) is given by: $z = A(\omega) \sin[\omega t - \alpha(\omega)]$
- $A(\omega)$ is the steady-state amplitude and $\alpha(\omega)$ is the phase lag; both are dependent on ω.

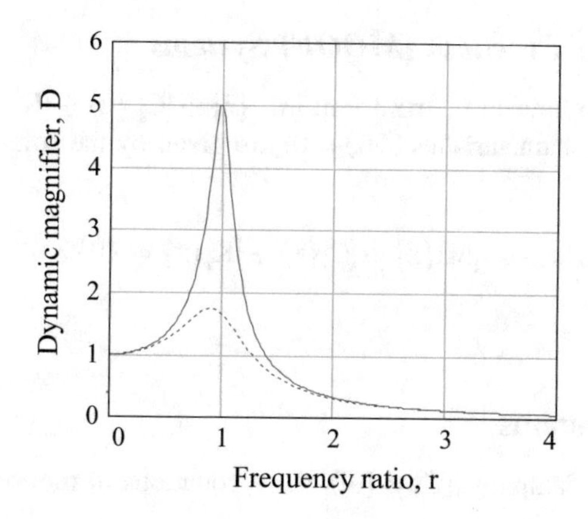

Fig. A.4 Amplitude response D in terms of frequency ratio r

- It may be shown that:

$$A(\omega) = F_0\,|H(\omega)| = F_0\left|\frac{1}{(k - m\,\omega^2) + (c\,\omega)\,i}\right| = \frac{F_0}{\sqrt{(k - m\,\omega^2)^2 + (c\,\omega)^2}}$$

- $H(\omega)$ is called the *frequency response function* and is generally *complex* (real and imaginary parts); it relates the input (excitation) to the output (response) in the frequency domain.
- The amplitude response can be presented in dimensionless form in terms of the dynamic magnifier $D = k\,A/F_0$ and frequency ratio $r = \omega/\omega_n$
- It may be shown that $D = \frac{1}{\sqrt{\left[1-\left(\frac{\omega}{\omega_n}\right)^2 + \left(2\zeta\frac{\omega}{\omega_n}\right)^2\right]}}$
- The variations of D with frequency ratio r for two different values of ζ are shown in Fig. A.4:

- It can be concluded from Fig. A.4 that:

(a) Maximum amplitude occurs at resonance when $\omega \approx \omega_n$

(b) The amplitude is strongly dependent on the damping in the system when $\omega \approx \omega_n$

(c) When ω and ω_n are very dissimilar, damping has very little effect on response amplitude—an important point when considering the use of damping to control vibration levels.

Multi-degree of Freedom (MDOF) Systems

- Equations of motion in matrix form: $[M]\,\{\ddot{z}\} + [C]\,\{\dot{z}\} + [K]\,\{z\} = \{F(t)\}$
- Free vibration characteristics (F(t) = 0) are given by the solution of the matrix equation:

$$[M]\{\ddot{z}\} + [C]\{\dot{z}\} + [K]\{z\} = \{0\}$$

Undamped Systems

- For negligible damping ([C] = [0]), these equations of motion become:

$$[M]\,\{\ddot{z}\} + [K]\,\{z\} = \{0\}$$

- Assuming solutions of the form:

$$\{z\} = \{A\}e^{st}$$

  leads to a set of homogeneous equations:

$$\left([M]\,s^2 + [K]\right)\{A\} = \{0\}$$

- The non-trivial solution can be written in the form of a *characteristic equation*:

$$\left|[M]\,s^2 + [K]\right| = 0$$

  or *eigenvalue equation*:

$$\lambda\,[M]\,\{u\} = [K]\,\{u\}$$

- This leads to a set of real roots, typically $s_i^2 = -\lambda_i = -\omega_i^2$, where λ_i is the i-th eigenvalue and ω_i the i-th natural frequency.
- For each eigenvalue λ_i, there is an eigenvector $\{u\}_i$, which relates the relative amplitudes at each of the degrees of freedom.
- Vibration in the i-th mode then can be described by: $\{z\}_i = \{u\}_i A_i \sin(\omega_i t + \alpha_i)$
- The general free vibration of the system is a combination of all modes:

$$\{z\} = \sum_{i=1}^{n} \{z\}_i = [u]\,\{q(t)\},$$

where

$[u] = \left[\{u\}_1\, \{u\}_2 \ldots\ldots \{u\}_n\right]$ is called the *modal matrix* and $\{q(t)\} =$

$$\begin{Bmatrix} A_1 \sin(\omega_1 t + \alpha_1) \\ \vdots \\ A_n \sin(\omega_n t + \alpha_n) \end{Bmatrix}$$ is a vector of *modal (principal) coordinates.*

- [u] represents a linear transformation between generalised coordinates {z} and modal coordinates {q}.

Lightly Damped Systems

For a lightly damped system ($C \neq 0$), the characteristic equation is given by:

$$\left|[M]\ s^2 + [C]\ s + [K]\right| = 0$$

- This equation has a set of complex conjugate roots having negative real parts which provide information about the frequency and damping associated with each mode of vibration
- For harmonic excitation of a damped MDOF system:

  (a) The system vibrates (in its steady state) at the same frequency as the excitation
  (b) The dynamic displacement at each DOF lags behind the excitation
  (c) The displacement amplitudes at each of the degrees of freedom are dependent on the frequency of excitation

- In MDOF systems, the excitation can be applied simultaneously at any of the DOFs
- For a linear system, the response at any of the DOFs is the sum of the responses due to each excitation force
- When the excitation is applied simultaneously at all DOFs, F(t) can be written as:

$$\{F(t)\} = \{F\}f(t) = \{F\}\sin(\omega t)$$

in which case $[M]\ \{\ddot{z}\} + [C]\ \{\dot{z}\} + [K]\ \{z\} = \{F\}\ f(t)$

- Taking Laplace transforms of both sides with zero initial conditions, replacing s with iω and pre-multiply both sides by $(-\omega^2[M] + i\,\omega\,[C] + K)^{-1}$ gives:

$$\{H_z(\omega)\} = [H(\omega)]\,\{F\}$$

where $\{H_z(\omega)\}$ is a vector of frequency responses at the DOFs

and $[H(\omega)] = \begin{bmatrix} H_{11} & H_{12} & \cdots & H_{1n} \\ H_{21} & H_{22} & \cdots & H_{2n} \\ \vdots & \vdots & \vdots & \vdots \\ H_{n1} & \cdots & \cdots & H_{nn} \end{bmatrix}$ is a matrix of frequency response functions such that H_{ij} = frequency response at i due to unit amplitude excitation at j

- The frequency response at DOF i is given by:

$$H_{zi}(\omega) = H_{i1}\,F_1 + H_{i2}\,F_2 + \;\cdots\; H_{in}\,F_n = \sum_{j=1}^{n} H_{ij}\,F_j$$

- The response at a particular DOF is made up of contributions from excitations at all the various DOFs in the systems
- FRFs and hence responses are complex functions (real and imaginary parts) when damping is included; the amplitude and phase at a given DOF are given by the modulus and argument of the complex frequency response function.

Bibliography

Balkwill J (2017) Performance vehicle dynamics. Butterworth-Heinemann

Brown JC, Robertson AJ, Serpento ST (2002) Motor vehicle structures: concepts and fundamentals. Butterworth-Heinemann

Crolla D (ed) (2009) Automotive engineering: powertrain, chassis system and vehicle body. Butterworth-Heinemann

Crolla D, Foster DE, Kobayashi T, Vaughan N (eds) (2014) Encyclopedia of automotive engineering. Wiley Online. https://doi.org/10.1002/9781118354179

Davies G (2003) Materials for automotive bodies. Elsevier

Elmarakbi A (ed) (2014) Advanced composite materials for automotive applications. Wiley

Hapian-Smith J (2002) An introduction to modern vehicle design. Butterworth-Heinemann

Heinz H (2002) Advanced vehicle technology, 2nd edn. Elsevier

Hillier A (2012) Hillier's fundamentals of motor vehicle technology, 6th edn. Nelson Thornes

Milliken WF, Milliken DL (2002) Chassis design: principles and analysis. Professional Engineering Publishing

Reimpell J, Stoll H, Betzler J (2001) The automotive chassis: engineering principles, 2nd edn. Butterworth-Heinemann

Seward D (2014) Race car design. Palgrave

Wong JY (2001) Theory of ground vehicles, 3rd edn. Wiley-Interscience

© Springer International Publishing AG 2018

D. C. Barton and J. D. Fieldhouse, *Automotive Chassis Engineering*,
https://doi.org/10.1007/978-3-319-72437-9

Printed in the United States
By Bookmasters